Environmental Contamination and Remediation Practices at Former and Present Military Bases

NATO Science Series

A Series presenting the results of activities sponsored by the NATO Science Committee. The Series is published by IOS Press and Kluwer Academic Publishers, in conjunction with the NATO Scientific Affairs Division.

General Sub-Series

A. Life Sciences	IOS Press
B. Physics	Kluwer Academic Publishers
C. Mathematical and Physical Sciences	Kluwer Academic Publishers
D. Behavioural and Social Sciences	Kluwer Academic Publishers
E. Applied Sciences	Kluwer Academic Publishers
F. Computer and Systems Sciences	IOS Press

Partnership Sub-Series

1. Disarmament Technologies	Kluwer Academic Publishers
2. Environmental Security	Kluwer Academic Publishers
3. High Technology	Kluwer Academic Publishers
4. Science and Technology Policy	IOS Press
5. Computer Networking	IOS Press

The Partnership Sub-Series incorporates activities undertaken in collaboration with NATO's Partners in the Euro-Atlantic Partnership Council – countries of the CIS and Central and Eastern Europe – in Priority Areas of concern to those countries.

NATO-PCO-DATA BASE

The NATO Science Series continues the series of books published formerly in the NATO ASI Series. An electronic index to the NATO ASI Series provides full bibliographical references (with keywords and/or abstracts) to more than 50000 contributions from international scientists published in all sections of the NATO ASI Series.
Access to the NATO-PCO-DATA BASE is possible via CD-ROM "NATO-PCO-DATA BASE" with user-friendly retrieval software in English, French and German (© WTV GmbH and DATAWARE Technologies Inc. 1989).

The CD-ROM of the NATO ASI Series can be ordered from: PCO, Overijse, Belgium.

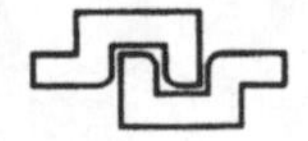

Series 2: Environmental Security – Vol. 48

Environmental Contamination and Remediation Practices at Former and Present Military Bases

edited by

F. Fonnum

VISTA and Norwegian Defence Research Establishment,
Kjeller, Norway

B. Paukštys

Hydrogeological Company "Grota",
Vilnius, Lithuania

B. A. Zeeb and K. J. Reimer

Environmental Sciences Group,
Royal Military College of Canada,
Kingston, Ontario, Canada

Kluwer Academic Publishers

Dordrecht / Boston / London

Published in cooperation with NATO Scientific Affairs Division

Proceedings of the NATO Advanced Research Workshop on
Environmental Contamination and Remediation Practices at
Former and Present Military Bases
Vilnius, Lithuania
October 12–17, 1997

A C.I.P. Catalogue record for this book is available from the Library of Congress

ISBN 0-7923-5247-5

Published by Kluwer Academic Publishers,
P.O. Box 17, 3300 AA Dordrecht, The Netherlands.

Sold and distributed in North, Central and South America
by Kluwer Academic Publishers,
101 Philip Drive, Norwell, MA 02061, U.S.A.

In all other countries, sold and distributed
by Kluwer Academic Publishers,
P.O. Box 322, 3300 AH Dordrecht, The Netherlands.

Printed on acid-free paper

Printed in the Netherlands

CONTENTS

REMEDIATION TECHNIQUES

PREFACE

NATO Advanced Research Workshop on "Environmental Contamination and Remediation Practices at Former and Present Military Bases was held on 12-17 October 1997 at the "Villon" hotel in Vilnius, Lithuania. The ARW was co-directed by professor, dr. Frode Fonnum, Norwegian Defence Research Establishment and dr. Bernardas Paukštys, Hydrogeological Company "Grota"/Vilnius University, Lithuania.

45 scientists from 10 NATO, 8 Co-operation Partner countries and Sweden participated at the ARW. 19 scientists made 17 keynote, 1 hour presentations*, 17 participants presented short, 15 minute papers and video material on military contamination in Latvia and Estonia was shown. The president of the Academy of Sciences of Lithuania, Member of Parliament and representatives from the ministries of Defence, Environmental Protection and Foreign Affairs extended their welcome greetings at the Opening Session.

Four main topics were discussed at the scientific sessions: i) general aspects of military contamination (5 keynote presentations and 4 participant contributions), ii) risk assessment and modelling (3 keynote and 3 participant papers), iii) modern clean-up methods and technologies (3 keynote and 5 participant contributions) and iiii) regional case studies (6 keynote and 5 participant papers). Characteristic feature of scientific sessions was an active participation of attendees in discussions after each presentation, willingness to share available knowledge and to learn more from each other. The papers from the majority of the ARW participants were collected before the meeting, bind into two volumes of the Workshop proceedings and included into information package. Papers brought by the attendees to the meeting, were copied and distributed to all participants.

A field trip was organised on 15 October. Participants visited former military fuel base of Valciunai and Vilnius Oil Storage. The remediation of contaminated groundwater is being carried out at both sites by the "Grota" company. More than 250 tons of liquid oil products was recovered in 9 months from Valciunai site and 2300 tons of oil was extracted at Vilnius oil base since 1989. The ARW participants were able to observe the remediation procedure, discuss its advantages and drawbacks. The site visit helped to more clearly understand the extent of military contamination.

It was agreed by all the ARW attendees that the meeting was a success as a lot of new scientific and practical issues were presented. An atmosphere for active professional discussions was created which developed new ideas. Clear proposals for further co-operation was established and included into Recommendations, and also personal contacts between the scientists were made which, hopefully, will lead to mutually beneficial co-operation in future.

ACKNOWLEDGEMENT

We would like to thank NATO Science Committee and its Programme Director L Veiga da Cunha for the support which made this meeting possible. We also extend our thanks to the Norwegian Foreign Affairs for a generous economic support.

On behalf of all the participants we extend our thanks to our Lithuanian colleagues and to the Hydrological Company "Grota" for their superb effort in hosting the meeting. We also like to thank the other members of the organizing committee, Professor Vytautas Juodkazis, Lithuania, Dr Vaclav Eliaš, Czec Republic and Algomantas Kulanovas, Lithuania.

For administrative support we are indepted to Norwegian Defence Research Establishment and Lithuanian Academy of Sciences.

In preparation of this book we are grateful to Deborah Reimer, Kari Skovli, Rannveig Løken, Gro Tunhøvd and Bente Persen Steihaug for secretarial assistance.

WORKSHOP SUMMARY

In all post communist and ex-Warsaw pact countries serious environmental damage due to military activities has been observed. Such problems are also encountered in most NATO countries. Former Soviet military bases, however, created specific environmental problems.

Various military installations built up during the cold war such as : training areas, airfields and air bases, rocket fuel and chemical storage centres, tank regiments, military towns and naval bases, have caused extensive damage to the environment. Pollution can also affect human health. Soil and groundwater deterioration at military sites is the most common environmental problem in all countries of the Baltic surface watershed and groundwater catchment areas, and also the most expensive to rehabilitate.

The military sector has only recently become seriously engaged in environmental site investigation and remediation and is, at least, in cooperation partner countries, technologically behind the civilian sector.

The main goal of the Workshop was to discuss and to provide a critical assessment of the risk and adverse impacts of military activities on the environment; to evaluate effectiveness of low-cost technologies for reclamation of contaminated military sites and to evaluate technological gaps, remediation methods, monitoring and sampling techniques. The ARW drew on the experience of participants from 10 NATO, 8 Co-operation partner countries and also from Sweden. An important objective was to facilitate East-West collaboration on scientific and technological aspects with specific emphasis on the environmental needs of the Baltic region countries.

The Vilnius-ARW recognize the existence of and need for coordination with other NATO activities such as "Pilot study on environmental aspects of reuse of former military lands" and "Pilot Study on transboundary pollution".

After four days of scientific sessions and a one day of field trip the ARW participants came to the following conclusions:

1. Hydrocarbons (petrol, diesel, kerosene) are the dominant pollutants found at military bases. Other common contaminants are chlorinated hydrocarbons and heavy metals. The extent of some contaminants such as rocket fuel at former Soviet bases are still mostly unknown. NBC training areas are special sites with specific toxic chemical agents: tear gas, chloropicrin and small amounts of mustard gas and environmentally aggressive decontaminants. Radionuclide contamination is a problem of concern in some countries. In some NATO military bases PCB, and other chlorinated hydrocarbons are special problems.

2. Several countries follow a step by step approach to investigations of polluted military bases. This is recommended for other ARW participant countries as well. The approach comprises a desk study archival material (aerophotos are particularly useful when available), interview with local people, site visits and visual inspection of the area. After a preliminary inventory of the site, physical and chemical investigations should follow if needed. The purpose of these investigations is to

investigations should follow if needed. The purpose of these investigations is to assess risk to human health, water resources and other environmental media. If there is a serious environmental threat or a threat to human health, then site remediaton measures should be recommended. In other cases the contaminated area can be left for natural attenuation to provide the clean up. All investigation and remediation should be risk driven. Monitoring of pollution will always be necessary at those sites where residual contamination is a risk or where the efficacy of natural attenuation is in doubt.

3. For all post-communist countries and also for several developed western countries, standard air, soil and water sampling procedures and certification of laboratories are required. Standardized protocols such as these should not be limited to analytical procedures but should be applied to all steps in field procedures, well construction and transport of samples. The importance of quality assurance programmes was stressed by all the members of the ARW.

4. Modern soil and groundwater remediation methods have been described by several participants from different countries. Importance of bioremediation and other cost-effective technologies was stressed.

5. There are technology gaps in remediation of PCB and heavy metals in soil and sea sediment. Remediation methods for organic mercury and organic tin compounds in marine sediments must also be developed.

6. Basic understanding and sufficient knowledge on the fate and transport of contaminants in the subsurface systems are lacking in all CP and some NATO countries.

7. It was agreed by the ARW participants that acceptable concentrations of pollutants depend on a risk-oriented approach. This would include an evaluation of the source of pollution, possible migration to and exposure from neighbouring areas. Acceptable concentrations of contaminants will depend on the future use of the military site.

8. It was agreed that internationally acceptable methods for detection of all pollutants are available and should be distributed between the countries concerned through e-mail or internet.

The ARW participants recommend:

1. To support a demonstration project of new remediation techniques on a specific site in the Baltic countries. This would demonstrate the cost efficiency of the

method and act as a training site for participants from partner countries. The project should as much as possible involve local equipment and personal. Funds to the project should be sought from national and international sources. A project group would be set up from NATO and CP countries to implement the project.

2. Accepted contaminant limits in soil, water and marine sediments need to be compared with the efficiency of remediation methods. It is at present not clear whether these limits are attainable at a reasonable cost using the prescribed remediation methods.

3. Knowledge on natural attenuation process is still insuffient and controversial. For practical purposes (such as remediation) and for improving the efficiency of clean up operations it will be of utmost importance to develop methods that transfer detailed information on parameters of subsurface media (stratification, fracturing, chemical composition, etc) to bulk characteristics, representing the average values of the processes and mechanisms of the given (desirable) operational scale. Research should focus on filling gap in knowledge.

4. When evaluating environmental military impact, focus on issues of quality assurance and quality control of analytical and sampling procedures should be underlined. The members of the Vilnius ARW should encourage the military and environmental authorities in their respective countries to insist upon QA/QC protocols when letting site investigation/remediation contracts to consultants. Each country should have at least one certified laboratory specialised in detailed assessment of contaminated sites.

5. Groundwater models are indispensable for the risk assessment and vulnerability evaluation of aquifers. Attention should be paid quantifying to uncertainty in modelling techniques and risk assessment, issues of homogeneity, fracturing, discontinuity and scaling. Reliability of the parameters should be considered. An ARW on preditive tools is recommended if it can include a posteriori results.

6. Vilnius-ARW recommend to establish a task force of experts consisting of the scientists from NATO and Co-operation partner countries. This list of names should be made available through NATO channels. Members of the group could be called upon if special problems surface in a country. Any cost involved is the responsibility of the user. Such a group could improve effective cooperation between nations on investigation and remediation of military sites.

7. In many NATO countries and partner countries there are training of military personal on environmental and health hazards of pollutants. An exchange of such training programs between military forces should be organized by NATO.

8. **Prevention and protection measures should be implemented at existing military sites with future military use.**

9. **Information on the composition and properties of rocket fuel and its products and its effects on human health and the environment should be made available.**

ENVIRONMENTAL CONTAMINATION AND REMEDIATION PRACTICES AT FORMER AND ACTIVE MILITARY BASES

German Experiences

B.-A. SZELINSKI
Federal Ministry for the Environment, Nature Conservation and Nuclear Safety
Schiffbauerdamm 15
D-10117 Berlin

1. Background

Clean-up of contaminated military sites and their conversion for future civilian use will remain a task for almost every country for quite some time. Germany has been expending resources for nearly 7 years on the process of registration, investigation, assessment and clean-up of these sites. The goals of the German program are two-fold: first, to avoid environmental damage and to protect the population from negative effects caused by contamination of such sites; and secondly, to render military sites fit for subsequent commercial, recreational or other civilian uses.

The withdrawal of former Soviet troops from Central Europe associated with the democratic development in Russia and the establishment of new independent states in Eastern Europe has been a strong incentive to develop international cooperation in this field. All countries having closed or abandoned soviet military sites on their soil face comparable problems in connection with former military use of large areas of land by the Soviet troops. In addition, many other countries outside of Eastern Europe are confronted with similar challenges as their armed forces are reduced in general. There are many sites all around the world that are no longer needed for military purposes since the policy of confrontation has changed into one of cooperation. Many of these sites have been closed to environmental authorities for inspection for a long time.

It certainly would not be fair to blame the military alone for damages that have occurred to the environment during military use. We often face a similar or even worse situation at sites which have been in industrial use for a long time. The issue of negative effects of activities on soil and other media is much younger than the activities that caused them.

The damages that we now endeavor to explore and understand are the price our societies now have to pay for the state of confrontation the world lived in for almost 50 years . In addition, many of these sites have already been in military use for more than one century and have often - at least in Europe - therefore been subject to even older

F. Fonnum et al. (eds.),
***Environmental Contamination and Remediation Practices at Former and Present Military Bases*, 1–11.**

war-related damages.

It is encouraging to see that all countries as well as NATO have adopted a positive approach to find safe solutions to deal with the existing problems that are viable and sound both from an environmental and economic perspective and make best use of the opportunities associated with the conversion of military sites in the future.

2. The German Experience

Up until now, Germany has had no generally applicable strict limits or action values which determine the level at which remedial action should take place and the extent to which contamination clean-up must be done. Although certain standards are applied by the Federal States, the system is still flexible enough to allow case-by-case decisions whereby the actual situation of the contaminated site - the actual risk associated with it and the need for clean-up in relation to the envisioned use of the site - considered from a risk-oriented approach. This allows one to differentiate according to the clean-up needs in a given situation. For example, a children's playground requires a different level of clean up than a parking lot or railway track. The proposed German Soil Protection Act, which is currently in parliamentary discussion, will also allow this kind of approach in the future.

2.1 THE GERMAN SYSTEM FOR ASSESSMENT AND CLASSIFICATION

In its position at the border between the Eastern and Western blocs, Germany has had to deal with all aspects of demilitarization of sites: the withdrawal of Soviet troops from former East Germany (GDR) as well as the reduction of the armed forces of Western Allies which has led to the closure of NATO sites on German territory.

Finally, the German armed forces (Bundeswehr) is also decreasing its manpower and will hand over many installations. For example the greater part of those installations that were former used by the GDR "Volksarmee" have already been demilitarized.

Table 1 gives an overview of those sites which were taken out of military use between 1990 and mid 1996 and have been handed back as part of general federal property.

Table 1: Sites Taken Over by the Federal Finance Administration

	BW*	NVA**	NATO	WGT***	Total
Total Number of Sites	1.215	8.953	4.043	1.968	16.179
Total Area of Sites (ha)	17.012	101.419	26.281	226.387	371.099
Number of Housing Areas	22	1.927	2.734	756	5.439
Number of Housing Units	152	66.155	36.390	49.522	152.219
Number of Barracks/ Garrisons	255	705	297	429	1.686
Number of Training Areas	86	164	70	169	489
	- Total Area approx. 217.839 hectares-				
Number of Airfields	21	46	45	37	149
Number of Agricultural/ Forestry Areas	213	2.024	29	95	2.361
	- Total Area approx.. 67.193 hectares-				

* Bundeswehr (Federal German Armed Forces)
** Nationale Volksarmee (GDR-National Peoples Army)
*** Western Group of the Soviet Forces

The German Federal Government started early with a comprehensive system to register, investigate, assess and clean up the sites left by the Soviet troops. The system Germany applied to deal with these sites was based on the following subsystems:

ALADIN : Altlasten - Dateninformationssystem (Residual Pollution Data Information System)

MEMURA : Modell zur Erstbewertung von militärischen- und Rüstungs-Altlastverdachtsflächen (Model for the Initial Assessment of Suspected Residual Pollution Areas)

MAGMA : Modell zur Abschätzung der Gefährdung durch militärische- und Rüstungsaltlasten (Model for the Risk Assessment of Military and Armament Contaminated Sites) and

KOSAL : System zur Kostenermittlung bei der Altlastensanierung (System for Cost Assessment for Clean-up of Contaminated Sites).

These systems, although specifically developed for the withdrawal of the Soviet troops from Germany, are recognized and used for other sites as well. They stood the test in a very difficult phase and guaranteed an even approach to all sites, thus providing a sound basis for the setting of priorities. The use of these systems has substantially helped to avoid excessive costs for individual site characterizations and assessments.

Since the problems in all Central and Eastern European states are so similar, the German Government has made available the systems and the software to run them to all interested countries. Representatives from most of these countries have been invited to training seminars in Germany in which they are given detailed information on how to operate the system and how to work with its subsystems. The idea behind this transfer of knowledge was to enable other countries with similar problems to make good use of a cost-efficient methodology, since the technical capability to deal with the problems is available almost everywhere.

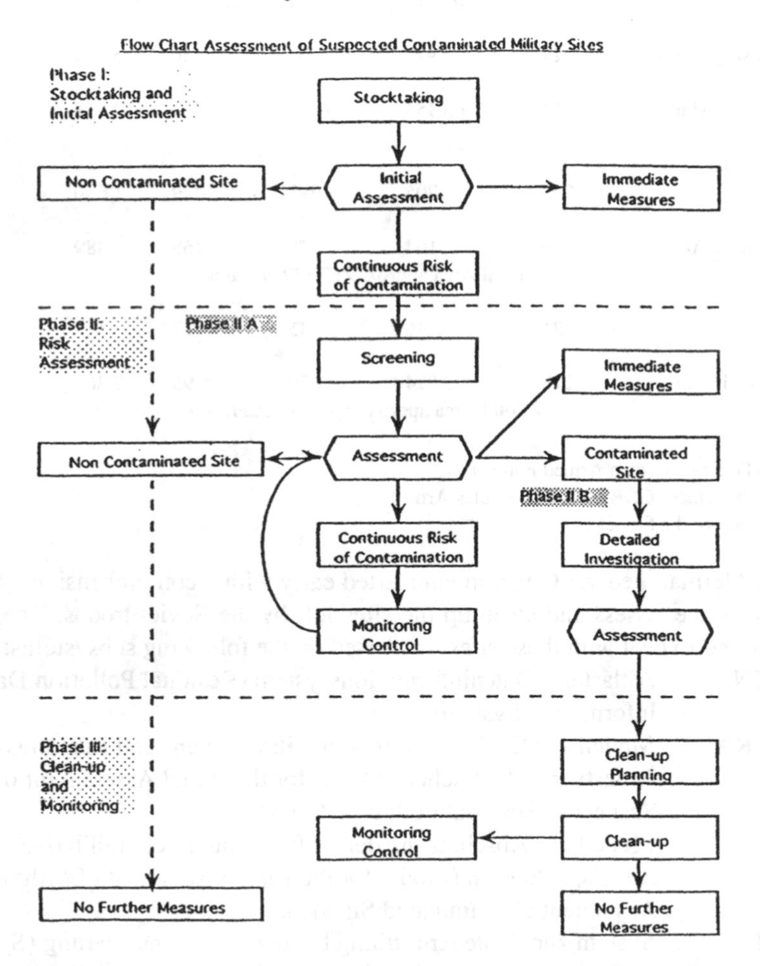

2.2 STATE OF ASSESSMENT AND CLASSIFICATION, RESULTS OF THE PROJECT

Risk assessment of the former military sites of the Soviet Army (WGT-West Group Troops) in Germany was concluded in 1996.

The extent of the problems in Germany is illustrated by figures showing what has been done and achieved following the withdrawal of the Soviet troops from Germany:

- 1052 installations have been subjected to the assessment procedure.
- Investigations covered an area of about 231000 hectares.
- The initial assessment documented 33000 suspected problematic sites.
- After risk assessment about half of these sites (19000) could be taken out of the investigation (risk free).
- At 10% of the sites (around 3.800) a potential risk existed that required immediate action, mostly gathering of ammunition, security measures, removal of wastes or scrap, removal of surface or soil contamination with hydrocarbons and the like.
- At about 400 sites (approximately 10 % of suspected risk sites and little more than 1% of the total of suspected problematic sites according to the initial assessment), additional investigations were necessary. Particularly this figure shows that the approach is a big money-saver as compared to individual investigation.
- Heavy environmental contamination was only found in a limited number of cases, but most frequently at air bases. Hydrocarbons were the most frequently detected pollutant. Although the number of heavily polluted sites was small the cost for remedial action at these sites can be extremely high.

It should be be noted here that the exercise of the WGT-Project was limited to the environmental aspects of the former military sites. It did not assess other damages (state of housing and facilities, etc). The Federal Government spent about 100 million DM on having the assessment done.

2.3 REUSE OF MILITARY SITES

What happened to these sites in the meantime? The conversion of military sites to subsequent civilian use is a very important task to keep costs for the public budgets as low as possible. The sale of former military sites to the private sector is one of the most important means to recover the costs incurred in the assessment and clean-up of the sites. Initially, all of the sites handed over by the military fell into the ownership of the Federal Republic of Germany. Most of the installations that were not needed for subsequent use by the Federal Government or its agencies have been handed over to the Federal States (Länder). Of the 1052 installations to which the Federal Government acquired ownership, 577 were transferred to the Federal States by 1996. Two of the Federal States have refused to accept sites in their territory, thus leaving the sites under the jurisdiction of the Federation.

All involved parties (States and Federal Government) are trying to sell these properties to the private sector, and here too a contamination assessment of the property is of prime importance. According to German law, the owner of a property must bear the risk of damages due to contamination and may be compelled to carry out clean up at his own expense. This situation makes it clear that property for which an assessment has not been made can hardly be sold. Even when an initial assessment has been made, investors will not take the risk unless substantial rebates are given to compensate for the residual risk - regardless how small this may be. The lack of an environmental assessment of military sites is thus an obstacle to attracting investments. We discovered that this is a quite universal phenomenon. It should, however, not be overestimated. Attractive sites for businesses or housing or other civil uses do find their market even when there is severe contamination. The environmental status of a military site to be put on the market is only one factor amongst others. The importance of this factor depends very much on the requirements for clean up and on who has to bear the cost. In Germany the Ministry of Finance follows a flexible policy when privatizing former military sites.

There is no sufficiently detailed overview available on the sales situation with regard to former military sites. Demand is strong in developed areas where the use of land offers good profits. The situation is quite different in more remote areas where most of the larger military facilities (training grounds, airfields etc.) are situated.

With respect to investments on former military land, the situation in Germany may not be comparable to other countries in Central or Eastern Europe. There, in some particular cases the specific infrastructure of former military sites may be viewed at as a bonus to attract investments. On the other hand, many countries are not densely populated. This aspect may have a negative influence on the demand for land in general and therefore also for former military sites. It certainly influences market prices. The German experience suggests that the geographical situation of the facility is the most significant factor. In this respect the infrastructure and economic strength of the surrounding area serves as an incentive for reuse. Even well-equipped former military sites which seem to be fit for reuse do not find a market in economically weak regions.

3. International Aspects

3.1 STUDY ON THE INTERNATIONAL STATE OF THE ART

International exchange and cooperation have been ongoing during the whole transition process, both on a bilateral and a multilateral level. This made it very important to compare the different approaches adopted by the different countries.

To obtain a reliable overview of the most important schemes used by other countries, the German government contracted a comparative study[1] on how countries dealing with the problems of contaminated military sites approach key issues such as:

- Definitions
- Legislative framework, administrative systems,
- responsibilities
- Financing
- Profile of contamination
- Registration of suspected contaminated sites

1 UBA Texte 5/97 "International Experiences and Expertise in Registration, Investigation, Assessment and Clean-Up of Contaminated Military Sites", study executed under the Environmental Research Plan of the Federal Ministry for the Environment, Nature Conservation and Nuclear Safety by Dames & Moore GmbH & Co, Berlin, 1997

• Methods for investigation of contaminated sites
• Clean-up attainment goals
• Clean up technologies
• Prioritizing of sites
• Health and safety aspects
• Cost estimates

as well as their current priorities for research and development. This study is made available free of charge upon request.

3.2 NATO-CCMS STUDY

In spring 1994, it was agreed to conduct a NATO-CCMS Pilot Study on the reuse of former military land. The study in which 19 countries participated (8 NATO-countries and 11 NACC-countries) is co-chaired by the United States and Germany. In the Pilot Study Group, military experts work together with experts from administrations for the environment. Phase I of the study was finalized in December 1996.
5 subgroups covered the following aspects:

• Identification and Assessment of Military Land for Reuse
• Remediation Strategies
• Analytical and Information Support
• Program Implementation
• Financial Resources.

The Pilot Study Report was finalized at the beginning of this year. It will be available soon after publication and can be ordered from NATO-CCMS.

In October 1997 NATO-CCMS decided that the study should be continued in a second phase. Here practical cases of conversion of military bases would be addressed. The initial meeting of Phase II was held on April 28 - May 2, 1997 in London. A number of military sites in NACC countries are currently investigated for their reuse potential.

Germany strongly believes that NATO-CCMS has to play an important role in exchanging the experiences of its members with respect to remediation and reuse of former military sites and in reaching out to distribute information on the technical, economic and management aspects of this task.

4. Need for Research, Open Questions

There are a number of issues which so far have not been fully addressed, either on the national or international level as well as a number of questions that cannot be fully answered due to a lack of sufficient knowledge.

4.1 CONSEQUENCES OF FINANCIAL CONSTRAINTS

The remediation of contaminated sites, either in the civilian or military sectors, faces increasing constraints nearly everywhere. Most of these constraints are of a financial nature. The ongoing development of economic globalization places national budgets under stress and has led to readjustments in government expenditures.

With regard to the remediation of contaminated military sites this will mean that even greater efforts must be made to arrive at a justifiable cost/benefit relationship. German efforts in the past have already had this goal in mind. The evaluation system which was presented above aimed at reducing the cost of site assessment by using a simple (common sense) approach to arrive at comparable data for all sites in question and to allow decisions on further activities (risk assessment and remedial action) to be made on a comparable data assessment.

Overall experiences in using this system are positive, and it works fairly well in practice. This is the reason why this system is accepted by all Federal or State authorities in Germany dealing with this matter. The simplified version used during the withdrawal of the Soviet troops has been refined since time is not that pressing any more. The phase of risk assessment has been split into two parts: The primary assessment following the scheme of MAGMA will be followed by a detailed investigation if the initial assessment shows that the site is contaminated. This secondary investigation will provide more information on the grade of contamination for subsequent clean-up steps which may be necessary. Remediation of contaminated sites can be a very costly operation. The costs incurred are basically a function of the following:

- The remediation goals (which limits have to be reached)
- the remediation methods chosen (on site/off site treatment, biological/ "technical" remediation)
- the duration of the remediation process.

Comparatively few military sites as compared to the overall figure are undergoing remediation[2] . In Germany for example only an estimated 10-15% of all problematic military sites are currently subjected to remediation processes. This is about 1 - 1.5% of all military sites which have been investigated.

There are multiple reasons for the relatively small number of remediation cases which are not specific to military sites. The situation is very much the same at contaminated industrial sites. The most obvious reason is the high cost of clean-up of contaminated sites. Additionally, unless there is an actual threat to human health or to the environment (namely groundwater resources requiring immediate action), there is no strict setting of priorities for clean-up procedures. Finally clean-up decisions are made on a case-by-case basis, i.e., cleanup is only likely to take place when it is a prerequisite for the reuse or sale of the property in question or when there is noticeable public pressure.

The technology for remedial action is under permanent development.

[2] Remediation as used here signifies soil and groundwater remediation from pollutants, as opposed to the removal of ammunition and small scale pollution (waste and scrap removal)

Uncertainty, next to the high cost of clean up operation, is the most obvious constraint. An emerging trend indicates that technological solutions (pumping and/or treatment in more or less sophisticated engineered facilities) should in the future not be the technology of choice primarily due to their very high costs. This is particularly true for off-site as compared to on-site treatment. The preferred choice of treatment clearly goes to on-site biological treatment wherever this option exists. The technologies used are developing rapidly. Subsequently, whenever a decision has to be made, it is advisable to check the latest state-of-the-art in on site remediation technologies.

The large number of contaminated former or active military sites works as a strong driving force to apply the phenomenon of "natural attenuation" since these sites are generally outside of heavily populated areas, often difficult to approach and sometimes not fit for reuse nor in any priority within reuse programs. In such cases, a "wait and see" decision often is the only viable option when there is no immediate threat to human health or the environment, in particular for water resources. This philosophy, however, is only justifiable when the contamination is being monitored. "See" is the important verb. In other words not taking concrete remedial action means to monitor the development at the site. This can also be very costly and may be necessary over long periods of time.

4.2 MAJOR FIELDS FOR RESEARCH

In the United States there is a growing interest in the "wait and see" option, i.e., ***remediation by natural attenuation***. This approach is also attracting more and more interest in Germany. The so called "intrinsic remediation" by "natural attenuation"[3] is neither new nor revolutionary. The process has been known for almost 50 years. The consequences of the existence of the phenomenon of natural attenuation are however difficult to assess since there are still a number of open scientific questions which require answers before this technology can be accepted as a sound and reliable method to deal with contaminated sites. Hence Germany has been thus far been very reluctant to base its official policy on an approach which *also* accepts natural attenuation or intrinsic remediation given that there is no sufficient operating history on using this method.

There is nothing wrong in taking natural processes into consideration for the clean-up of contaminated sites. It would in fact be an inexcusable mistake not to do it. However, the uncritical overestimation of attenuation processes ***without sufficient scientific evidence on their mechanisms*** is cause for concern. It is quite clear that lacking reliable knowledge of the processes which occur during the "wait and see" period, the utilization of this "method" could be running substantial risks, especially if the sites are not being regularly monitored.

With respect to military sites only a handful of relevant possible pollutants require special attention. This should make it feasible to elaborate an assessment scheme for long term transport of the pollutants in the environment (soil,

---

[3] This means natural contaminant degradation with no human intervention other than monitoring

groundwater). A scientific basis is the primary requirement to justify to the public that a proven pollution will not result in further damages in the future if active measures are not taken. To this end we feel that the development and/or refinement of prognostic tools to allow a forecasting of the behavior of pollutants on a scientifically sound basis should be particularly emphasized. Policy-oriented research to fill the gaps in knowledge commands a high priority. It will, however, not solve the *political issues* of intrinsic remediation which may ease the financial responsibility of those who carelessly pollute soil and groundwater. This strictly political issue should, however, not obstruct necessary research activities.

Answers are urgently necessary for the following questions:

1. On what substances (other than mineral oils and other hydrocarbons) can this "method" be used?
2. What are the process products, are there metabolic reactions, and is there a transformation into other substances which may be more hazardous than the contaminant itself?
3. What is the influence of the *concrete* natural conditions, i.e. properties of the medium and transport of pollutants vs. degradation rate.

If these questions can be answered in a satisfactory and scientifically sound manner, there should be an opening to use this "phenomenon" as a practical and very cost efficient *tool* for site remediation and in so far as possible, develop it into a *treatment method* using techniques that support the natural process.

Another important issue is the reuse of former military lands. The size reduction of armed forces throughout Europe has resulted in the closure of many military facilities, which, wherever possible should be recycled for civilian use. The burden of this task is manifold:

- The sale of military installations to private investors requires in most cases an assessment of the environmental condition of the respective facility since the owner will have to take over responsibility for the status of the site.
- Installations taken out out of service require permanent maintenance and care which can be very costly.
- Military installations are often "out in the middle of nowhere". This is frequently an obstacle for the sale of facilities and installations.

In the past, the closure of facilities came like a "snow slide". Too many facilities where released in too short time to allow proper management of sales. In the future this process will be slowing down. It would therefore be very helpful to develop a strategy for reusing former military sites, taking into account the experiences gained in the last few years. NATO-CCMS is investigating this matter with the pilot study on "Reuse of Former Military Lands". Furthermore additionally research may be necessary, particularly on the economics of the sale of former military installations and on the mechanisms pertaining to it and which should be considered in preparing such sales. It would be extremely helpful to those responsible to have a guidance document to properly assess not only the market value of property but also the obstacles and chances for sales of military lands comprehensively.

5. Outlook

The conversion of military sites for civil use and the environmental issues connected with this process will remain a high priority for the coming years. The assessment of the problems is almost complete. However, this does not mean that they will be solved soon.

High costs and conflicting interests, particularly budget priorities are limiting factors for fast and easy solutions. An international exchange of experience and international cooperation will therefore play an important role in the future - also as a means of avoiding unnecessary costs in "reinventing the wheel". With reduced budgets not everything that should be done can be done in practice. In the technical and scientific field there is a noticeable shift from "high tech solutions" to bioremediation and from remediation technology to safeguarding techniques. Open questions with regard to the safety and effectiveness of low-cost remediation options should be answered soon. It may be anticipated that only more cost-effective technologies both for remediation and for the safeguarding and containment of pollutants will become the long term technologies of choice. To this end, the distribution and the exchange of information and experience is of prime importance. The NATO-CCMS framework offers excellent conditions for this.

GENERAL ENVIRONMENTAL SITUATION, RISK ASSESSMENT AND REMEDIATION OF FORMER MILITARY SITES IN EAST GERMANY

V. ERMISCH, P. SCHULDT
HGN Hydrogeologie GmbH
Rothenburgstr. 10/11
D - 99734 Nordhausen

Abstract

About 1,500 contaminated military sites - of which about 1,100 belonged to the former Western Group of the Soviet Troops and about 400 of the Bundeswehr - have been identified in East Germany since 1991. The assessment and remediation of contaminated military sites includes the following steps: inventory, preliminary assessment, investigation, risk assessment, and remediation.

The contamination potential is dependent upon the use of the sites. The risk is greatest in the case of fuel depots, airfields, military training areas, and barracks.

Petrochemicals, scrap metal and mineral wastes are the dominating contaminants, with soil and groundwater being the most frequently affected. Stabilisation and averting dangers are at the centre of efforts for the remediation of contaminated sites. Further remediation measures depend on the subsequent use of the given site.

1. Introduction

As per 1993/94, about 192,000 non-military sites suspected of having been contaminated from previous uses - thereof about 64,000 in East Germany - had been recorded in Germany. Table 1 shows the distribution of those sites among the five new (East German) Laender of the Federal Republic of Germany (compiled acc. to [1]).

Identification of contamination from military uses on the estates of the former Western Group of the Soviet Troops (WGT) and of the Bundeswehr (formerly of the People's Army of the GDR) was started in East Germany in 1991.

On the 1,052 WGT estates covering an overall area of 231,595 hectares, the inventory and assessment of sites suspected of being contaminated from previous uses was under the direction of Bundesministerium für Umwelt, Naturschutz und Reaktorsicherheit (Federal Ministry for the Environment, Nature Conservation and Reactor Safety) - project management: Industrieanlagen-Betriebsgesellschaft mbH, Ottobrunn. Table 2 shows the distribution of those estates among the new (East German) Laender of the Federal Republic of Germany.

F. Fonnum et al. (eds.),
***Environmental Contamination and Remediation Practices at Former and Present Military Bases*, 13–32.**

TABLE 1. Non-military sites suspected of being contaminated from previous uses, by Laender

Land	Number (absolute)	Number (percent)
Mecklenburg-Western Pomerania	11,692	18.3
Brandenburg	13,565	21.3
Saxony-Anhalt	14,715	23.1
Free State of Saxony	18,168	28.5
Free State of Thuringia	5,587	8.8
Total	**63.727**	**100.0**

TABLE 2. Number and size of WGT estates in several Laender (acc. to [2])

Land	**Number (absolute)**	**Number (percent)**	**Area (ha, absolute)**	**Area (percent)**
Berlin	10	1.0	79	< 0.1
Brandenburg	379	36.0	119,551	51.8
Mecklenburg-Western Pomerania	118	11.2	14,909	6.5
Free State of Saxony	170	16.2	16,233	7.0
Saxony-Anhalt	245	23.3	60,406	26.2
Free State of Thuringia	130	12.4	19,617	8.5
Total	1,052	100.0	231,595	100.0

Inventory and preliminary assessment of those estates resulted in the listing of a total of 33,750 sites with an area of about 5,700 hectares under suspion of having been contaminated, [2].

Within the framework of the contaminated sites programme of the Bundeswehr (Federal Armed Forces), which has been carried out for the East German Laender on behalf of Bundesministerium der Verteidigung (Federal Ministry of Defence) under the direction of Oberfinanzdirektion Hannover (Hannover Regional Revenue Office), altogether 1,900 suspect sites were identified on 412 out of a total of 1,058 estates of the Bundeswehr (as per 1994, acc. to [1]).

Former military production sites were not included in these considerations.

2. Methods of investigation of sites contaminated from previous uses

There is no difference between the exploration of contaminated military sites and that of non-military sites. This is understandable, since there are no differences in contamination potential or their safety problems, except for the specific features of military-chemical contaminations and their safety problems. However, once the remains from ammunition and explosives have been cleared away, this concern also fales away.

The investigation methodology can be separated into three steps.

- Inventory and preliminary assessment
- Investigation and risk assessment
- Remediation planning, remediation and monitoring

A flow chart of the process of suspect site investigations is shown in Fig. 1.

2.1. INVENTORY AND PRELIMINARY ASSESSMENT

The step of inventory and preliminary assessment is also known as the "historical exploration" of a contaminated site.

Historical exploration includes the study of files and documents, inquiries with authorities, evaluation of maps and aerial photographs, and interviews. Inspection on site with photographic documentation is an indispensable element of investigations at this stage.

As a result of the inspection, the suspicion of contamination is either confirmed or not. In case there is contamination, the risk potential emanating from the contaminated site is preliminarily assessed and the need for action is determined from further investigation and/or monitoring.

2.2. SITE INVESTIGATION

The stage of site investigation includes "orientational investigation" and "detailed investigation".

Orientational investigations provide a survey of the size of the risk potential and of the spatial extent of contamination and of the goods/media to be protected,

If the orientational investigations have shown that it would be necessary to continue, further detailed investigations must provide comprehensive information about the kind and spatial extent of contamination on the site and in the goods/media to be protected.

The following steps help to obtain the required in-depth knowledge of a contaminated military site:

- Geophysical measurement
- Drilling of bore holes
- Setting up of observation wells
- Sampling and analysis of soil air, soil, groundwater and surface water
- Computer models simulating the transport of contaminants in soil and groundwater (only at the stage of detailed investigation and not in all cases)

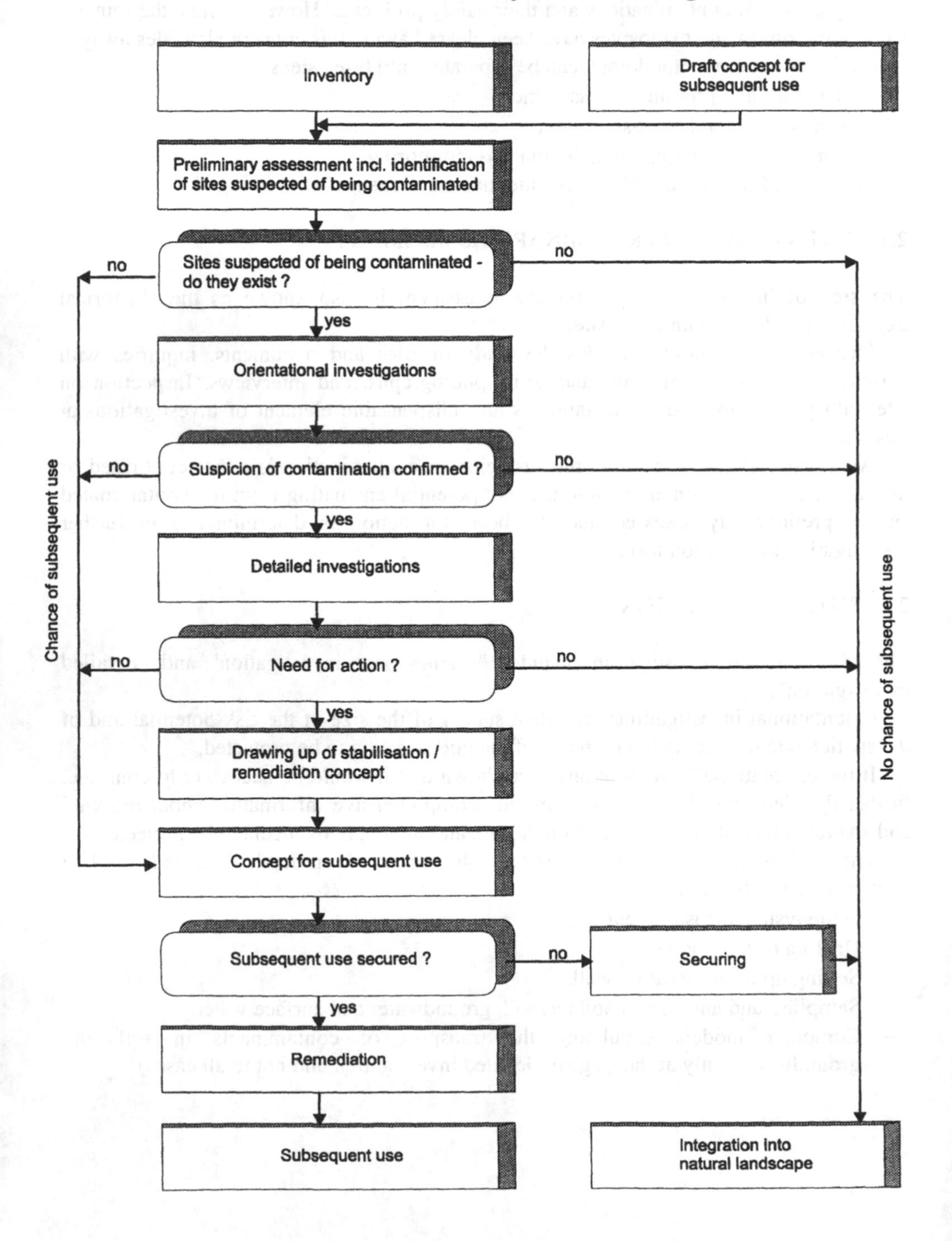

Figure 1. Flow chart of site investigations

2.3. RISK ASSESSMENT

Work at any step of the investigation of a contaminated military site (inventory, orientational investigation, detailed investigation) concludes with an assessment of the potential risks emanating from the contamination towards the factor to be protected. This assessment become the starting point for drawing up the necessary measures for monitoring, stabilisation and remediation.

The factors most at risk from contaminated military sites include:

- human health (top priority)
- groundwater
- soil

Toxicological assessments are the decisive element when it comes to assessing the hazards to human health that would emanate from a site suspected of being contaminated. Human health will be at risk if - under a realistically assumed use-related exposition - the permissible daily exposure to a harmful substance was continuously exceeded.

This, however, needs toxicologically substantiated criteria for assessment (orientational values).

As the toxicological effects of many of the substances contained in contaminations from previous uses of a site are known either insufficiently or not at all, certain safety factors have to be applied on risk assessment. Those specific features, however, are not discussed further here.

In the following, we shall go into more detail about the criteria to be applied when assessing the risks for groundwater and soil, the two media most frequently affected.

2.3.1. *Groundwater*

In Germany, about 70 percent of the drinking water comes from groundwater. Hence, the protection of groundwater against contamination is particularly important.

This has to be taken into account when assessing the contamination-borne risk to groundwater. The criteria underlying risk assessment are shown in Table 3.

As the many site-specific data are often of limited availability and their procurement would require extensive and costly investigations, in many cases the use of generalised parameters for assessment of the risks to groundwater has been the only recourse.
Such parameters include:

- Depth of groundwater table
- Thickness of aeration zone
- Shares of cohesive layers in the aeration zone
- Permeability of aeration zone and aquifer
- Slope of groundwater surface
- Pressure situation in the aquifer (confined or unconfined)

2.3.2. *Soil*

In Germany there are many lists of values for the assessment of soil contamination, that are used for risk assessment (Berliner Liste, Hamburger Liste, Niederländische Liste, etc.).

The problem with many of those lists is that the use- and good/medium-related values are not always comparable, and in some instances different numbers of parameters are used. This does not simplify risk assessment or to make it easier.

The dilemma shall be remedied by a Federal law on soil protection (Bundes-Bodenschutzgesetz, draft of 25 September 1996 [3]). This law establishes that the safeguarding of the soil is a necessity of life, as well as the future use of the soil is not impaired.

It is essential to make provisions ensuring that the soil's ecological efficiency not get overdone by material and physical effects, and there is the obligation to remedy if the soil holds hazards for man and environment.

There are plans for the enactment of sub-statutory rules in the form of a "Bodenschutz- und Altlastenverordnung" (decree on soil protection and handling of contaminated sites) providing a national standard for precautionary and remedial requirements based on use-specific soil data.

2.3.3. *Risk assessment for WGT estates*

With the large number of WGT estates having been abandoned in East Germany between 1991 and 1994, assessment in line with the Laender-specific regulations was for reasons of money and time not possible for the many places suspected of being contaminated.

With a view to ensuring a uniform approach founded on a comparable data base, a computer-aided formalised concept was developed - on behalf of Ministerium für Umwelt, Naturschutz und Reaktorsicherheit - for the preliminary assessment (MEMURA) and risk assessment (MAGMA) of contaminated military sites [2]. MEMURA (Fig.2) is a relative value method related to goods/media to be protected. The overall risk r4 express a measure for the need for action related to the assessed media to be protected.

MAGMA (Fig.3) is an absolute value method that only measurements processed. The risk assessment based on site investigations. The MAGMA value M describes the degree of risk by the contaminated site to the man.

2.4. REMEDIATION

Remediation of contaminated sites means the implementation of administrative and technical measures which ensure that, after specific measures have been applied, the site does not hold any hazards to human life and health or to the natural environment in connection with existing or planned uses (Der Sachverständigenrat für Umweltfragen, 1995, [1]).

TABLE 3. Criteria for determination of the potential risks to groundwater from a contaminated site

Contaminant-related criteria	Site-related criteria	
	Geometry of contamination	Site conditions
– State – Solubility – Mixability – Interaction with other substances – Viscosity – Density – Bond type – Biological and chemical degradability – Eco- and human-toxicological properties	– Punctiform – Linear – Areal limited diffuse	– Contaminant concentration – Amount of precipitation – Position to surface water – Position to groundwater (aeration zone, in groundwater surface zone, in aquifer) – Structure and properties of aeration zone and aquifer (homogeneity, permeability, particle size distribution, porosity, ion exchange capacity, clay minerals content, organic matter content, pH value, redox potential, temperature micro-organisms, etc.) – Groundwater flow velocity – Information about regional groundwater dynamics (recharge area, transit area, discharge area), distance from groundwater catchment facilities

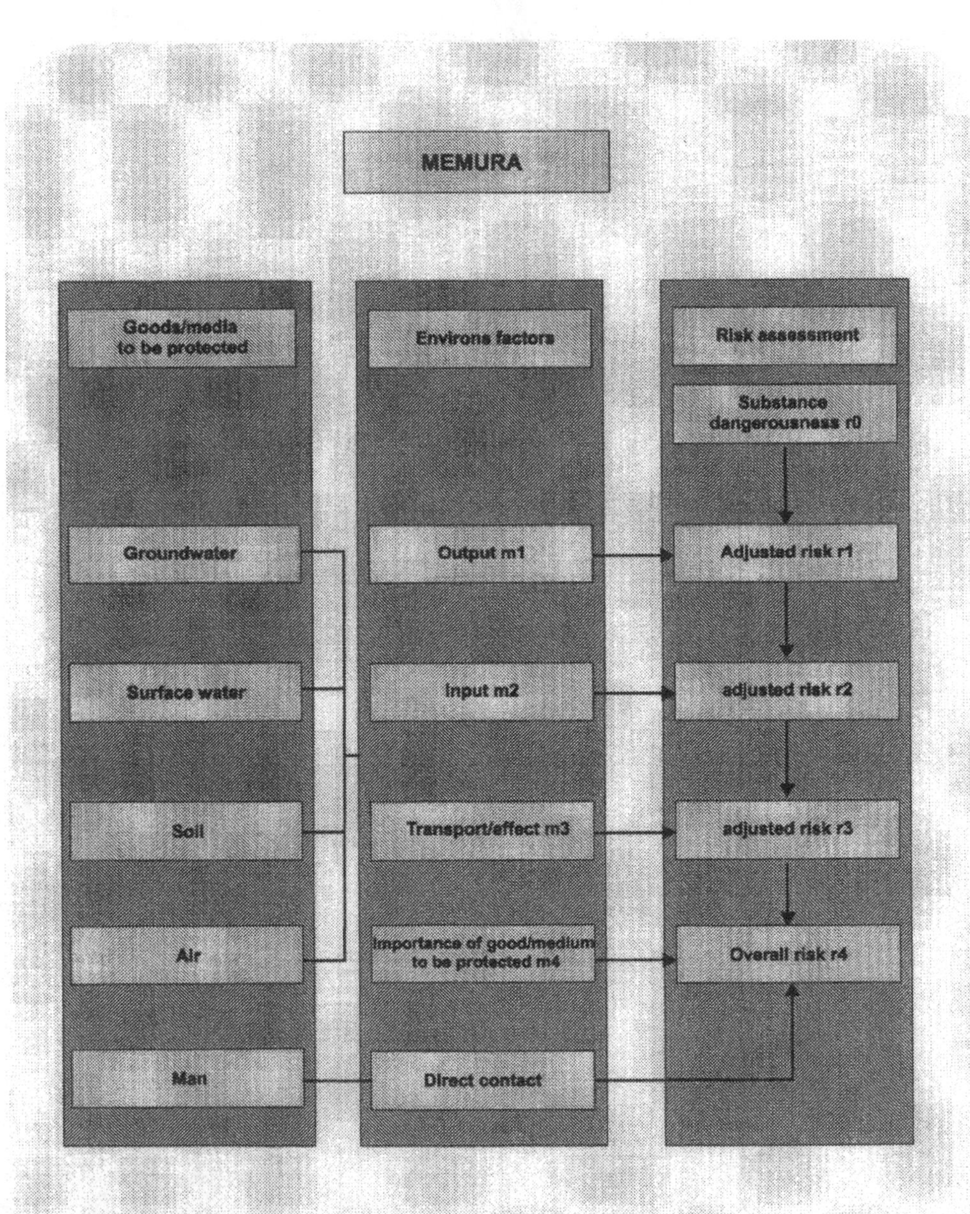

Figure 2. Structure of the model MEMURA acc. to [2]

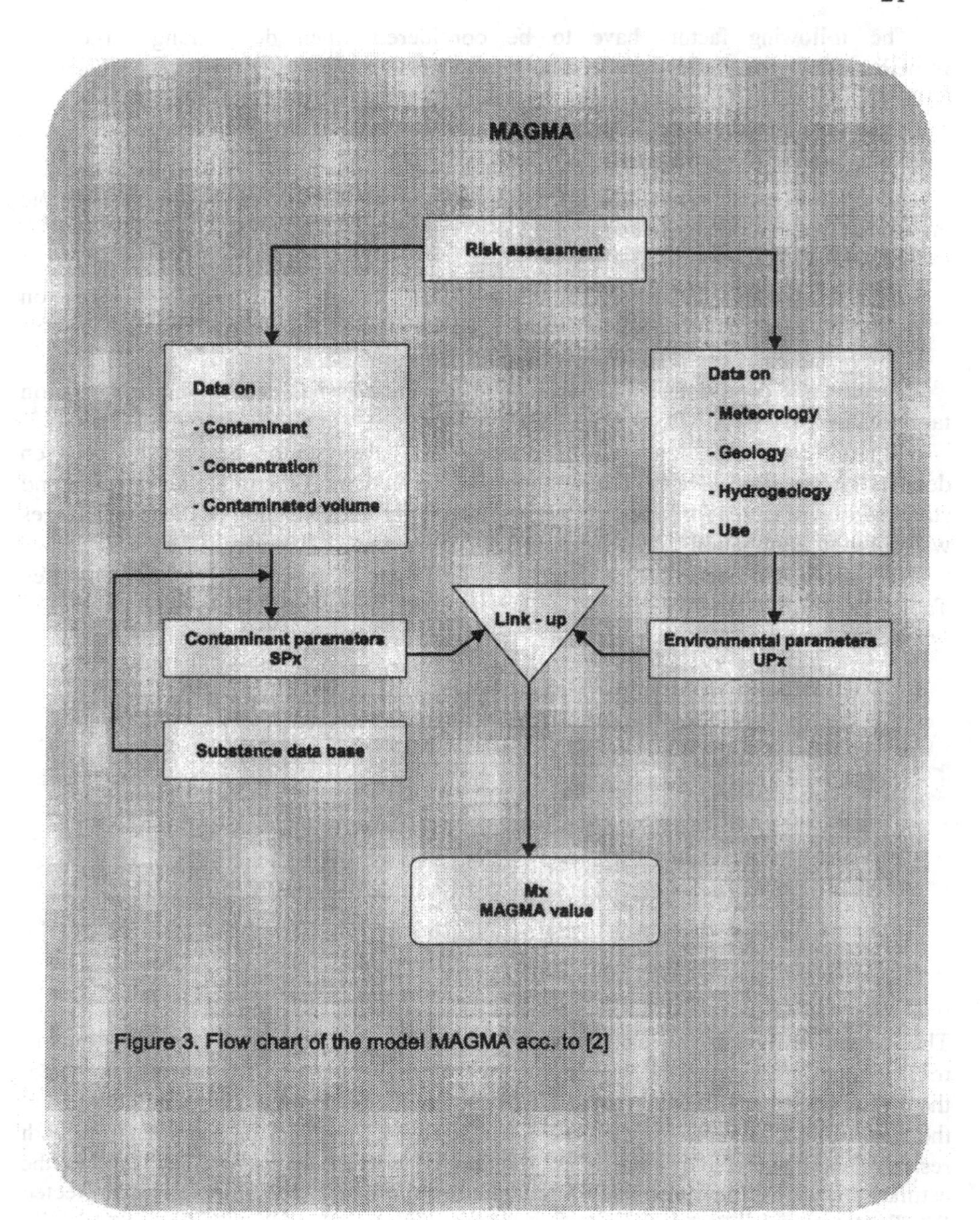

Figure 3. Flow chart of the model MAGMA acc. to [2]

The following factors have to be considered when determining targets of remediation:

- The goods/media to be protected and their functions on the given site
- Site use in line with planning permission
- Possibility of modifying remediation targets with advanced state of knowledge

The criteria for determination of the remediation target value by the responsible environmental authority [1], are as follows:

- Falling below the risk threshold
- Weighing up of technically suitable methods (maximum avoidance of risks), on the one hand, and economic efficiency, on the other
- Environmental compatibility of the remediation measure

Specific site conditions should always be considered when determining remediation target values.

Regarding remediation, a distinction has always to be made between decontamination measures for elimination or reduction of harmful substances, and stabilisation measures for long-term prevention of further spreading of those substances without their elimination.

These two approaches differ greatly in the financial expense of remediation (see Table 4).

TABLE 4. Cost estimates for remediation measures acc. To JESSBERGER [4]

Remediation measures	Remediation costs
Stabilisation measures (encapsulation)	
- surface sealing	DM 175.-/m²
- vertical sealing (15m deep)	DM 300.-/m²
Decontamination measures	
- "general" decontamination	DM 450.-/m³
- "specific" decontamination	DM 750.-/m³

The specific features of the remediation of contaminated military sites compared to non-military ones ensue from the large number of places suspected of being contaminated, the different uses and the different kinds of contamination that may be found on one and the same estate. In some places there is the additional problem of contamination with residual ammunition and, sometimes, remains of warfare agents. Table 5 shows the estimated costs for the remediation of contaminated military sites, illustrated by selected WGT estates.

Due to the large number of WGT estates abandoned in East Germany, not all of these estates can be transferred to economic uses. One option might be to make those places available for nature conservation (military training areas, in particular).

TABLE 5. Estimated costs for the remediation of select WGT estates, according to [5]

Estate	Size [ha]	Contaminants	Contaminated medium/type of waste	Quantity [m³]	Method 1)	Costs 2) [DM]
Hospital	11	HC*, lead	Soil Old deposits Building rubble	60 4,100 100	Landfill Microbial On-site soil cleaning	697,000.-
Fuel depot	10	HC	Soil Groundwater Building rubble	6.900 100	Soil air extraction Microbial on-site soil cleaning Percolation drainage	6,328,500.-
Garrison	32	HC,CHC,*PA HC*, heavy metals	Soil Building rubble Old deposits	6,700 9,700 77,000	Landfill Soil washing	29,800,00.-
Firing range	12	HC, CHC, heavy metals, phenols	Soil Old deposits	375 1,500	Landfill Microbial on-site soil cleaning	451,500.-
Barrack	50	HC, CHC, PAHC, heavy metals, phenols	Soil Old deposits	130 50	Soil washing Landfill	98,000.-
Garrison	14	HC, heavy metals	Soil Building rubble Scrap metal	11,000 2,000 6,000	Landfill Microbial on-site soil cleaning	11,289.800.-

1) The remediation methods were selected with the following aspects in mind: feasibility, present condition, time, legal measures, costs.

2) Cost-affecting factors considered: establishment of remediation site, site equipment, reduction of emission, safety provisions for workers and pollution control, demolition work, earthwork, drilling work, de-watering, protective and restriction measures, safeguards to interrupt contamination paths, de-contamination of soil and/or groundwater, treatment of seepage, process water, etc., treatment of exhaust air, indoor air, etc., disposal and depositing including transport and conditioning, accompanying analysis, restoration of terrain surface (re-naturalisation), after-care and monitoring.

* HC - hydrocarbons, CHC - chlorinated hydrocarbons, PAHC - polycyclic aromatic hydrocarbons

3. Results

3.1. WGT ESTATES

The following results of the various steps of the investigation of WGT estates were taken from the final documents on the project "Ermittlung von Altlastenverdachtsflächen auf den Liegenschaften der Westgruppe der sowjetischen Truppen (WGT)" (Determination of sites suspected to be contaminated on estates of the Western Group of the Soviet Troops) [2]. Investigations carried out by our company (HGN) within the framework of that project are included in [2].

Altogether 33,750 suspect sites with 5,708 contaminated hectares were identified within the framework of inventory and preliminary assessment. Military training areas and firing ranges accounted for 35%, and barracks for 34% of the sites suspected of being contaminated. With reference to the size of the contaminated area, military training areas accounted for 69%.

Altogether 18,888 suspect sites (56%) can be cleared or are of minor relevance to the environment (MEMURA rating < 3).

On the other hand, 14,862 sites were found to require further investigations and/or remediation (MEMURA rating ≥ 3; 44%).

Of those sites suspected to be contaminated, a total of 4,048 sites (total area: 2,021 hectares) were identified (MEMURA rating ≥ 10) as needing urgent action (consideration of measures to ward off hazards, further investigations urgently required). About fifty percent (1,999) thereof were military training areas and firing ranges, accounting for 70% (1,406 hectares) of the area.

Regarding the main sources of contamination (see Table 6), disorderly depositions rank first (13,358 = 40%), with handling of fuels and lubricants ranking second (7,771 = 23%).

Suspect sites where ammunition and explosives had been handled account for only 3% (948) by number, but for 35% (2,022 hectares) in terms of area.

TABLE 6. Suspect sites and contaminated areas by groups, acc. to [2]

Group	Suspect sites		Contaminated areas	
	Absolute	Relative	Absolute	Relative
	[-]	[%]	[ha]	[%]
Handling of fuels and lubricants	7,771	23	531.5	9
Handling of other materials	1,075	3	14.4	< .5
Handling of explosives and ammunition	948	3	2,021.8	35
Buried materials and scraped grounds	2,573	8	650.1	11
Disorderly depositions	13,358	40	1,338.6	23
Scrap yards	2,668	8	743.0	13
Residues from incineration	2,230	7	106.8	2
Sewage, sewage sludge and fecal matter	1,249	4	115.7	2
Other suspect sites	1,878	6	185.7	3
Total	33,750	100	5,707.5	100

A special point should be made of the fact that groundwater is endangered or affected in 32.8% (11,079 suspect sites) of the cases. Moreover, 53.8% of the suspect sites are situated within, or immediately adjacent to, drinking water protection zones or reserve zones.

The former WGT estates were found to hold an estimated 2.9 million tonnes of waste (Table 7). Almost half of it (1.4 million tonnes) account for mineral wastes (ashes, building rubble, excavated soil, coal dust, asbestos-containing wastes) which, however, are of minor importance as to their environmental relevance. This applies also to the about 750,000 tonnes of scrap metal identified on those estates.

Residential wastes (456,000 tonnes) are a greater hazard to both soil and groundwater.

Petrochemicals, organic chemicals, acids, lyes, salts, ammunition and explosives are found in much smaller quantities. Nevertheless, their potential hazard to the environment is significantly higher.

TABLE 7. Amount of contaminants on WGT estates (estimates), acc. to [2]

Kinds of waste material	Amount of contaminants [t]	Percent in total amount [%]
Mineral wastes	1,370,313	46.5
Waste metal	749,907	25.5
Residential wastes	456,414	15.5
Petrochemical wastes	126,266	4.3
Organic wastes	100,394	3.4
Change of terrain (buried materials, etc.)	83,602	2.8
Plastics and textile wastes	23,265	0.8
Explosives and ammunition	19,823	0.7
Organic chemicals	8,176	0.3
Acids, lyes, salts	6,126	0.2
Total	2,944,286	100.0

Immediate measures

Under the WGT estates project, immediate measures had to be taken

- when there was an immediate risk to the general public (people who might come into contact with the danger spot).
- when a substantial amounting of contamination was spreading so quickly such that there was the risk of immediate danger in the not too distant future.
- when this was required for reasons of local (Laender) law.

Against this background, a total of 16,932 immediate measures were taken on 9,862 suspect sites for warding off direct or indirect hazards and/or as precautionary measures against such hazards.

Some 91% of the immediate measures taken were non-structural ones for warding off direct danger to man. A further 9% involved structural activities, mainly to prevent groundwater contamination or to eliminate existing groundwater contamination. Tables 8 and 9 summarise the kinds of immediate measures taken along with the kinds of environmental hazard involved.

TABLE 8. Types of immediate measures, according to [2]

Types of immediate measures	Number of immediate measures
Non-structural	14,882
• Protective and restrictive measures	4,851
Recovery, de-activation and disposal of ammunition	1,126
Disposal of contaminants	7,228
Clearing	1,677
Structural	1,480
• Continuing investigations	1,162
Stabilisation	107
Decontamination	86
Translocation	125
Total	16,362

TABLE 9. Types of acute environmental hazards, according to [2]

Kinds of hazard	Number of suspect sites concerned		
	Non-structural immediate measures	Structural immediate measures	Total of immediate measures
Danger of explosion	1,782	183	1,965
Fire hazard	359	27	386
Risk of acid burn/poisoning	1,029	37	1,066
Danger to groundwater	214	385	599
Danger to soil/groundwater	4,163	351	4,514
Danger of fall/collapse	692	17	709
Danger of contamination	235	2	237
Miscellaneous	226	160	386
Total	8,700	1,162	9,862

3.2. BUNDESWEHR ESTATES

The contaminated sites programme of the Bundeswehr with its three steps
– inventory and preliminary assessment
– investigation and risk assessment
– remediation planning and remediation

is still at the stage of implementation. Inventory and preliminary assessment are most advanced on the estates in East Germany. In 1994, out of the total of 1,900 suspect sites

396 had been preliminarily assessed, 133 had been investigated and risk-assessed, and remediation measures and immediate action had been taken for 24 of those sites [1]. Unlike the WGT estates, however, no summary assessments are available so far.

According to estimations based on work (inventory and preliminary assessment, investigation, risk assessment) of our company within the framework of that contaminated sites programme, all in all the environmental hazards from those estates are less serious than those emanating from the WGT estates.
The average number of suspect sites per estate is significantly smaller, and the amounts of waste materials found are much smaller, too.
Most at risk are soil and groundwater.

The vast majority of sites suspected to be contaminated from previous uses comes under:

- fuel depots and filling stations
- heating oil depots
- paint, dye and chemicals depots
- car wash stations
- vehicle parking grounds
- repair and maintenance facilities

Hydrocarbons are the predominant contaminants.
Remediation planning and remediation proper are decisively influenced by the fact that most of the estates will continue to be used for military purposes.

3.3. CASE STUDIES

3.3.1. *Garrison (barracks of motorised unit)*

Description of estate:

- Edge-of-town location with adjacent residential quarter and commercial areas
- Size: about 15 hectares
- Military use since 1935, used by WGT from 1945 to 1993.
- Earmarked for subsequent use by police

Investigations made:

- Inventory and preliminary assessment
- Orientational and detailed investigations including risk assessment
    - Setting up of seven groundwater observation wells
    - 110 bore probings
    - Collection and analysis of 14 groundwater samples, 340 soil samples and 12
      samples from residues in the sewer system
    - Organoleptic assessment of the structures

Results of risk assessment:

- Thirteen suspect sites
- Contaminants found: hydrocarbons, heavy metals (mercury, cadmium, lead, zinc)
- Affected medium: soil (close to surface), amount of contaminated soil: 10,770 m³ contaminated with hydrocarbons and 3,430 m³ contaminated with heavy metals
- No hazard to groundwater catchment or drinking water protection zones
- Partial contamination of structures
- Contaminated solid residues in the sewer system

Suggested remediation:

- Parts of the contaminated soil with IR-determined hydrocarbon concentration > 1,500 mg/kg dry matter to be cleaned by micro-biological degradation of contaminants, using specific facilities (on-site or off-site).
- If de-contaminated soil is re-introduced on site, a residual IR-hydrocarbon concentration of ≤ 500 mg/kg dry matter is tolerable.
- Contaminated soil with heavy metal concentrations higher than the C-value of the Niederländische Liste gets remedied (off-site by soil washing) or stabilised (depending on the specific use intended for the given site).
- Residues in the sewer system get disposed in an off-site unit.
- Scrap metal gets disposed in an off-site unit.
- Contaminated building rubble gets disposed of off-site during reconstruction work.
- Estimated cost of remediation: DM 1.2 to 1.6 million.

3.3.2. *Airfield*

Description of estate:

- Forest, farmland and - in the southern part - surface water form the boundary of the airfield.
- Size: about 520 hectares
- Military use since 1936; used as military airfield by WGT from 1945 to 1993
- Subsequent use: not yet decided

Investigations made:

- Inventory and preliminary assessment
- Orientational investigations
  - 130 bore probings
  - 35 machine bores for setting up 55 groundwater observation wells
  - Soil and groundwater sampling and analysis for hydrocarbons and BTEX
  - Soil air analysis

- Detailed investigation of a partial area
    - 70 bore probings and setting up of 60 groundwater observation wells
    - Groundwater sampling and analysis for hydrocarbons and BTEX

Results of risk assessment

- 99 suspect sites with very different types of contamination from previous uses and very different risk potentials
- Main sources of contamination: kerosene handling station (emptying of tank wagons), main fuel depot, pre-takeoff line (fuelling of aircraft) with about 20 hectares of contaminated ground
- Contaminants: hydrocarbons, BTEX, polycyclic aromatic hydrocarbons (to a very small extent)
- Affected goods/media
    - Soil
    - Groundwater (floating kerosene, depth of layer up to 4m), no immediate danger for drinking water catchments
    - Surface waters (kerosene entering the water bank sector)

Remediation measures

- Hydraulic measures for groundwater remediation in the range of the kerosene handling station
- Sinking of 21 wells to extract the kerosene phase on the groundwater (ca. 11,000 m^2)
- Erection of a hydraulic collector channel to prevent more kerosene entering the surface water
- With these measures, 390 m^3 of kerosene were retrieved in the course of six years
- Since the termination of hydraulic measures in 1996, a groundwater monitoring scheme with regular sampling has been going on; if required, hydraulic measures can be resumed
- Expenditure on remediation so far: ca. DM 3.2 million
- Detailed investigations still have to be made for the other main suspect sites (main fuel depot, pre-takeoff line)

3.3.3. *Military training area*

Description of estate

- Heathland, some parts with forest, watercourses running through, marsh and swampland biotopes, directly adjacent to barracks
- Size: about 7,500 hectares
- Military use since 1904; from 1945 to 1992 used by WGT

Investigations made
- Inventory and preliminary assessment

Results of preliminary assessment
- More than 1,000 suspect sites
- Many depositions of numerous kinds, buried materials, fuel depot
- Environmentally relevant contaminants: hydrocarbons, BTEX
- 60% of suspect sites of minor environmental impatence
- Affected media: soil, groundwater (in most parts, however, only temporary groundwater flow), surface waters

Concept for remediation/subsequent use
- A development concept is being prepared for the military training area. It is planned to clear the fuel depot and the structures and to integrate the estate as a conservation area into the natural landscape.

4. Prospects

The inventory and preliminary assessment of suspect sites on the estates formerly used by the WGT has been completed. The work is well advanced on the estates of the Bundeswehr. A number of orientating investigations have already been done. Detailed investigations are still in their beginnings, but investigations will have to be forced in order to get more reliable information about the real risk potential that exists.

It turned out that no specific methods are required for the assessment of contaminations from previous military uses. The only difference between military and non-military uses consists in the specific features of military chemicals.

The volume of formalised assessment methods has become quite obvious, constituting a uniform and comparable basis for risk assessment. However, they cannot replace risk assessment which includes all the site-specific peculiarities as the basis for remediation planning. This holds true above all in view of the fact that remediation targets must always be derived to match the specific case, the good/medium involved, and the intended future use of the site.

Protection and restriction have been priority aspects so far in the remediation of contaminated military sites. Those measures, however, cannot be substitutes for sustainable remediation.

Co-ordinated remediation planning, regional planning and development planning would provide a good solution for contaminated military sites that will no longer be used for military purposes in future any more, as this would provide a sound basis for land recycling.

The abandoned WGT estates seem to be particularly suitable to this end. In Brandenburg, Saxony and Thuringia all those estates were taken over by the Land and transferred to "Landesentwicklungsgesellschaften" (regional development corporations). This should provide favourable conditions for land recycling at least in those Laender.

References

1. Der Rat von Sachverständigen für Umweltfragen (1995) *Sondergutachten Altlasten II*, Metzler-Poeschel, Stuttgart.
2. Industrieanlagen-Betriebsgesellschaft mbH, Ottobrunn (1995) *Ermittlung von Altlastverdachtsflächen auf den Liegenschaften der Westgruppe der sowjetischen Truppen (WGT), Abschlußdokumentation - Zusammenfassung*, Ottobrunn, 20.11.1995.
3. *Gestz zum Schutz vor schädlichen Bodenveränderungen und zur Sanierung von Altlasten* (Bundes-Bodenschutzgesetz), Entwurf vom 25.09.1996.
4. Jessberger, H. L. (Hrsg), (1993) *Sicherung von Altlasten*, Rotterdam, Balkema, S.IXXII.
5. Franzius, V., (1994) *Kosten und Finanzierungsbedarf der Altlastensanierung in den neuen Bundesländern*, Zeitschrift für angewandte Umweltforschung, Sonderheft 5 Altlastensanierung, Berlin, Analytica, S. 21 - 35.

CURRENT AND FUTURE RESEARCH AND DEVELOPMENT PROJECTS OF THE FEDERAL MINISTRY FOR THE ENVIRONMENT ON CONTAMINATED MILITARY SITES

SILVIA REPPE
Federal Environmental Agency
Bismarckplatz 1
14193 Berlin
Germany

As a result of the political changes in Europe and the concomitant disarmament process, military use of about 50% of what once were 1 million hectares of land used for military purposes has ceased since 1991. This process includes the withdrawal of the Western Group of the former Soviet Forces (WGT) as well as sites of the former GDR's People's Army, the six western Allies, the NATO and the German Federal Armed Forces.

At a multitude of these sites, environmental damage has resulted not only from military use after 1945, but also from military operations prior to and during the Second World War. Before military sites can be converted to civilian uses, it is necessary to determine and evaluate the status of contamination, take action to avert imminent dangers to human beings, soil and the groundwater, undertake in-depth risk assessment and, based on this, perform remedial measures.

Due to the dimension of the problem, the recording, evaluation, investigation and cleanup of contaminated military sites is accorded high political priority in Germany. An expression of this is that the discontinuation of military use has been accompanied by comprehensive programmes of the Federal Government and the federal Länder aimed at compiling a record and evaluating the status of contamination of these sites. Measures to avert imminent hazards as well as the recording of military sites throughout Germany was largely completed in 1995, including the development of general procedures and instruments for the treatment these sites. Future tasks comprise detailed hazard assessment and the containment/decontamination of contaminated sites as an integral part of the activities to convert former military sites to civilian uses.

In the work under those cleanup programmes, it became apparent that there was a need for the provision of technical and scientific guidance and for the creation of a sound methodological basis.

The first step towards this end took the form of an international review of the state of the art. Under the Federal Environment Ministry's Environmental Research Plan, the Federal Environmental Agency in 1996 commissioned a study to ascertain the current level of knowledge in 21 countries in the field of contaminated military sites. Based on the overview

F. Fonnum et al. (eds.),
Environmental Contamination and Remediation Practices at Former and Present Military Bases, 33–40.

thus obtained of different approaches and levels of technological development, research an
development needs in the Federal Republic of Germany were identified.

The complete study (country reports on 21 countries / synopsis / research an
development needs) has been published in German (TEXT 4/97) and English (TEXT 5/97
under the Federal Environmental Agency's TEXTE publication series.

Based on the findings obtained from the Federal Government's contaminated site
programmes and the results of the study referred to above, a number of research project
have been launched on the subject of contaminated military sites:

- Investigations into the degradation and migration behaviour of mineral oil, aromati substances and chlorinated hydrocarbons in soil and groundwater, using the WGT sit Vogelsang (fuel storage depot) as an example;
- decision-making aids for containment and decontamination concepts for military fu storage depots;
- mobility, transfer and degradation behaviour of contaminants from military sites;
- establishment of a substance database on substances of relevance to contaminated sit (military and armament-contaminated sites)
- establishment of a contamination profiles database for military and armamen contaminated sites;
- technical requirements for contaminated site cleanup within the framework subsequent-use planning and alternative solutions to financing cleanup (civilian sites an military sites to be converted to civilian uses);
- subsequent uses of former military sites;
- aerophysical investigation of contaminated sites - innovative use of different airbor sensors.

A brief outline is given of the aims of these projects. All are still in progress. The resul
will be in hand in 1998 / 99in the form of research projects or databases and will
available both nationally and internationally through the Federal Environmental Agency.

Topic 1

Study on degradation and migration behaviour of mineral oils, aromatics and chlorinated hydrocarbons in soil and ground water

Background

- Characteristic for former military sites, especially fuel storages: *outdoor* location, contamination of soil and ground water with fuels and lubricants, lack of future use concepts
- Search for cost-saving treating methods to clean up such sites against the background of increasing public budget restrictions and simultaneous prevention liability for environmental receptors; time frame is of minor priority

Objective:

- Study of the long-time behaviour of TPH, BTEX, VHHC
- Development of innovative, cost-saving treatment technologies for these hazard substances
- Elaboration of an action concept for the model site "Fuel storage Vogelsang" (former WGT)
- Derivation of a generalised action concept for similar sites

Work programme:

- Risk assessment 1992, *follow-up* analyses of soil and ground water 1995, 1996, 1997
- Specific laboratory tests to stimulate degradation activity of *autochthon* microorganisms in soil and ground water Determination of site-specific parameters: soil permeability, hydrogeology, microbial degradation potential
- Soil pilot test to check ~ applicability of biotechnological methods on the model site
- Evidence of the activation of natural degradation of TPH in soil by soil fertilisation (liquid manure), breaking-up and irrigation of soil in pilot test

Topic 2

Decision fundamentals for containment and decontamination concepts for military fuel storages

Background:

As a result of inventorying military contaminated sites it was realised that relating to a risk of~the environmental receptors soil and groundwater fuel storages are of special concern. Considerable *inputs (discharges*) of *operation (*process) *agents (substances)* (fuels, lubricants, detergents) as a result of handling losses in transport, decanting and cleaning processes as well as leakages of tanks and pipelines. Discharges of *operation agents* are in general of high environmental relevance. Mainly contaminations by TPH, BTEX, PAH, lead and VHHC occur. From the specificity of military contaminated sites it should be taken into consideration that fuel storages often are situated in forests without infrastructure. This leads to considerable problems for developing civil purpose use concepts and reclamation of investors. Often recultivation and integration in forestry is the only future use concept. At present in Germany nation-wide several projects and activities for risk assessment are under way, safety/decontamination, *pulling down,* recultivation and utilisation of military fuel storages. Up to now specific, technologically harmonised and general approaches do not exist.

Objective:

- For former military fuel storages under specific cost aspects decision fundamentals for safety/decontamination concepts have to be developed in connection with strategies for future use and utilisation taking into account specific parameters (size, *outdoor* site). These concepts shall enable authorities and site owners efficiently to restore fuel storages taking into account technological requirements *and* cost-saving aspects in connection with utilisation concepts.

Work programme:

- Analysis of 10 representative fuel storage projects on the basis of a unified matrix which takes into consideration all relevant parameters (legal, site-specific, technological, economic, costspecific).

- Elaboration of decision fundamentals based on the experience of model projects.

- Derivation of still remaining research demand for the field "Military fuel storages".

Topic 3

Mobility, transfer and degradation behaviour of pollutants from military contaminated sites

Background:

- Pollutants and mixtures of them from civil and military contaminated sites are a considerable risk potential for ground water. This is of high importance for water resources policy, especially concerning pollution of ground water resources used for drinking water supply. Particularly on wide-area sites used for military purposes with fuel storages and technical departments considerable amounts of hydrocarbons from mineral oils in the form of different fuels with specific additives for military requirements were handled and the soil was contaminated on a large area. To give preconditions for a unified assessment of consequences of existing soil contaminations for ground water resources for restoration measures, bases of case-to-case assessments shall be continued.

Objective:

- Assessment criteria for pollutants (substances contained in water) have to be developed with the help of which effects of anthropogenic substances on the environment, their biological degradation and their transformation and degradation products can be quantified. Therefore it is necessary to further study the connections between chemical-physical properties and biological effects for further investigation of the synergistic combination of different substances.

Work programme:

- Investigation of substance-specific data for relevant pollutants from military contaminated sites concerning mobility, degradation and transformation behaviour (metabolisation) in soil and ground water

- Data assessment incl. toxicological parameters concerning future ground water utilisation

- Evaluation of standardised statements concerning action demand

- Control of the verification of the results by the example of military sites with known soil contamination profiles

Topic 4/5

4 Data bank for substances relevant to contaminated sites (Military and sites contaminated due to armament production)
5 Data bank "Contamination profilea" for military and sites contaminated due to armament production

Background:

- Data relevant to contaminated sites are available in different sources and data banks.

Objective:

- A complex INFORMATION SYSTEM ON CONTAMINATED SITES is intended to collect all data relevant to contaminated sites and make them available in a data bank. Suitably this system will be developed from single modules, which can access to one another. At present the data bank on substances relevant to contaminated sites and the data bank "Contamination profiles" are being developed as the first components. Later the development of further modules on clean-up technologies, models on risk assessment and cost evaluation for cleaning up contaminated sites is planned.

Work programme:

- Data bank on substances: Development of a computer-based data base for the assessment of hazardous substances relevant to contaminated sites conceming their physical-chemical substance behaviour, their bioavailability and their environmental impact in the media soil, soil air, water (surface and ground water) for the time being. for 47 substances relevant for contaminated sites.

- Data bank"Contamination profiles": The inventorying of military contaminated sites in Germany and the studies on environmental importance connected to this lead to the elaboration of contamination profiles for types of military sites (e.g. air force base, training ground, barrack) and their specific fields of use. The military use caused contaminated sites here is given a typical stock of pollutants/groups of pollutants and a potential of hazardous substances belonging to it.

- These data are at present available as tables and in writing now will be made available as a data ban The access system allows to get data on relevant pollutants from the data bank "Contamination profiles" and data to assess these substances from the data bank on substances provided that the site was used for military purposes.

Topic 6

Study "Follow-up use of former military sites"

Background:

- The conversion of former military sites affects a lot of aspects which, on the one hand, are merely related to environmental protection, on the other hand involve complex socioeconomic and legal problems as well as questions of urban and landscape planning. Whereas for the assessment of environmental questions knowledge and approaches are available, a general view considering all criteria relevant for a follow-up use has not been made sufficiently up to now. As the conversion of military sites is a current process at present no compilation or analytical assessment of the experience made exist. An analysis how methods and scenarios of area recultivation from the field of civil contaminated sites can be applied to (former military) sites to be converted has not been carried out yet.

Objective:

- The study intends to make a typisation of criter}~ levant for an efficient follow-up use of military sites and to prepare them in the form of a ;~; . Experience from conversion projects shall be used to find a optimal choice of alternatives for future conversion measures considering the cost-benefit-aspect, to avoid misinvestments and to optimise the use of special financial funds.

Work programme:

- Analysis of conversion projects taking into account criteria relevant for decisions for investors/site owners:

- Decision processes and criteria during the planning phase, assessment if financing models, importance of environmental loads for conversion methods, information of the public, gainingof investors, cost-benefit-analysis,

- Compilation of a checklist containing the characteristic site criteria which can serve as basis to control conversion options.

- Development of assessment standards for individual criteria

Topic 7

Aerophysical investigations of contaminated sites - innovative use of different air-bas sensors

Background:

- Characteristic for former military sites, especially training grounds: situated outdoc often in forests, wide areas, buried subjects/shifted areas, loads by ammunitio underground facilities (e.g. tanks, shelters, canalisation, pipeline systems), contaminatio of soil and ground water with TPH.

- Difficult preconditions for extensive investigation.

Objective:

- Development of scientifically reliable statements on technological possibilities and limi of application of air-based sensors for the registration of suspected contaminated sites

- Magnetic, electromagnetic and optical methods are applied

- Harmonisation of evaluation and interpretation methods

Work programme:

- Selection of appropriate sites considering aspects of sensors, aeronautics a contaminated sites (training grounds)

- Experimental phase: development of a test field and characterisation of anomalies conditions characteristic for contaminated sites, helicopter flights, additional input typical objects to be observed

- Combination of different sensors verification phase with test on a great area a cost-benefit-study.

ENVIRONMENTAL RISK ASSESSMENT AND MANAGEMENT IN EVALUATION OF CONTAMINATED MILITARY SITES

A Canadian Approach

K.J. REIMER AND B.A. ZEEB
Environmental Sciences Group
Royal Military College of Canada
Box 17000 Stn Forces
Kingston, Ontario K7K 7B4
CANADA

1. The Canadian Remediation Program

The National Contaminated Sites Remediation Program (NCSRP) was established in Canada in 1991 to ensure a coordinated, nationally consistent approach to the identification, assessment, and remediation of contaminated sites in Canada [1, 2]. This program addresses sites that currently are impacting, as well as those that have the potential to impact on human health or on the environment. Special rules apply when an industrial site or a military base is involved. In these cases, the site must be 'decommissioned' prior to selling the land or using it for other purposes. Decommissioning refers to the process of closing; it includes the removal of equipment, the dismantling of buildings and structures, the remediation of any soil contamination, and the reclamation of land to render the property suitable for other purposes [3]. It may apply to all or part of the facility, but it aims to ensure that the end product is a site that is:

- not a risk to human health and safety;
- not the cause of unacceptable effects on the environment;
- in compliance with all applicable laws and regulations;
- suitable for the proposed new land use;
- not a liability to current and future owners; and
- aesthetically acceptable.

Decommissioning projects conducted over the past decade have shown that a phased approach is the most practical and cost-effective. There are six phases in the decommissioning of a Canadian industrial or military site.

F. Fonnum et al. (eds.),
Environmental Contamination and Remediation Practices at Former and Present Military Bases, 41–56.

Phase I: Site Information Assessment

This phase consists of a review of all available information relating to the site. Its primary objective is to determine whether practices on the site have resulted in environmental contamination or unsafe conditions. It also assesses the need for immediate interim actions to ensure the protection of human health and safety, and to prevent the spread of any contamination in the environment. The evaluation takes into consideration the anticipated future land use for the site. Phase I is largely research-oriented, but it does include a physical inspection of the site by trained specialists. In some cases it may be difficult to access pertinent historical records. This places additional emphasis on Phase II.

Phase II: Reconnaissance Testing Program

The main objective of Phase II is to identify the types and concentrations of contaminants present in various media on the site, and to specify the locations of contaminant problems. During this phase, known and suspected areas of contamination that have been identified in Phase I, as well as areas believed to be relatively unaffected by site operations are targeted. Background concentrations of chemical contaminants in soil, subsurface materials, surface water, groundwater, and air are also determined. Results from Phase II indicate whether further detailed testing is required in specific areas.

Phase III: Detailed Testing Program

During Phase III, contaminated areas identified in Phase II are targeted, and a sample grid system is laid out to delineate the boundaries of contamination. Computer spatial analysis programs can be used to interpolate chemical concentrations between sample points, and to ensure that a statistically significant amount of data is available. The data is used to outline the areas, volumes, and concentrations of contaminants throughout the site. From this information, technological options for cleaning up the site are selected.

Phase IV: Preparation of Decommissioning and Cleanup Plans

This phase involves a feasibility study to evaluate remedial alternatives if site cleanup is necessary. Preparation of the Decommissioning and Cleanup Plans may have been initiated as early as Phase I in order to assess all of the options. The final plan must include the actual design and procedures for implementing the recommended cleanup plan. The cleanup of contaminated soil must be logically integrated with the demolition, removal, and/or stabilization of buildings and structures on the site. The plan must also address occupational health and safety issues to ensure that the on-site cleanup staff as well as the public are adequately protected from exposure to contaminants.

Phase V: Implementation of Decommissioning and Cleanup
Depending on site conditions, Phase V may be relatively simple or quite complex. It is recommended that the work be carried out by contractors experienced in demolition, hazardous toxic waste cleanup, and waste treatment to ensure successful completion and to minimize the potential for accidents or inadvertent release of contaminants into the environment. Phase V may include any or all of the following: worker safety and health monitoring, construction of on-site containment facilities, handling of wastewater and surface drainage, control of fugitive atmospheric emissions, removal and disposal of materials, residues, equipment, and buried services, cleaning and dismantling of buildings, excavation of contaminated soils and sediments, remediation, and monitoring.

Phase VI: Confirmatory Sampling/Completion Reporting
This is the final phase in decommissioning and cleanup. Its first objective is to confirm through testing that contamination has been removed or effectively stabilized on site, and that the cleaned up site meets the required standards or criteria. On completion of the project, a report documenting the six-phase process must be submitted to the regulatory agency for review and acceptance that the site is suitable for the proposed future land use.

It is recognized that this process is, in its entirety, extremely costly. The Canadian military is currently following more environmentally sound practices for product and waste management so as to avoid excessive costs and liabilities in future decommissioning projects. A legacy of abandoned sites, however, remains. The Environmental Sciences Group (ESG) at the Royal Military College of Canada (RMC) has completed assessments at many of these sites and has made recommendations for their cleanup. Two distinct approaches have been used by ESG to evaluate the status of a site, and to assess the risk it poses to environmental and human health.

2. Assessing Environmental Risk

2.1. THE ABSOLUTE APPROACH

An 'absolute' or criteria-based approach defines numerical values or limits that can be compared with measured environmental levels. It establishes baseline concentrations for various substances below which the impact on the environment is assumed to be minimal. Environmental samples containing concentrations above this limit may indicate the need for further investigation and/or remediation of the sampling area, depending on the extent to which the level is exceeded.

2.1.1. Canadian Criteria

In 1983, the Netherlands Ministry of Housing, Planning and the Environment issued the frequently cited 'ABC' numerical guidelines for soil and water quality to be used in assessing the severity of contamination and the urgency for further investigation or remediation [4]. The Canadian Council of Ministers of the Environment (CCME) Interim Canadian Environmental Quality Criteria for Contaminated Sites were published in 1991 and updated in 1997 [1, 5]. These criteria are numerical values for the assessment and remediation of water and soil. Assessment criteria are either approximate background concentrations or approximate analytical detection limits that provide a starting point for examining data. Remediation criteria are applied as guidelines for cleaning up contaminated sites, depending on whether the intended land use is agricultural, residential/parkland, or commercial /industrial.

The application of absolute criteria has certain advantages such as relative ease of administration. This approach also forms a common base for communication between interested parties, it is not arbitrary, and it reduces confusion. However, the assumption that the criteria apply equally in all circumstances is a limitation. Although the criteria are sometimes referred to as 'generic' because they were developed for the generic Canadian climate and do not take into account any unique characteristics of a particular site, the CCME documentation does allow for the development and application of such criteria [6, 7]. Guidelines that have been adjusted to reflect the characteristics of a specific site are referred to as site-specific criteria, and are an essential component of a modified, semi-quantitative approach to risk assessment, discussed further in Section 2.2.

2.1.2. Quantitative Risk Assessment (QRA)

In order to manage risk, risk must first be assessed. Quantitative methods provide discrete values or a distribution of values for the components of the risk assessment. In a QRA, mathematical models built on the parameters for a specific site are used to estimate exposures or dose, and the results are then combined with toxicological information to obtain a numerical value representing the risk assessed for the site in question. In the field of contaminated site remediation, detailed quantitative models do exist, but they must be applied correctly.

Quantitative risk assessment has some inherent difficulties such as the quantification of uncertainty, or the defining of acceptable risk levels. Acceptable risk has been defined by some agencies as a one-in-a-million chance of an occurrence of an adverse effect, but public acceptance of any risk can be difficult to achieve. In addition, more- and less-conservative definitions of acceptable risk have been employed. In some situations, an absence of available toxicological data forces the use of approximations or assumptions, eroding the usefulness of a QRA.

3. A Modified Approach

3.1. THE SEMI-QUANTITATIVE RISK ASSESSMENT

Extremes of climate have created diverse environments in Canada, and a corresponding need for alternative approaches to ensuring their protection. The

Canadian Arctic, for example, supports a uniquely fragile ecosystem in a harsh and remote environment. There are many reasons why a quantitative approach to environmental assessment may not be appropriate. This type of assessment of a typical Arctic site does not account for the fact that the impact of chemical contaminants that enter the foodchain may be far more profound than in more moderate environments. Conversely, a contaminant immobilized in permafrost will have less impact than it would have in an environment that promotes contaminant mobility. A consideration when assessing risk in the Arctic or other sensitive environments is that the physical impact resulting from the removal of contamination may be more damaging than the contamination itself. To address these challenges, an alternative or semi-quantitative approach to risk assessment has been developed that focuses on the potential for the movement of contaminants from the substrate in which they are contained (source), to humans or other parts of the environment (receptors). This approach evaluates risk according to existing or potential environmental impact: where contaminant movement is demonstrated, a risk exists; where there is no risk of contaminant movement, there is a greatly reduced environmental risk. It is therefore necessary to examine the pathways for this movement in a site-specific context, and to assess the impact that this will produce.

This approach is appropriate in a number of circumstances. It can be used:

- when national criteria do not exist for a contaminant;
- where cleanup to criteria-based levels is not feasible for the targeted land use;
- where criteria-based objectives do not seem appropriate given the site-specific exposure conditions;
- where significant or sensitive receptors of concern have been identified; or
- where there is a significant public concern.

Remediation addresses the elimination of pathways to environmental receptors. The question to be answered in choosing remedial solutions is not How great is the risk of adverse affect? or How toxic is this source? but rather, Is there a link between the source and any environmental receptors? When the answer is Yes, then there is a risk, which must be managed by breaking the link; the amount of risk is related to the strength of that link. This approach can be more effective than a quantitative risk assessment because it is easy for the public to conceptualize, and it does not rely on estimates of toxicological effects.

When remediation is based heavily on economic considerations (which it usually is), this semi-quantitative approach may be the most practical. When the costs of meeting existing environmental criteria are extremely high, priorities must be established to focus remediation efforts and to evaluate potential impacts of remediation. The same applies when the contaminated area is very large. Remedial solutions are most successfully approached with the goal of managing risk where it exists, rather than through a quantified assessment.

3.2. TARGETED SAMPLING AND THE DEVELOPMENT OF SITE-SPECIFIC CRITERIA

The same concentration of a contaminant in soil can display very different effects under different circumstances, depending on the soil type and depth, ecosystem features etc. It is therefore necessary to determine the impact of contaminants on a site-to-site basis, which means that site-specific criteria must be established. To achieve this, samples are collected to include a variety of media. The choice of sampling locations is guided by obvious indications of site use, such as building foundations, dumps and storage areas, as well as by previous experience with patterns of chemical contamination at former military sites to include sewage disposal areas, building doorways, landfills, etc. The objective is to identify all potential sources of concern. Each distinct operational area of the site is therefore targeted.

Samples are also collected from drainage pathways and from potential receptors. Samples in these categories provide information to determine whether contaminants are migrating and whether they are having a negative impact on the environment. Samples are therefore collected in drainage pathways leading away from potential contaminant sources. Special efforts are made to identify and sample receptors. These include valued ecosystem components (VECs) or environmental features that:

- are important to human populations;
- have economic and/or social value;
- have intrinsic ecological significance; and/or
- serve as a baseline from which the impacts of development can be evaluated.

3.3. THE ROLE OF RECEPTORS IN RISK ANALYSIS

Potential receptors such as plants are collected at the same locations as some of the soil samples. These can be used as a direct measure of the entry of contaminants into the food chain, resulting in a measured rather than modelled estimate of risk. Species such as willows (Salix sp.) and oleasters (Shepherdia sp.) which have importance in the food chain are collected. Willow, for example, demonstrates the ability to accumulate and, in some cases, concentrate one or more of the parameters analyzed from the soil or sediment [8]. The analysis of plant material is not intended to preclude the interpretation of soil and water data. Instead, plant analysis provides a different perspective on the environmental status of the site, helps to identify substances with particular biological activity, and assists in the selection of remedial options.

Plants vary greatly in their ability to accumulate contaminants, and some naturally accumulate substances that are toxic over certain concentrations. Although elevated concentrations of contaminants in plants demonstrate a measurable movement of contaminants into the foodchain, the specific concentrations at which they become a threat to the ecosystem are often unknown. Impact is therefore measured by

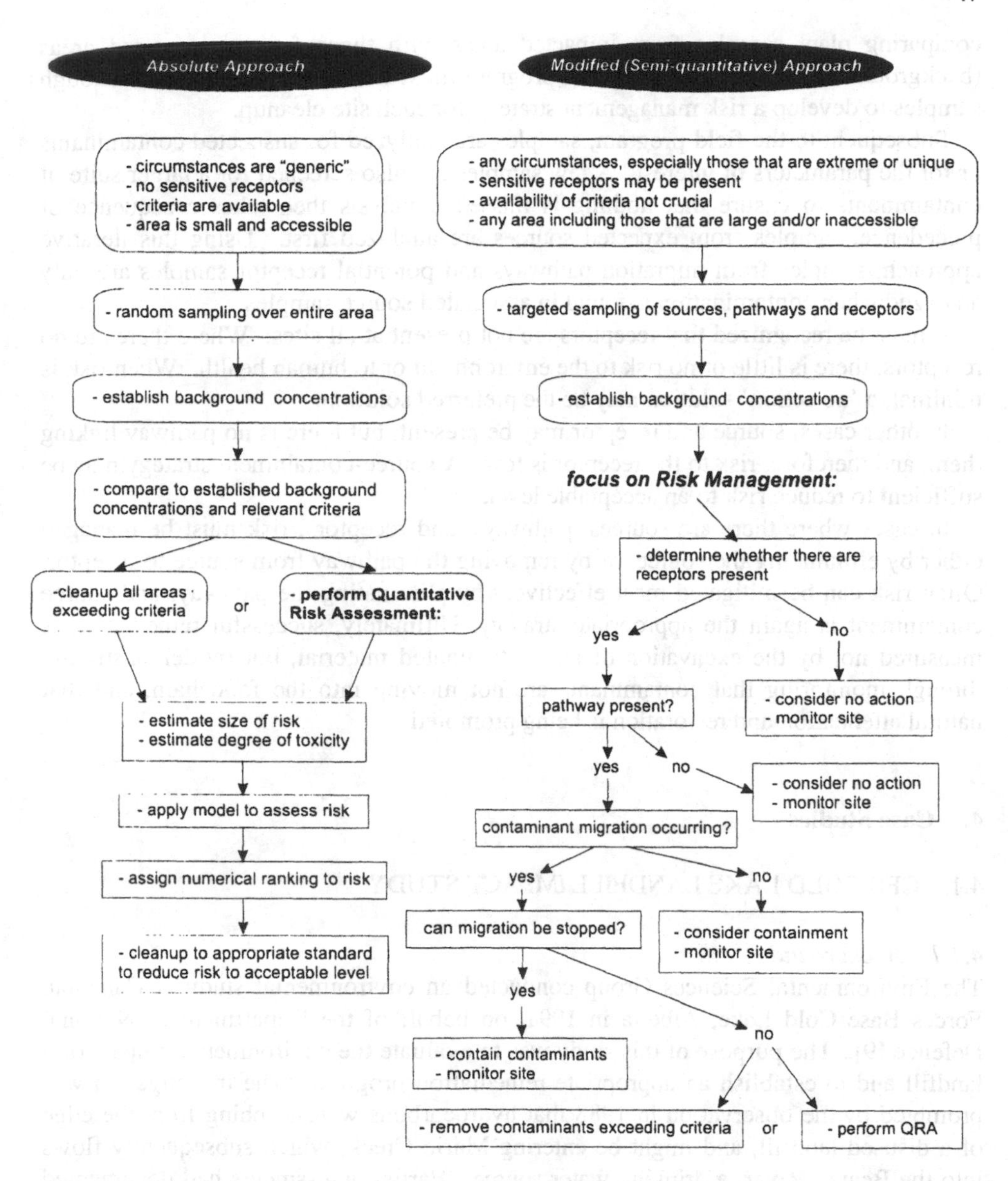

Figure 1. Flowchart showing approaches in assessing environmental test.

comparing plant samples from impacted areas with those from unimpacted areas (background samples). The sampling program must ensure the collection of enough samples to develop a risk management strategy for each site cleanup.

Subsequent to the field program, samples are analyzed for suspected contaminants or for the parameters of interest. A few samples are also screened for a larger suite of contaminants to ensure that nothing is missed. Analysis then takes a sequence of precedence: samples from expected sources are analyzed first. Using this iterative approach, samples from migration pathways and potential receptor samples are only analyzed when contamination is found in associated source samples.

It must be recognized that receptors are not present at all sites. Where there are no receptors, there is little or no risk to the environment or to human health. When risk is minimal, a 'no action' solution may be the preferred solution.

In other cases, source and receptor may be present, but there is no pathway linking them, and therefore, risk to the receptor is low. A source-containment strategy may be sufficient to reduce risk to an acceptable level.

In cases where there are sources, pathways, and receptors, risk must be managed either by eliminating the source, or by removing the pathway from source to receptor. Often risk can be mitigated most effectively by eliminating the pathway, and source containment is again the appropriate strategy. Ultimately, successful remediation is measured not by the excavation of all contaminated material, but by demonstrating through monitoring that contaminants are not moving into the foodchain, and that natural attenuation and restoration is being promoted.

4. Case Studies

4.1. CFB COLD LAKE LANDFILL IMPACT STUDY

4.1.1. Background

The Environmental Sciences Group conducted an environmental study at Canadian Forces Base Cold Lake, Alberta in 1991, on behalf of the Department of National Defence [9]. The purpose of this study was to evaluate the environmental impact of a landfill and to establish an appropriate remediation program. The investigation was prompted by the observation in 1988 that hydrocarbons were leaching from the edge of a disused landfill, and might be entering Marie Creek, which subsequently flows into the Beaver River, a drinking water source. Earlier assessments had documented the local hydrogeology and identified contaminated areas. Although there was no evidence of an immediate threat to human or environmental health, it was recommended that further action be taken to define the contaminant zone and to promote rehabilitation.

4.1.2. Implementation

Cold Lake was visited on three occasions by members of ESG. An initial reconnaissance was undertaken during which interviews were conducted with Base personnel, and existing monitoring wells were inspected. Water samples were collected from the wells, and from several surface locations; these samples were subjected to a complete set of chemical analyses. The data from this first set of samples confirmed suspicions that contamination was confined mainly to a bog at the edge of the landfill. There were several possible sources for the contamination, since the bog was also receiving discharge from a runway storm sewer, and as well, groundwater from an extensive surrounding region was surfacing within the bog. The identification of the major contaminants as organic compounds consistent with the landfilling of lubricants, fuels, and degreasers, eliminated the runway as a source. A review of geological and site maps narrowed the possible source locations to within the general landfill area, and the most likely source was finally pinpointed after an examination of historical aerial photographs, which clearly showed a pond of some kind within the landfill in a flat area to the north of a salt/sand storage building. Further enquiries revealed that an open oil pit created from waste oil disposal had existed at that location. This practice of oil disposal was subsequently discontinued, and the pit was no longer visible at the time of the ESG investigation. Shortly after the first visit, and in response to a request by Air Command, an electromagnetic induction (EM) survey was conducted under the supervision of ESG. The survey mapped a plume of high conductivity originating at the salt/sand storage building, and spreading northeast towards the bog. Although the plume masked the leachate from the oil pit, which might otherwise have been apparent as a region of relatively low conductivity, the results did provide confirmation of the direction of groundwater flow. It is interesting to note that, without additional data from well sampling, this plume might have been identified as the contaminant source, rather than the source of a benign sodium chloride discharge.

On the third visit to the site, several new monitoring wells were installed. The extent of contamination was defined by locating wells in and around the suspected area. The analytical results confirmed both the identity and the location of the source. In order to determine the impact of contamination from surfacing groundwater, water and soil/sediment samples were collected from shallow pits dug in the bog, and down-slope in the wooded area between the bog and Marie Creek. Elevated levels of several organic contaminants were found in the bog, but no substances of environmental concern were found in samples collected in Marie Creek itself.

Seasonal variations in contaminant loadings were examined by means of an additional set of water samples obtained from selected wells by Base personnel later the same year. Good agreement was found with the results of the previous sampling programs.

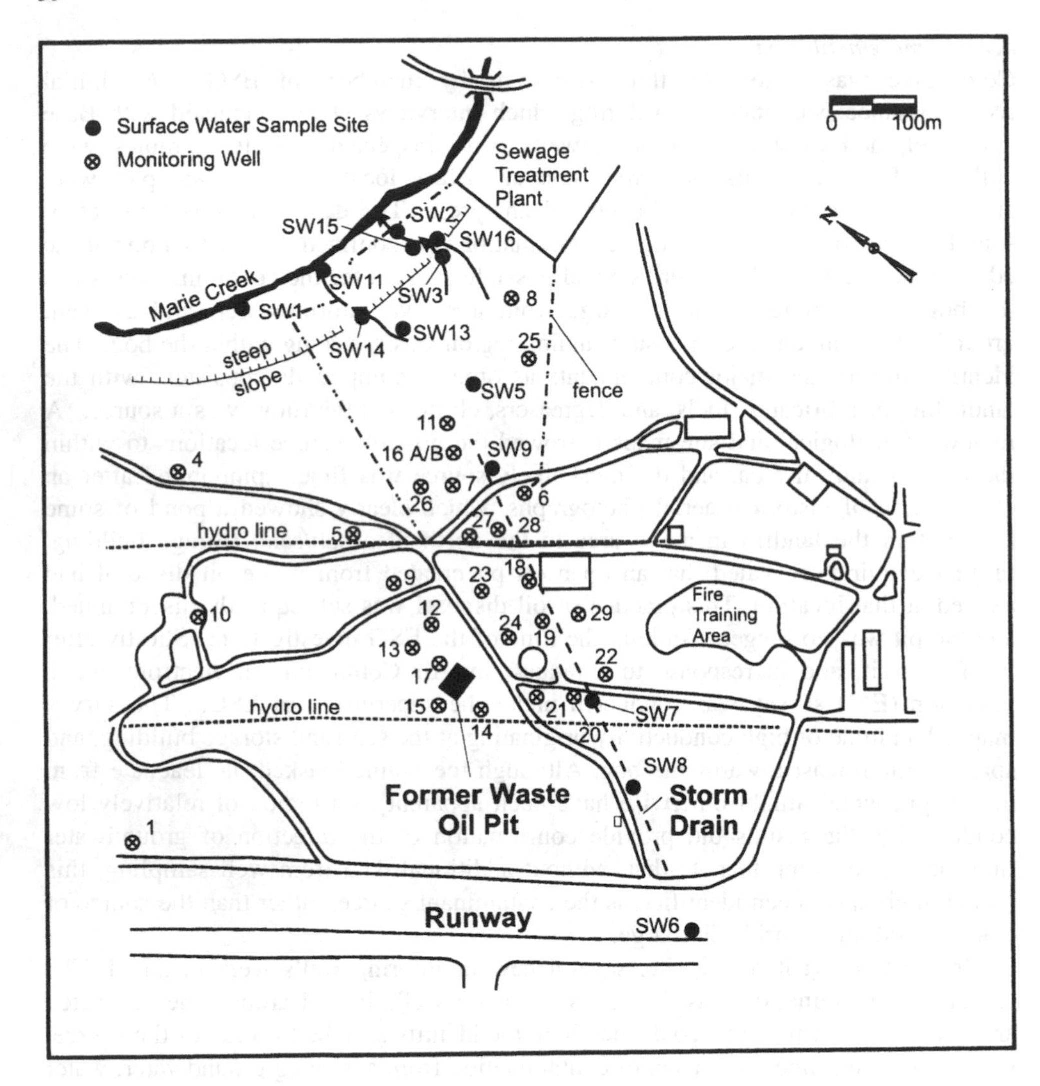

Figure 2. Layout of the Cold Lake Landfill Site showing the location of the contamination source (abandones oil pit), receptor (Marie Creek) and water sampling stations.

4.1.3 Analytical Results

The principal compounds of environmental concern were determined to be benzene, toluene, xylenes, ethylbenzene, phenols, 1,1-dichloroethane, and 1,1,1-trichloroethane. These compounds were present in the groundwater, and in surface water samples in the bog at the point where the groundwater surfaces. Only two contaminants (1,1-dichloroethane and 1,1,1-trichloroethane) were detected in the wooded area adjacent to Marie Creek, and these were at very low concentrations. It should also be noted that water flow in this area was very limited. The contaminants detected were consistent with contamination emanating from a now buried oil disposal pit located to the north of the salt/sand storage building.

4.1.4. Implications

Contaminant levels in the groundwater plume exceeded many current environmental quality criteria. Groundwater surfacing at the edge of the landfill contained elevated levels of organic compounds, but the concentrations were found to be significantly reduced closer to Marie Creek. Had contaminants been found in Marie Creek, the environmental impact would have been of far more immediate concern. However, the bog was shown to be acting as an effective barrier to contaminant migration to Marie Creek, and natural attenuation was in progress within the bog.

It may be argued that significant quantities of contaminants might enter the Creek at some future time. This would be a possibility if, at the time of the assessment, the concentration of contamination at the bog had not yet reached the maximum, or if the filtering capacity of the bog were to become saturated. Remedial actions were therefore recommended to preclude these possibilities.

4.1.5. Recommendations

The following recommendations were made by ESG:

- The area of surface contamination was to be fenced and posted.
- A lagoon was to be constructed to receive both surfacing groundwater and the discharge from the runway storm sewer. This would permit volatilization and photolytic degradation of the compounds of concern. It would also encourage phytolytical/microbiological degradation. Outflow from the lagoon would be dispersed over a larger area of the bog, and would further reduce the already slow seepage of water to the wooded area below.
- Continued monitoring of ground and surface water was required.

Consideration was also given to the installation of two large-diameter purge wells within the area of contaminated groundwater. These would permit the interception and further reduction of contaminant loadings by air sparging. This option was not considered to be essential at the time, but it was recommended as an action that would reduce the duration and cost of the monitoring program.

The Cold Lake Landfill Impact Study is an example in which a risk-based approach was applied. In this case, the most feasible solution was to manage risk by assisting the natural attenuation process occurring within the bog. The final cost of managing risk was in the order of $10,000, whereas a source removal solution would have cost at least an order of magnitude more. Continued monitoring of water in the bog and in the creek will ensure that neither the health of Marie Creek nor of humans in the area will be compromised.

4.2. THE DEW LINE CLEANUP PROJECT: MANAGING CONTAMINATION IN THE ARCTIC

4.2.1. Background

The Distant Early Warning (DEW) Line was a surveillance system built by the United States across the North American Arctic in the 1950s. Over the years, some of the stations were decommissioned or abandoned, and by 1989, with the installation of the new North Warning System, many more became redundant. In 1993 the old DEW Line was officially closed. Environmental investigations commissioned by the Canadian government found evidence for significant physical and chemical impact at the DEW Line sites, and pointed to a need for environmental standards and cleanup procedures appropriate for the Arctic. The Environmental Sciences Group, working on behalf of the Canadian government, visited all of the 42 closed DEW Line sites in the Canadian Arctic, and over a period of five years, completed environmental assessments at every site [10]. In the course of these investigations, a practical strategy for cleanup was developed that would leave the sites in an environmentally safe condition. This approach includes a set of numerical guidelines for contaminant tolerances in soil, and lays out general recommendations for dealing with contaminated soils, landfills, and site debris.

The DEW Line sites are spread over a vast area, and many are in remote settings with very limited access. The field season in the Arctic is extremely short, and the environment can be hostile. For these reasons, remedial and cleanup programs are, under any circumstances, enormously expensive. An additional challenge in developing such a program is that the Arctic ecosystem is far more sensitive to environmental influences than is a more temperate system. Chemical contaminants have a more profound impact on the foodchain because of the limited biological diversity, and physical disturbances are likely to disrupt the permafrost, with repercussions to all parts of the ecosystem. Restoration of normal conditions may take many years. The cleanup strategy takes into consideration all these factors: it places special emphasis on preventing the movement of chemical contaminants from sources at the DEW Line Sites into other parts of the Arctic ecosystem; and in the interests of practicality and cost-effectiveness, it does not call for the restoration of sites to a pristine condition, but to a state that allows natural attenuation to take place. An accompanying engineering design includes plans for the removal of hazardous materials from the Arctic, for the containment of substances, which might threaten the environment, and for the construction of special facilities for the containment of

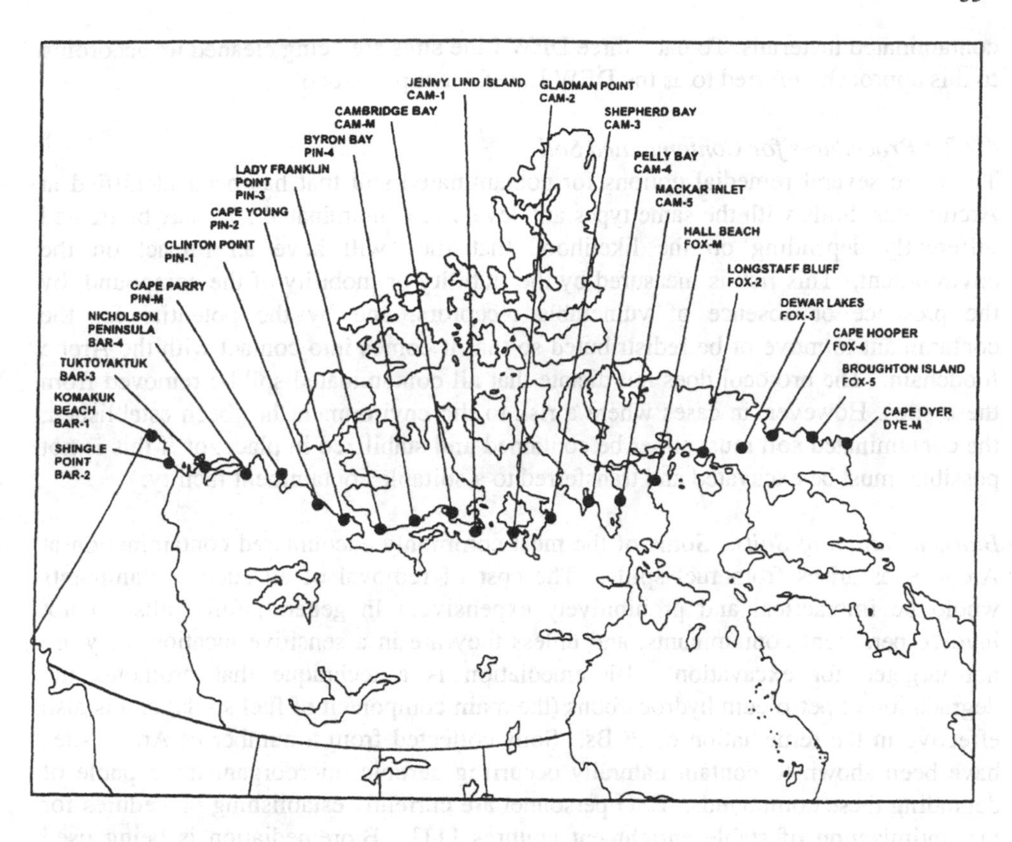

Figure 3. Location of old Distant Early Warning Sites across the Canadian Arctic.

contaminated materials. To date, three DEW Line sites are being cleaned up according to this approach, referred to as the DEW Line Cleanup Protocol.

4.2.2. Procedures for Contaminated Soil

There are several remedial options for contaminated soil that has been identified at Arctic sites. Soils with the same types and levels of contaminated soil may be treated differently depending on the likelihood that they will have an impact on the environment. This risk is measured by the stability or mobility of the compound, by the presence or absence of vulnerable receptors, and by the potential for the contaminant to move or be redistributed so that it comes into contact with the Arctic foodchain. The protocol does not dictate that all contaminated soil be removed from these sites. However, in cases where a risk to the environment has been established, the contaminated soil must either be contained and stabilized in place, or if this is not possible, must be excavated and transferred to a suitable containment facility.

Bioremediation of Soils. Some of the most commonly encountered contamination at Arctic sites arises from fuel spills. The cost of removal of all such contaminants would be impractical and prohibitively expensive. In general, full spills do not involve persistent contaminants, and unless they are in a sensitive location, they are not targeted for excavation. Bioremediation is a technique that promotes the degradation of petroleum hydrocarbons (the main component of fuel spills) and is also effective in the remediation of PCBs. Soils collected from a number of Arctic sites have been shown to contain naturally occurring aerobic microorganisms capable of degrading these compounds. ESG personnel are currently establishing procedures for the optimization of stable enrichment cultures [11]. Bioremediation is being used with the other technologies to promote natural attenuation, and also to reduce the volume of contaminated material that is destined for containment – for example in a landfill or a containment facility - as described in the following sections.

Landfill Remediation. The method of waste disposal during the operational years of the DEW Line was with landfills. Waste of all kinds, including contaminated material, was deposited in a designated area and generally buried, but sometimes just piled up. All of the sites contain landfills of some description, and a large part of the cleanup strategy is devoted to their remediation. It is considered both impractical and environmentally damaging to attempt to remove landfills in permafrost soils. Therefore, the remediation choice for most of the DEW Line landfills involves leaving them intact, but with the application of additional clean overburden material in order to ensure that they will become part of the permafrost rather than part of the active layer. This is possible when the landfill is in a stable area not subject to erosion. If there is evidence that contaminants are leaching, or the possibility that they might leach in the future, then measures must be taken to contain the leachate. The proposed design uses a low-permeability synthetic barrier that is incorporated with the permafrost by covering it with frozen saturated soil, then covering all with granular fill to a specified depth.

The On-Site Disposal Facility. In cases where contaminants in soil are threatening the environment and cannot be contained where they are – for example, where contaminated soil is eroding and cannot be stabilized – the soil must be excavated. The concept for a special disposal facility for contaminated Arctic soils was developed by UMA Engineering of Alberta, as a solution for reducing costs and risks associated with the transportation of soil to an existing southern facility in situations where soils cannot be contained in place. Its function is to provide a secure facility in the North for soil contaminated to a level that threatens the environment and that cannot be stabilized in its present location.

The facility is designed with a double containment system: the primary containment is the natural permafrost, which provides a thick barrier of extremely low permeability surrounding the facility; the secondary containment is provided by a synthetic liner placed on the inside of the face of the perimeter slope of the facility. The cover of the facility is also lined to maximize run-off from snow and heavy rains. Movement of the contaminated material is restricted further by the fact that it is maintained in a permanently frozen state.

4.2.3. *Monitoring Programs*

When contaminated soil is confined using leachate containment, or transferred to a containment facility, a monitoring program is essential. A three to five year program has been suggested for landfills requiring leachate containment in permafrost; this is the time required to attain thermal equilibrium. The program should consist of visual, thermal, and surface water or active layer monitoring. Water quality should be monitored using wells to within a specified distance from the structure. Baseline water quality should be established before any waste is placed in the landfill.

5. Summary

The two case studies described here represent two very different remediation strategies. In the first case, the contaminant source and receptors were identified, and it was demonstrated that no pathway connected them. The solution was to promote natural attenuation. The second case proposed approaches for managing environmental risk when there is a pathway between source and receptor. It was also shown that risk can be mitigated by eliminating the pathway. When this pathway cannot be eliminated, the risk must be managed by removing the source. In all cases, a monitoring program that continues beyond the cleanup process is an essential part of remediation. These case studies demonstrate that a semi-quantitative approach to risk management presents a variety of effective solutions, some of which might otherwise be missed. Often the solution is surprisingly simple, but in all cases the end result is that contaminant movement into the surrounding ecosystem is precluded, and natural restoration is promoted.

6. Acknowledgements

We would like to thank Marjorie Cahill for extensive editing of this manuscript. In addition, we would like to thank all members of the Environmental Sciences Group who contributed extensively to the development of the ideas and concepts presented in this paper.

7. References

1. Canadian Council of Ministers of the Environment. (September 1991) *Interim Canadian Environmental Quality Criteria for Contaminated Sites.* 20 pp.
2. Canadian Council of Ministers of the Environment. (1994) *A Framework for Ecological Risk Assessment at Contaminated Sites in Canada: Review and Recommendations.* C. Gaudet, EVS Environment Consultants, and Environmental and Social Systems Analysts. 108 pp.
3. Canadian Council of Ministers of the Environment. (March 1991) *National Guidelines for Decommissioning Industrial Sites.* Monenco Consultants Ltd. 98 pp.
4. Moen, J.E.T. (1988) *Soil Protection in the Netherlands.* In Contaminated Soil '88. Edited by K.Wolf, J. van den Brink, and F.J. Colon. Kluwer Academic Publishers. pp 1495-1503.
5. Canadian Council of Ministers of the Environment. (March 1997) *Recommended Canadian Soil Quality Guidelines.* 185 pp.
6. Canadian Council of Ministers of the Environment. (March 1996a) *A Framework for Ecological Risk Assessment: General Guidance.* Then National Contaminated Sites Remediation Program. 32 pp.
7. Canadian Council of Ministers of the Environment. (March 1996b) *Guidance Manual for Developing Site-specific Soil Quality Remediation Objectives for Contaminated Sites in Canada.* 45 pp.
8. Environmental Sciences Group. (February 1993) *The Environmental Impact of the DEW Line on the Canadian Arctic* – Volumes 1-3. Department of National Defence. 488 pp.
9. Environmental Sciences Group. (1992) *CFB Cold Lake Landfill Impact Study.* Department of National Defence.
10. Environmental Sciences Group and UMA Engineering Ltd. (1995) *DEW Line Cleanup Scientific and Engineering Summary Report.* Department of National Defence.
11. Mohn, W.W., Westerberg, K., Cullen, W.R., and Reimer, K.J. (1997) Aerobic biodegradation of biphenyl and polychlorinated biphenyls by Arctic soil microorganisms, *Applied and Environmental Microbiology* 63, 3378-3384.

SWEDISH GUIDANCE ON THE PROCEDURE FOR THE INVESTIGATION AND RISK CLASSIFICATION OF URBAN AND INDUSTRIALLY CONTAMINATED AREAS

DAG FREDRIKSSON
Geological Survey of Sweden
Box 670
Uppsala
S-751 28 Sweden

1. Introduction

In Sweden, as well as in other countries many land and water areas have been polluted by past industrial activities or in other ways. There are thousands of sites where soil, groundwater or water sediments have been polluted to such an extent that risks to human health or to the environment have to be considered. The pollutants include heavy metals, polycyclic aromatic hydrocarbon solvents, pesticides and persistent organic compounds such as PCBs and dioxins. A preliminary inventory (BKL) of these sites was accomplished during the years 1992-1994. This inventory has resulted in the forecast that eventually about 7,000 sites will be classified as contaminated. At present, about 1,200 sites are registered as suspected contaminated sites.

In order to achieve a satisfactory level of knowledge for risk assessment and to decide the order of priority between sites, a systematic inventory has to be fulfilled. For this reason the Swedish Environmental Protection Agency together with the Geological Survey of Sweden, the Institute for applied Environmental Research at the Stockholm University and the Institute of Environmental Medicine at Karolinska Institutet have recently developed a guidance manual for inventories with regard to contaminated sites. This guidance is aimed to be a standard for the Swedish authorities in preparing and execution of future inventories on contaminated sites.

2. Strategy

The new inventories will be based on site and industrial information obtained from BKL. This information will then be supplemented with new desk and field information. The

F. Fonnum et al. (eds.),
Environmental Contamination and Remediation Practices at Former and Present Military Bases, 57–61.

purpose is to achieve a comparable level of knowledge about all sites, and also to identify previously unknown sites.

The guidance manual consists of a two step strategy (*Fig.1*) of investigation similar to the "Guidance on the procedure for the investigation of urban and industrial sites with regard to soil contamination" (1).

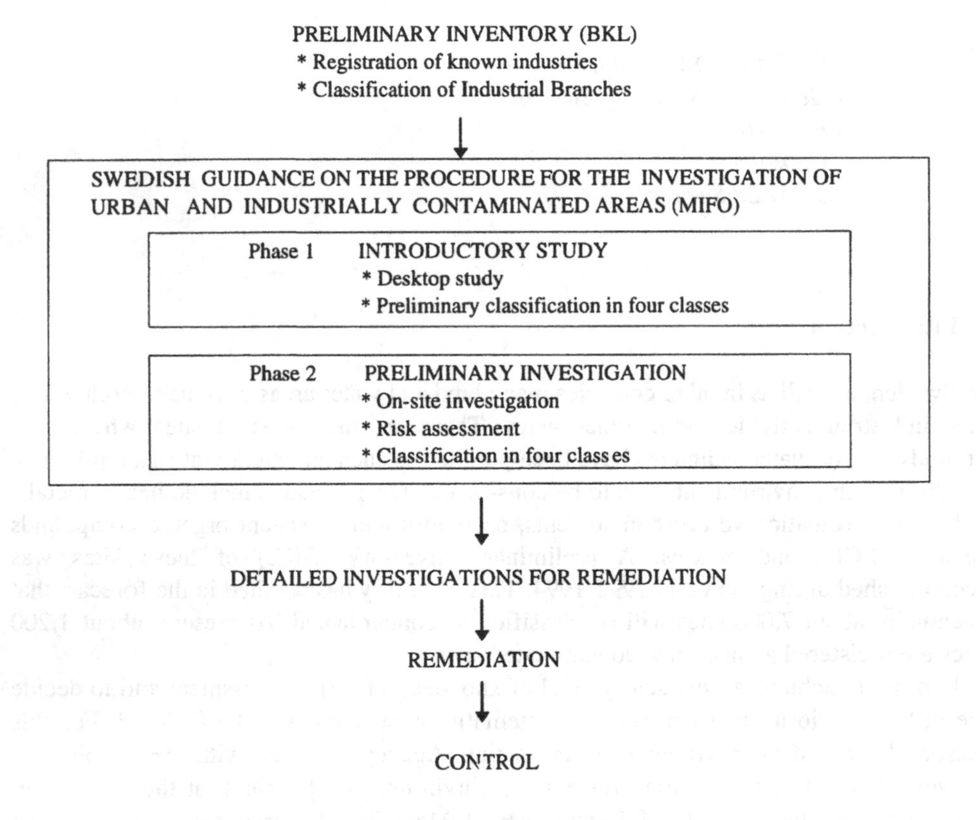

Figure 1. Main strategy for inventory and investigation

The introductory study (Phase 1) is mainly a desk study, which include identification and historical research, past and present usage of the site, drawing- and map studies and soil- and hydrogeological properties from available geological maps. This Phase 1 study normally also includes a short site reconnaissance. From this study the possibility of contamination can be deduced and a preliminary classification and risk assessment can be carried out into four classes.

The risk assessment consists of an evaluation of four different factors:

* Chemical risks of handled chemicals

* Level of contamination if known
* Prediction of mobility and spread of contaminants in soil and groundwater
* Human health risks and ecological values of the area.

The preliminary classification is finally made in four classes.

The second step, the preliminary investigation (Phase 2), consists of an on-site investigation which includes soil sampling, groundwater sampling, in special cases sampling of surface water and if possible sampling of sediments from receptors. The design of sampling strategy will derive from the introductory study, and will consist of a very limited number of samples.

Sampling medium:

* Soil
* Groundwater
* Surface water, in special cases
* Sediment, if available

The prime purpose with the sampling normally is to verify the level of contamination, but in some cases also samples are taken to get local background values or information on the distribution of the contaminants.

The proposed analytical programme will vary with sampling medium, soil, groundwater, surfacewater or sediment. It is based on a standards suite of chemical- and physical analysis and biological tests (TABLE 1 and 2). In some cases when the character of the contamination is known, or when the pollution is typical for a specific branch, the analytical programme could be reduced to specific substances or elements.

The evaluation of the level of contamination is based on guidance values if possible, but if such values are lacking, then it is based on local or regional background values. The chemical risks of the contaminants are adopted from official lists and textbooks.

The prediction of migration of contaminants is made from a very simplified calculation of groundwater velocity, which is adjusted due to influence of for example, decomposition and absorption of substances, organic content in the soil, observed ditching, dry fissures in clay, distance to groundwater, known migration of contaminants, impermeable layers, water holding zones etc.

The final classification is then based on the same principles as those used in the introductory study, and is made in four classes (*Fig. 2*). A data base for collection of data will be set up on a regional basis.

TABLE 1. Chemical- and physical analysis

Basic programme	Soil	Ground-water	Surface water	Sediment
pH		x	x	
Temperature		x	x	
Conductivity		x	x	
Colour		x	x	
Turbidity		x	x	
Oxygen			x	
Tot-N		x	x	
Tot-P		x	x	
Chlorine		x	x	
Metals Ag, Al, V, Cr, Mn, Fe, Co, Ni, Cu, Zn, As, Cd, Pb (+Hg for sediment and soil)	x	x	x	x
EGOM+GC-MS	x	x		x
SPME		x		
Dry substances	x			x
Loss of ignition	x			x
Complementary programme				
AOX		x	x	
EOX	x	x		x
HEGOM	x	x		x
PBS	x	x		x
Polarised and nonpolarised hydrocarbons		x		
SPOT-test	x	x		x
TOC		x	x	
Branch specific analyses	x	x	x	x

TABLE 2. Biological analysis

Basic programme	Soil	Ground-water	Surface water	Sediment
Microtox	x	x	x	x
Complementary programme				
Algae test	x	x		
Clam test	x	x		x
Cell-test EROD			x	x
Umu-C test	x		x	x

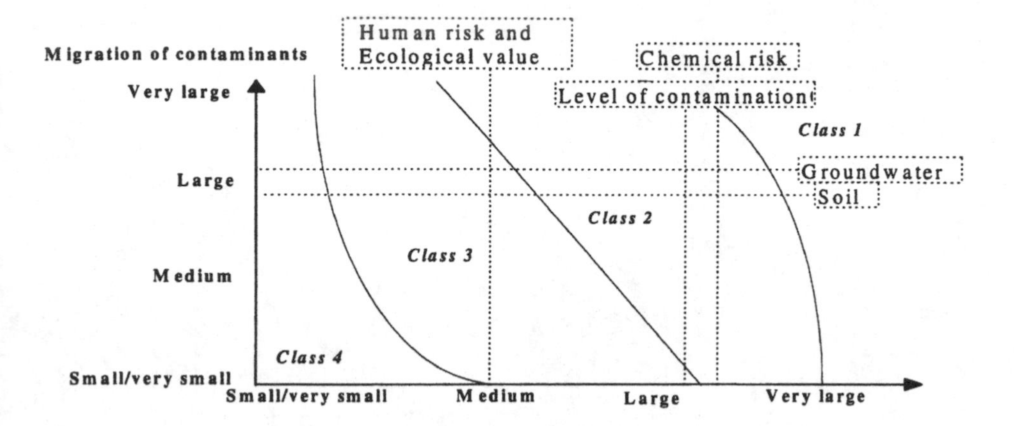

Figure 2. Example of assessment of risks and classification (Class 2, large risk) of an contaminated area.

3. References

1. ISO/TC 190 / SC2 N 78 vs.8. Guidance on the procedure for the investigation of urban and industrial sites with regard to soil contamination, 10381-5, Soil Quality- Sampling.

ENVIRONMENTAL SITUATION AND REMEDIATION PLANS OF MILITARY SITES IN LATVIA

DZIDRA HADONINA
Ministry of Environmental Protection and Regional Development of the Republic of Latvia
Head of Division of State Cadastres and Natural Resources
Peldu iela 25
Riga, LV 1494, Latvia

1. Introduction

Considering the fact that Latvia's administrative system, national economy and society are in a transition period, clear environmental requirements should be included in the implementation of large scale state administration reforms and in the drafting of new legislation. A strategy should be devised for implementation of relevant economic projects and environmental protection as for any branch of the national economy. The National Environmental Policy Plan offers such a strategy.

Currently Latvia faces two types of environmental problems of military origin:

A. The first part may focus on the potential impacts on the environment from the newly-formed Latvian National Defence Forces installations' and activities';

B. The second major part could focus on the environmental damage caused by former Soviet troops in Latvia over the past 50 years.

The first type is due to the grim heritage left by the totalitarian period: characterized by inefficient, environmentally hostile, resources-squandering state sector industry; the restructuring of agriculture, energy and transportation; underdevelopment of the environmental and utilities sectors; former Soviet military territories that became polluted during the occupation period and still have not been cleaned up; increased emergency risk and insufficient environmental quality in several regions.

The second category of problems can be seen from the restructuring of the national economy into a market economy. The amount of investment currently under way is insufficient to bring about a rapid improvement in Latvia. A preventive environmental policy is hindered by an ill defined economic policy and an adverse legal system.

Majority of environmental problems in Latvia are concentrated in the so-called "hot spots" - the largest industrial centers, transportation crossroads or in territories abandoned by the Russian army (the former Soviet Army).

F. Fonnum et al. (eds.),
Environmental Contamination and Remediation Practices at Former and Present Military Bases, 63–69.

2. Environmental issues and Latvia's newly-formed national defence forces

The basic tasks of the newly-formed National Armed Forces (NAF) are to maintain the peace, security and stability of the Latvian state and its population and to eliminate any (not only a military) threat.

The NAF consists of Ground Forces, Border Guards, the Air Forces and Naval Forces. The Ground Forces, which consists of regular, territorial and support troops, is the dominant forces.

Latvia places great importance on military cooperation with Estonia and Lithuania, not only as a means of improving defence capabilities, but as an opportunity to demonstrate our common interests and willingness to cooperate to solve serious problems. For this reason, we greatly appreciate the formation of the UN Baltic Peacekeeping Battalion (BaltBat).

The Nordic and Baltic states' defence ministers have signed a memorandum of understanding that pledges cooperation in the development of the Baltic Battalion. The Battalion receives language training, technical and equipment and basic infantry instruction from NATO countries.

The NAF occupies training bases established by the former Soviet Army. The NAF's troop deployments and training bases do not pose any environmental burdens or threats. The necessary review of conditions and sanitation procedures was carried out during the Latvian take-over of the former Soviet facilities. Specific examples will be reviewed in the context of descriptions of environmental problems the current inherited from the Soviet Army.

For example, the NAF's basic training center occupies the former Soviet Army ground forces' training grounds in Âdaþi. The NAF's infantry and the Baltic Battalion forces train at this base. The Âdaþi training grounds and the training activities conducted here, conform to NATO standards, both militarily and environmentally.

Main environmental problems at Latvian Armed Forces units are resolved taking into consideration local management policies and according to Law on Environmental Protection and other environmental protection laws and regulations within allocated common budget.

"Infrastruktura AM" - Infrastructure MOD (InfMOD) is nonprofit state organization which is responsible for dealing with all environmental protection issues involving Ministry of Defence. Responsibilities of Infrastructure MOD include:

A. Repair and replacement of sewage and water supply system to prevent contamination of all water bodies and ground waters in areas on, or near military facilities.
B. Protection of soil in areas near military refueling facilities.
C. Control of air pollution around boiler houses at military bases and installation of gas operated equipment in facilities where solid fuel is used today.

InfMOD Construction and Maintenance Department has prepared a prioritized environmental protection program which has been accomplished only partially due to budget shortages. The following projects have been accomplished:

- Soil analysis at Lielvârde Air Base;

- Repair sewage and water supply systems at Southern Region Navy base in Liepâja;
- Repair of waterlines at Liepâja Navy Training Center;
- Repair of waterlines at National Armed Forces HQ Battalion installation in Rîga;
- Due to a malfunction of a water treatment plant at Suþi Airborne Reconnaissance Battalion base near lake Íîðezers sewage, spillage contaminated the lake. Therefore, a preliminary project for reconstruction and replacement of the sewage treatment plant at Suþi base was developed.

Hitherto MOD of the Republic of Latvia is not receiving financial assistance from various programs which are designed to finance environmental protection projects. Latvian military would benefit from participation in such programs since there are many environmental problems at the MOD facilities that the Latvian military cannot solve due to financial problems.

Structures of the MOD and MEPRD of the Republic of Latvia consider the information gained via the Mil-to-Mil environmental activities essential and useful. Particlarly the information concerned reuse of former military sites has found more wide application in Latvia.

3. Withdrawal of the armed forces of the Russian Federation from the territory of Latvia

On August 31, 1994, all Russian forces withdrew from Latvia in accordance with the "Agreement between the Republic of Latvia and the Russian Federation on the Conditions, Terms, and order of a Complete Withdrawal of the Armed Forces of the Russian Federation from the Territory of the Republic of Latvia and Their Legal Status During the Period of Withdrawal" signed on April 30, 1994.

One of Latvia's specific problems is the ballistic missile early warning radar station in Skrunda in the Kuldîga district. In accordance with the "Agreement" this radar will continue to function for up to four years (until August 31, 1998). Dismantlement of the radar system must start on September 1, 1998 and be completed no later than February 28, 2000. Latvia views the continued existence of this system as, not only an environmental problem, but a social and psychological one as well.

This site had two main structures - an old radar station (Dnepr), which will continue to function for up to two years (until August 31, 1998), and new unfinished one (Darjal), consisting of a 19-story receiver building and 8-story transmitter building, which was dismantlement by implosion and mechanical demolition. This was possible thanks to assistance of the United States Government. Demolition was directed by the Maryland based company ***Controlled Demolition Inc. (CDI)*** and many Latvian subcontractors took part in demolition as well.

Dismantlement of the new radar station and putting in order the place cost more than 6 mill. USD. The project had been successfully completed within 9 months. At the present moment there is a clean area prepared for planting a forest.

4. Environmental damage and problems caused by the Former Soviet Army in Latvia

The former Soviet Army's activities, installations and military bases, and its confiscation of land from Latvian citizens for military use, caused serious damage to Latvia's environment and natural resources. The conversion of these abandoned sites and installations to civilian use, have high economic, environmental, health and political priority.

The territories occupied by the Soviet Army, and the location of its bases in Latvia, were chosen not only for the country's advantageous geographical location but because it is the strategic center of the Baltic region. The USSR's Baltic Military District, later, the Russian Northwest Group's army headquarters, Air Force headquarters and border guard headquarters were located in Rîga. The reserve command centers and training areas or grounds for these headquarters were also located in this region. Large fuel depots, chemical and chemical warfare depots, munitions, and military-technical support supply centers were located in Latvia to ensure the availability of centralized supplies for the entire Baltic Military District. The region also contained large military technical repair depots. The military naval bases were located in Rîga and Liepâja. According to our data, former Soviet Army military units and bases of differing scales and purposes, occupied approximately 100,000 hectares or 1.5% of Latvia's territory.

Examination of environmental pollution caused by the Russian Army was initiated in 1992. Two military territories (Suþi and the Spilve fuel depot) were investigated in detail. Primary observations (without chemical analysis) and an initial assessment of former Russian Army territories have been completed. Three hundred military sites, occupying approximately 96,000 hectares were studied. MEPRD conducted these investigations. It should be noted that Latvia lacks experience in conducting such studies, lacks many of the financial and technological resources to do so, and relies heavily on other countries' s assistance.

The most serious environmental and economic damage was caused by the former Soviet Army's military firing grounds, airfields, rocket bases, fillingstations, fuel depots and naval ports. The Russian Army had firing grounds for every kind of weapon in Latvian territory. Many of the buildings in these areas are not suitable for conversion to civil use.

The environmental problems caused by the former Soviet Army are similar to those found in many post-socialist Eastern European countries. Only the order of priorities to deal with them differs. The problems posed by restoration of the military harbour in Liepâja for civilian use offers one characteristic example of the environmental challenges Latvia faces.

Liepâja is the third largest city in Latvia situated on the Western coast of our country some 210 km South-West of Riga with a population of approximately 115,000 inhabitants. From the end of World War II Liepâja has served as Soviet (Russian) naval base, and was the second largest of its kind in the Baltic. It means that Liepâja was a closed or semi-closed military territory excluded from normal economic development and the ice-free port was closed to all commercial activities. After the withdrawal of

Russian Navy we have received a city with undeveloped infrastructure and very polluted harbour with a lot of sunken military ships. The harbour needs very serious reparation works and improvement, dredging and disposal of approximately 5 mill.m^3 of contaminated and uncontaminated sediments.

In addition the Northern part of the city named Karosta which is one third of the territory of Liepâja was used only for military purposes. This district had historically developed as *"a state within a state"*, separated from the rest of the city. There are side by side buildings with unique architectural value in a well planned structure from the beginning of the century, and buildings with extremely bad quality from times of Soviet Army presence. Many buildings are in a bad state. Several of them after the withdrawal of Russian Army had been left unattended and without any owner. Many of the buildings in their present condition cause a real threat to the public. In the district of Karosta there are 498 buildings of various character (apartment houses, barracks, auxiliary buildings, bomb-shelters, ammunition warehouses, garages, clubs, canteens, manufacturing facilities etc.), 401 of which are not suitable for any further use, and they have to be demolished. This will take considerable resources.

In the past, the Soviet military firing range and aviation targets covered 24,500 hectares of farm land and forests at Zvârde in the Saldus district. Mechanical and chemical pollutants such as aviation bomb splinters and undetonated bombs have defaced the terrain. Diffuse contaminants such as aircraft fuel, burning wastes and explosives have rendered the soil unusable. Undetonated bombs must be disposed of and soil samples analyzed so that remediation measures can be identified before the land is to be reclaimed for cultivation.

5. The assistance in environmental programs within the military offered by other countries

Thanks to the Government of Germany and its Federal Ministry for Environment, Nature Conservation and Nuclear Safety, since 1992, we have had excellent cooperation from the Industrieanlagen-Betriebsesellschaft mbH (IABG) company. Latvian specialists had the opportunity to familiarize themselves with environmental problems in former Soviet military territories in Germany. They have participated in the applications seminar: "Data Processing Programs - ALADIN, MEMURA, MAGMA" and received copies of software. In 1994 the joint Latvian-German project "Contaminated Military Airfield Lielvârde Environmental Impact Assessment" to investigate the Lielvârde airfield using German methodology has been completed. This project - the first joint project focusing on Latvian territory formerly occupied by the Soviet Army - was conducted to identify measures needed to prevent the diffusion of ground water pollution. The airfield of Lielvârde is situated 60 km South-East of Riga. The area includes 657 ha and is situated 50 - 60 m above the sea level.

In 1994 Latvia began a three-year joint project with Canada to conduct a demonstration project to remediate rocket fuel-polluted soil in the rocket bases Tâôi and Bârta in the Liepâja region. This project also calls for training of Latvian specialists in Canada.

Proponents of the project are:

A. Environment Canada-Emergencies Engineering Division;
B. Gartner-Lee International Inc.;
C. Canadian-Latvian Community;
D. Riga Technical University;
E. Health Protection Branch, Health Canada.

The Latvian Ministry of Environmental Protection and Regional Development strongly supports this project.

The Tukums bulk fuel terminal site, located 70 km West of Riga is one of the oldest in Latvia and was at one time crossed by a Soviet Army aviation fuel pipeline. Over the years, routine operations and numerous leakage from pipeline resulted in severe ground water contamination. Some of the pipeline releases reportedly flooded a nearby stream.

American and Latvian joint-venture "Baltec associates, inc." performed a subsurface investigation of the site in early 1992. The investigation included excavation of exploratory pits and installation and sampling of ground water monitoring wells. Laboratory analysis of ground water samples revealed the presence of fuel hydrocarbons at concentrations exceeding drinking water standards. Aviation fuel was identified in the samples and observed floating on the water table to thickness of one-half meter in one of the on-site wells.

Based on these findings, Baltec constructed a system of three ground water recovery trenches around the periphery and one infiltration trench in the center of the site. Fuel and contaminated ground water is pumped from the trenches to an aboveground treatment system consisting of oil-water separation, aeration, and biological treatment, and treated water is subsequently reintroduced into the zone of contamination to enhance in-situ bioremediation. It is estimated that within three to five years, the floating fuel will have been removed and the concentration of hydrocarbons dissolved in ground water will be significantly reduced. The remediation system began operation in late 1993. Already, more than $280 000 has been invested in the project. It is estimated that and additional $50 000 per year will be invested for the next five years. These funds will be applied toward the development and installation of a biological filter for processing treatment system vapor emissions and for overall system maintenance.

Danish company "Kruger Int. Consult A/S" in cooperation with the State Geological Survey of Latvia and MEPRD has prepared the project proposal "Investigation of Proposed Sites for Groundwater Development and Remediation of Former Soviet Military Site".

Under an agreement on environmental cooperation between Latvia and Denmark, the Danish Environmental Protection Agency is presently funding a study on ground water management for the Riga Water Supply. Two sites of particular interest for the Riga Water Supply with respect to contamination are: Vangaþi Military Site (fuel storage) and Adaþi Military Site.

In our country up to now we have not carried out the all-embracing evaluation of environmental damage in all former Russian military sites according to one common methodology. There are several investigations, but those are made according to different methodologies and different degree of details. Since they do not cover all sites, it is difficult to obtain a complete picture of the situation. However, in 1996 such investigations was started according to the standard methodology in the framework of Latvian-Norwegian two-years cooperation project "Investigation of the Former Soviet Army Bases in Latvia and Investigation of Environmental Damage and Problems". The investigations carried out by the Norwegian Defence Research Establishment and by Geological Surveys of Latvia and Norway.

The project has comprised three parts:

A. Introducing the Norwegian-developed "Waste" Database to Latvia;
B. Carrying out the first-step audit and hazard ranking of contamination at known military bases in Latvia;
C. Developing methodologies for more detailed assessments of contaminated bases. This have been demonstrated at two military bases in Riga - fuel storage depot at Viestura prospects and armoured vehicle repair workshop at Valmieras street.

These two sites have been chosen, not because they represent the worst contaminated military bases in Latvia, but because they are typical of a number of registered sites, are easily accessible and have both organic (Viestura prospect) and inorganic (Valmieras street) contamination.

6. The most serious overall constraints

The most serious constraints in Latvia are:

- lack of financial resources;
- lack of common methodology;
- special training courses for our specialists in European countries where laboratories are already working on the determination of soil and groundwater contaminants from an environmental standpoint;
- technological resources are scarce.

7. References

1. Hadonina, D. (1996) Country Report Latvia, in E.A. McBean et al. (eds.), *Remediation of Soil and Groundwater,* Kluwer Academic Publishers, Dortrecht, pp. 59-77.
2. Hadonina, D. (1995) *Environment and Defence Issues for Latvia,* 13-15 September 1995, CCMS Report No. 211, pp. 268-275.
3. Plaudis, A. (1995) *State of the Environment and Defence Environmental Issues for Latvia,* Defence Environmental Conference '95, Garmisch, Germany, 7-13 May 1995, Proceedings, pp.E-1-E-29.
4. Strauss, I., Plaudis, A. (1996) *Latvian Experience in Reusing Former Russian Military Bases for Solution of the Environmental Protection and Economical problems,* presentation in Central/Eastern European Military-to-Military Environmental Engineering Conference, Garmisch, Germany, May 1996.
5. Strauss, I., Skultans, V., Plaudis, A. (1997) *Short Overview of the Environmental Policy of the Republic of Latvia and State of the Environmental Protection Issues in the System of the Ministry of Defence of Latvia,* presentation in Central/Eastern European Military-to-Military Environmental Conference, Berlin, Germany, 12-16 May 1997.

STRATEGY FOR INVESTIGATIONS AND REMEDIATION OF A POLLUTED NORWEGIAN NAVAL BASE

A. JOHNSEN
Norwegian Defence Research Establishment
Division for Environmental Toxicology
Postboks 25, N-2007 Kjeller, Norway

1. Introduction

In 1990-91, the Norwegian Defence Construction Service in co-operation with the State Pollution Control began a survey of military facilities on behalf of the Norwegian Defence Forces. The purpose of this survey was to assess the kinds of hazardous waste produced by various military units, and to locate waste disposal sites and areas of polluted ground.

All information collected in this survey has been stored in a special database and will serve as input for future planning operations and utilisation of areas where hazardous waste has been shown to exist. The database is continuously updated as new information and new registrations are compiled.

About 310 sites have been registered so far. At 19 sites immediate investigation or remedial action was deemed necessary in light of the amount and type of hazardous waste disposed of or spilled, and potential conflict with the surrounding environment.

1.1. CRITERIA FOR RANKING SITES

Landfills and polluted grounds have been ranked in four priority groups according to the need for further investigations or remedial actions. The criteria for ranking are:

- information on type of waste and producer
- location of sites with respect to residential development areas and water supplies
- existing and planned utilisation of sites and water supplies
- recipient conditions
- geological conditions, when reliable data exist for evaluation

The ranking has been based solely on the danger of pollution and conflict arising from the presence of hazardous waste. Ranking has been based on the following criteria:

Priority group 1: Immediate investigation or remedial action required

F. Fonnum et al. (eds.),
Environmental Contamination and Remediation Practices at Former and Present Military Bases, 71–81.

- Reliable information exists about hazardous waste in landfills or leakage of hazardous waste and/or dangerous chemicals to the ground

- Types and amounts of waste and location of site are such that it may lead to serious pollution or can harm humans or animals.

Priority group 2: Further investigation required

- Sufficient information exists to suspect that hazardous waste or dangerous chemicals are present in the ground and this can lead to serious pollution or harm to humans and animals

- Reliable information exists about hazardous waste and/or dangerous chemicals in the ground.

Priority group 3: Further investigation required in cases of altered use of land or water recipient

- Reliable information or suspicion exists concerning the presence of hazardous waste and/or dangerous chemicals in the ground

- The location and usage of the site and recipients will not lead to serious pollution or harm to humans or animals.

Priority group 4: No further investigation required

- Survey evaluation of the site has not revealed the presence of hazardous waste nor any reasons for suspecting the presence of significant amounts of disposed hazardous waste.

1.2. RESULTS OF THE SURVEY

Table 1 show the results after ranking the sites in each priority group. Most of the sites in priority group 1 involve conflicts with drinking water or groundwater interests near the site.

TABLE 1. Numbers of sites in each priority group and site types

Site/Ranking	Group 1	Group 2	Group 3	Group 4	Total
Landfills/dumpsites	9	38	76	38	161
Polluted ground	8	24	105	0	137
Landfills/dumpsites and polluted ground	2	4	5	0	11
Total	19	66	186	38	309

In the past, it was commonly accepted that waste, now defined as hazardous waste, could be disposed of along with other types of refuse. Most military units have their own dump site for heavy refuse, scrapped material etc. Waste was often incinerated at military dump sites along with the remains of paint, glue and varnish. In some cases, oil and solvents were included. Some landfills were also used as fire training areas.

Many of the registered dump sites were established either by the Germans during World War II or in connection with post-war clean-up operations. Three such wartime sites have been registered in priority group 1. It is difficult to obtain information about the content of these wartime dump sites. Since proper control was lacking, many types of hazardous waste may have been dumped: paints, oil products, heavy metals, products containing PCB (Polychlorinated Biphenyls), medical wastes and ammunition.

Polluted ground is most often caused by leakage or spillage of oil products from oil tanks or pipelines. Some former fire training sites also contain solvents as well as remains of paint, glue and varnish. Ground pollution has also been registered in locations where hazardous waste have been released into sewage systems. Over long periods of time activities at these sites have resulted in ground pollution. This has probably led to the contamination of groundwater aquifers with the resulting negative consequences for drinking water supplies.

2. Contamination at Haakonsvern naval base

On the basis of information received from the survey of military facilities, Haakonsvern naval base was ranked in group 1. Further investigation uncovered several areas with contaminated ground. At the fire training area, large amounts of hydrocarbons were found both in soil and water. Several small areas were also found to have been contaminated with hydrocarbons, while several other areas were proven to be contaminated with PCB. Investigation of sea sediments showed that the sediments were polluted with hydrocarbons, PCB, heavy metals, and PAH (Polycyclic Aromatic Hydrocarbons). In an area near the boat yard extremely high concentrations of organic tin compounds were found. High levels of PCB detected in fish and shellfish caught in the area around the naval base resulted in restrictions on the consumption of fish and on commercial fishing by the State Food Control.

3. Remedial action at the fire training area

Chemical analysis has confirmed that the ground at the fire training area is highly polluted with hydrocarbons; furthermore, because the area is situated close to the seashore, hydrocarbons are dispersed into water and transported into the sea. The fire training area was closed in 1993 immediately after it became clear that the area was heavily polluted with hydrocarbons.

The remaining pollution by hydrocarbons in the fire training area was estimated to be 5000 - 20000 litres in a soil volume of 5000 m^3. This pollution is partly decomposed as a result of biological activity.

It was decided to start a project in which methods of cleaning up the fire training area were evaluated. Two main methods were evaluated, *ex situ* remediation and *in situ* remediation. The cost for *ex situ* remediation was several times higher than for *in situ* remediation, due to high excavation costs. Hence the decision was made to commence a pilot study on *in situ* remediation for the purposes of determining the rate of degradation and to discover whether the addition of fertilisers (nitrogen and phosphorus) increased the rate of degradation. This pilot study would also assist in estimating how much time must have elapsed before the contamination fell to acceptable levels.

Based on the information emerged from the pilot study it was decided to clean up the area utilising *in situ* biological remediation. To increase the rate of degradation, the addition of fertilisers and the injection of air were recommended. This remedial action started in 1996 and is planned to take about three years. The costs for this project are estimated at 0.3 mill $.

The criteria for terminating the *in situ* remediation were:

- remaining concentrations of hydrocarbons in the soil shall not necessitate restrictions in using the area as a parking place or storage area.

- the concentration of hydrocarbons in the groundwater shall not exceed 0.5 mg/l.

The final concentration of hydrocarbons in the soil has been estimated to 1500 - 2000 mg/kg following the termination of *in situ* remediation. It is planned to use the remediated area as parking place.

Figure 1 shows a map of the remediation area. Air is injected at a depth of five meters. Groundwater is pumped up and fertiliser added automatically before the water is infiltrated back again on the top soil layer. Several wells are installed in the treatment area in order to get groundwater samples for chemical analysis. The groundwater is sampled approximately four times a year. Figure 2 shows the concentration of hydrocarbons in four groundwater wells from the beginning of the remediation until the present.

So far this project shows that *in situ* remediation of hydrocarbon-polluted soil gives promising results. The hydrocarbon content in groundwater has decreased to a level below that which is requested by the State Pollution Control. Gas chromatograms verify that a degradation in the level of hydrocarbon pollution has occurred.

A new fire training area is under construction in a new area at the navel base. This fire training facility will utilise simulation methods and will utilise gas as fuel.

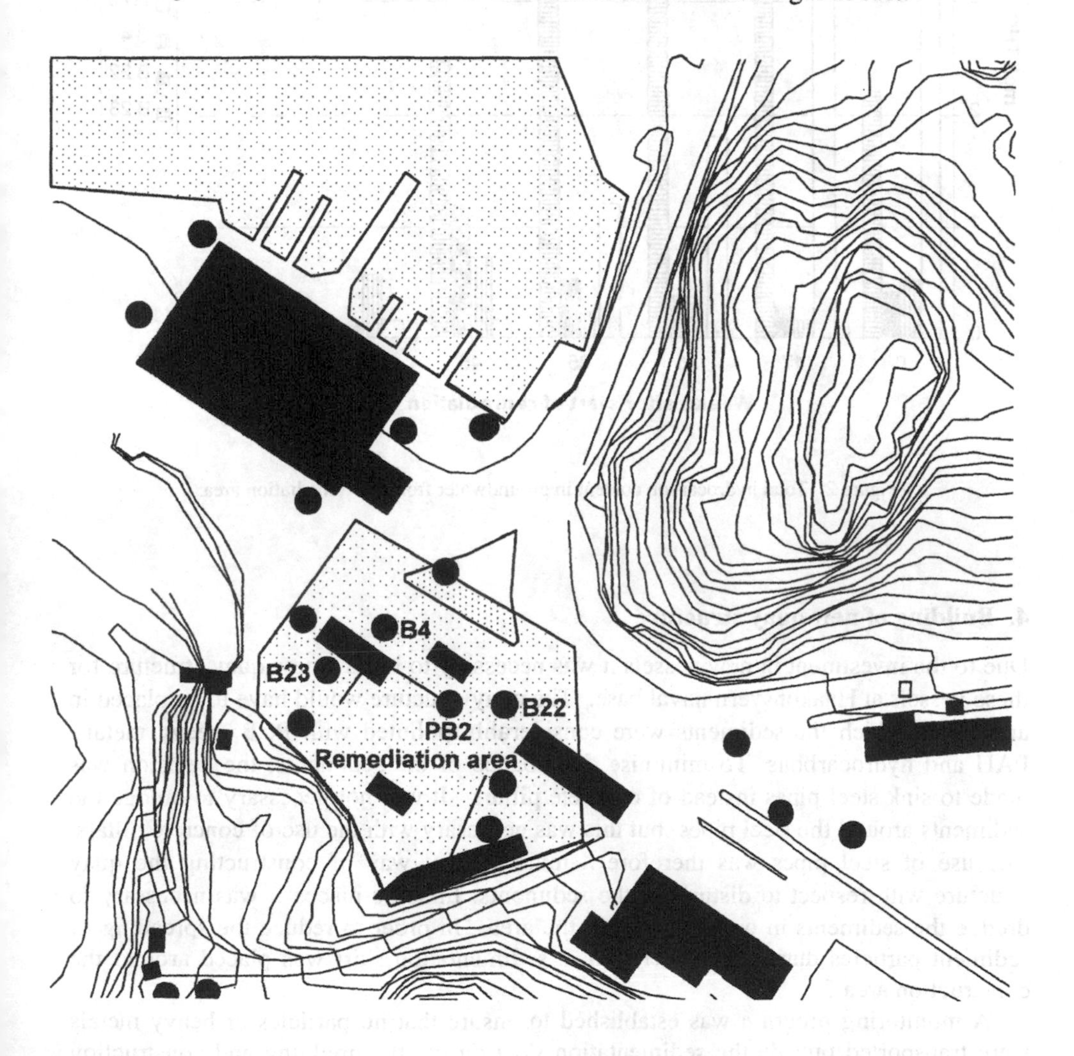

Figure 1. Map of the remediation area indicating the location of groundwater wells.

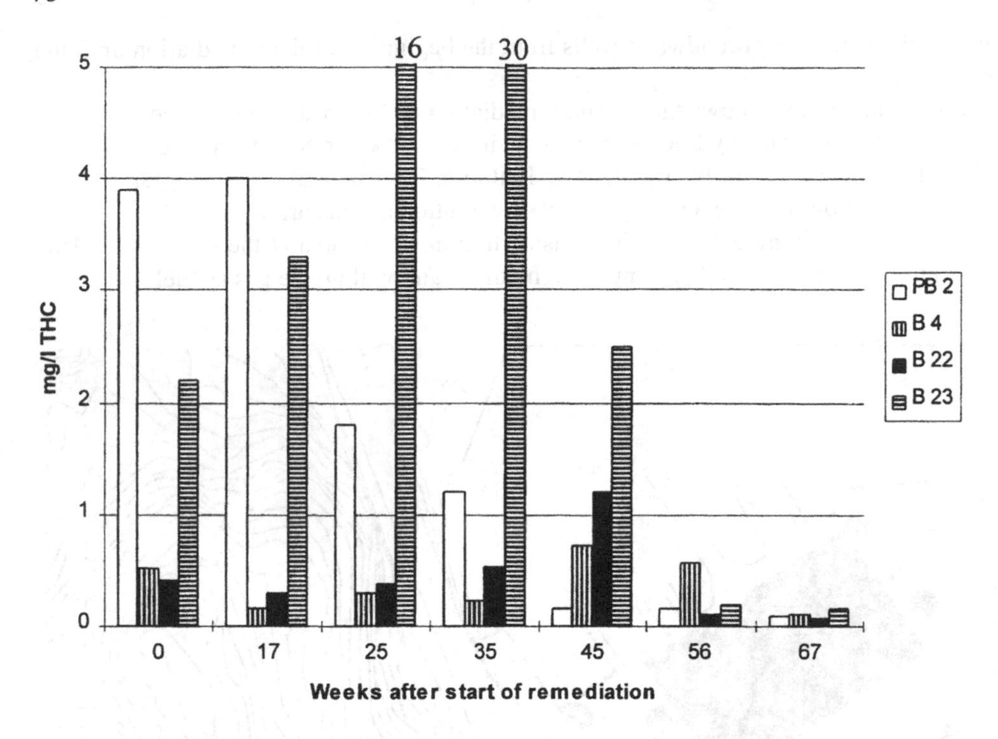

Figure 2. Total hydrocarbon content in groundwater from the remediation area.

4. Building of new quay structure

Due to the investment in new vessels it was necessary to build a new quay structure for these vessels at Haakonsvern naval base. The quay structure would have to be placed in an area in which the sediments were considerably polluted with PCB, heavy metals, PAH and hydrocarbons. To minimise disturbance of the sediments, the decision was made to sink steel pipes instead of concrete pillars. It was not necessary to dredge the sediments around the steel pipes, but this was necessary with the use of concrete pillars. The use of steel pipes was therefore a more careful way of constructing the quay structure with respect to disturbing the sediments. In some places it was necessary to dredge the sediments in order to deepen the area. In order to reduce the spreading of sediment particles during this operation, a sedimentation skirt was placed around the construction area.

A monitoring program was established to ensure that no particles or heavy metals were transported outside the sedimentation skirt during the dredging and construction work. Turbidity in sea water 20 m outside the sedimentation skirt was measured continuously at two sites. A reference station for turbidity was also established. In the event that the turbidity outside the sedimentation skirt exceeded the level at the reference station by a factor of 2, the work was halted and the State Pollution Control informed. The concentrations of copper, lead and zinc were measured three times each week at four stations outside the sedimentation skirt. The action limit for these metals was 6.0 μg/l for copper, 1.3 μg/l for lead and 20 μg/l for zinc. If the concentration exceeded these action limits, work had to be stopped and the State Pollution Control informed. Figure 3 shows the working area for the quay structure, monitoring stations, placement of the sedimentation skirt and the dredging area.

In the dredging area, pollution was found down to a depth of approximately 50 cm in the sediments. Therefore it was decided to carefully dredge the top layer of the sediments and to use conventional dredging techniques to get the right depth. A special washing device was constructed which fractionated the sediment into in four different grain sizes. The grain sizes in the fractions were 8-60 mm, 2-8 mm, 0.2-2 mm and 0.03-0.2 mm. In the first fraction stones, seaweed, shell and other big particles were collected. This fraction was discharged inside the sedimentation skirt where it was deep enough. The last fraction was collected in a filtering cloth.

The top layer of the sediment was pumped in small segments to the washing device in order to avoid disturbing the sediments. All of the fractions except the first one above 60 mm were so contaminated with PCB that they had to be stored in a sealed concrete silo. Prior to the dredging operation the volume of the dredged material was estimated to be 400 m^3, but only 21 m^3 were dredged mainly due to high water content in the sediments.

The level of turbidity, copper, lead and zinc outside the sedimentation skirt did not exceed action limits during the dredging activity. Figures 4 and 5 show the turbidity and copper, lead and zinc concentrations in sea water 20 m outside the sedimentation skirt during the dredging and working period in the sediments. The turbidity inside the sedimentation skirt during the careful dredging activity was at the same level as the level at the reference station. The concentration of metals in sea water inside the sedimentation skirt did not exceed the action limits during careful dredging.

While conventional dredging was underway, the turbidity increased dramatically to a level which exceeded the measuring range of the instrumentation (> 1000 NTU) inside the sedimentation skirt. The concentration of copper, lead and zinc in filtered (0.45 μm) sea water from this area does not exceed the action limits. Twenty meters outside the sedimentation skirt the level of turbidity was at the same level as the reference station, and the concentrations of copper, lead and zinc were below the action limits.

Samples taken from the sediment after the quay structure was finished indicate that the concentration of PCB in the sediment has decreased, but is still high. This may be ascribed to the conventional dredging activity which disturbed the sediment. It is possible that not all polluted sediments were removed during the careful dredging. If there were any polluted sediments left, this pollution was spread over the whole area during the conventional dredging. PCB and metals are absorbed to small particles and these particles have a low sedimentation rate and will sink against the top layer sediments. Hence the top layer, which was sampled, may have been polluted in this way.

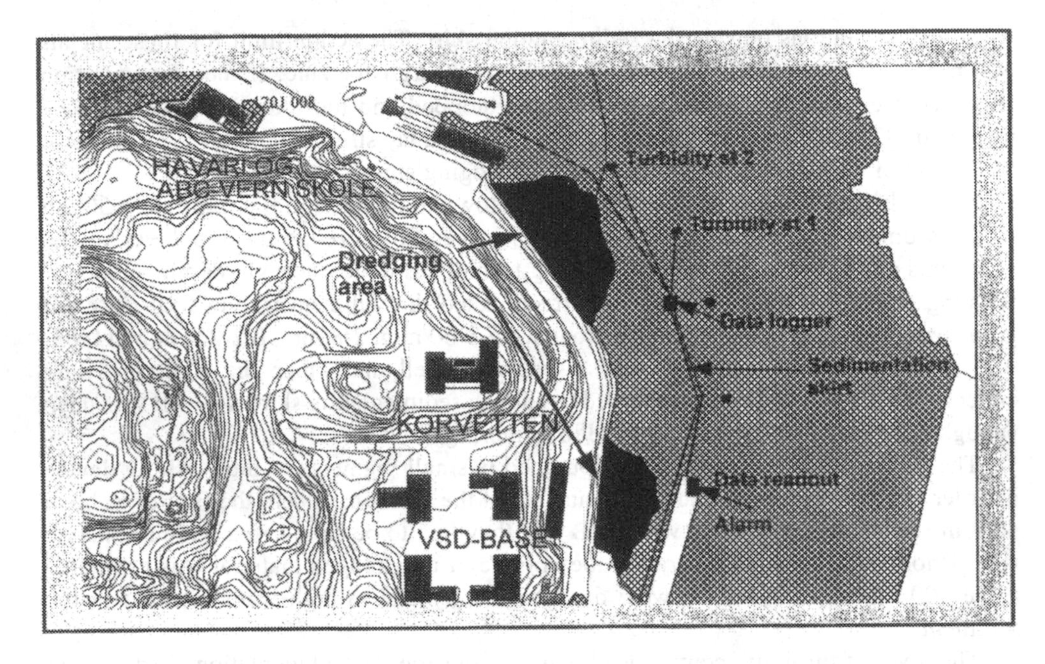

Figure 3. The working area for the quay structure, monitoring stations, placement of the sedimentation skirt and the dredging area.

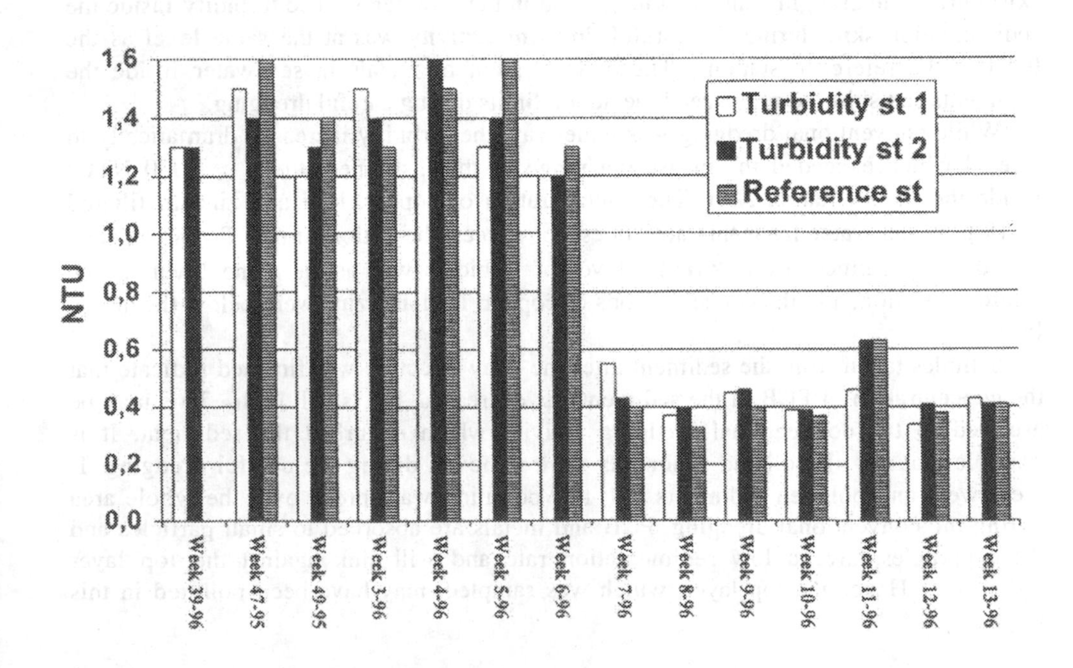

Figure 4. Turbidity measurements outside the sedimentation skirt during the period of work on the seabed.

This project demonstrates that it is possible to carry out a careful dredging without disturbing the sediment. It also shows that the pollution is absorbed very strongly to the sediment particles and very little is desorbed into the sea water during dredging. The use of a sedimentation skirt prevented the transportation of particles outside the dredging area. If the sediment is to compact, it is very difficult to use the careful dredging method, for almost no sediment was pumped up. It is very important to ensure that all polluted sediments are removed during the careful dredging operation; otherwise the remaining pollution could be mobilised and spread during vessel activity or any other activity that can influence the seabed.

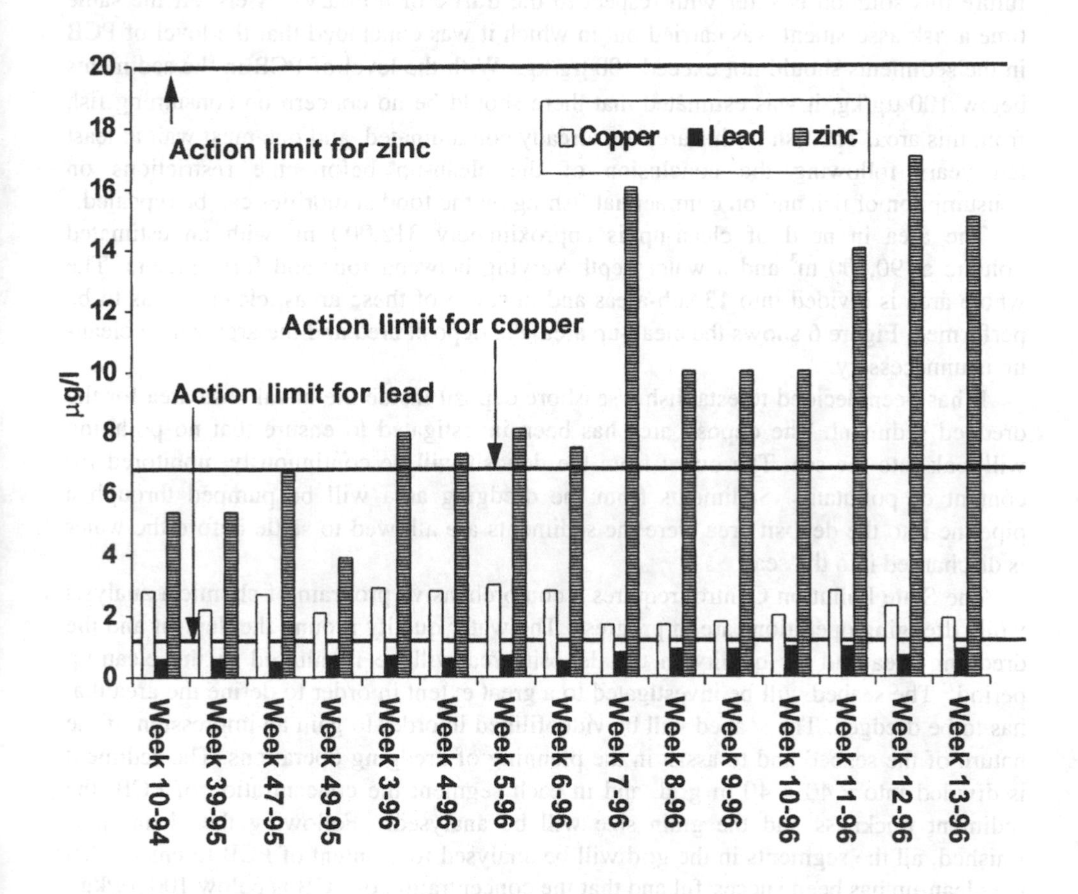

Figure 5 Measurement of copper, lead and zinc in seawater outside the sedimentation skirt during the period of work on the seabed.

5. Clean-up of the sediment around the naval base

Investigation of the sea sediments around Haakonsvern naval base reveals that the concentration of PCB is very high and in certain locales up to 1,000 times above background level. The PCB concentration in cod liver and fillet is about 20-30 times greater than the assumed background level. In mussel and crabs the concentration is more moderate.

A technical solution for the clean-up of the sediment has been studied. The conclusion from this was that the sediments could be dredged or they could be covered *in situ* by a geotextile. Dredging the sediments was the cheapest solution and for the future this solution is safer with respect to the traffic of military vessels. At the same time a risk assessment was carried out in which it was concluded that the level of PCB in the sediments should not exceed 100 µg/kg. With the level of PCB in the sediments below 100 µg/kg, it was estimated that there should be no concern on consuming fish from this area. The fish in this area are already contaminated, and one must wait at least ten years following the conclusion of the clean-up before the restrictions on consumption of fish and on commercial fishing by the food authorities can be repealed.

The area in need of clean-up is approximately 312,000 m^2 with an estimated volume at 90,000 m^3 and a water depth varying between four and forty meters. The whole area is divided into 13 sub-areas and in seven of these areas, clean-up has to be performed. Figure 6 shows the clean-up area, the deposit area and the area which clean-up is unnecessary.

It has been decided to establish a seashore deposit inside the naval base area for the dredged sediment. The deposit area has been investigated to ensure that no pollution will leak into the sea. The outlet from the deposit will be continuously monitored for content of pollutants. Sediments from the dredging area will be pumped through a pipeline into the deposit area were the sediments are allowed to settle before the water is discharged into the sea.

The State Pollution Control requires a comprehensive program of chemical analysis while dredging operations are in progress. The water quality around the deposit and the dredging area and air quality in the deposit area will be monitored in the clean-up period. The seabed will be investigated to a great extent in order to define the area that has to be dredged. The seabed will be videofilmed in order to gain an impression of the nature of the seabed and to assist in the planning of dredging operations. The sediment is divided into a 40 x 40 m grid, and in each segment the concentration of PCB, the sediment thickness and the grain size will be analysed. Following the clean-up is finished, all the segments in the grid will be analysed for content of PCB to ensure that the clean-up has been successful and that the concentration of PCB is below 100 µg/kg.

The construction of the deposit area will be completed in the beginning of January 1998. Dredging operations will commence in the middle of January 1998 and will be finished at the end of 1998. Clean-up of the sediment is estimated to cost about 14 mill US dollars.

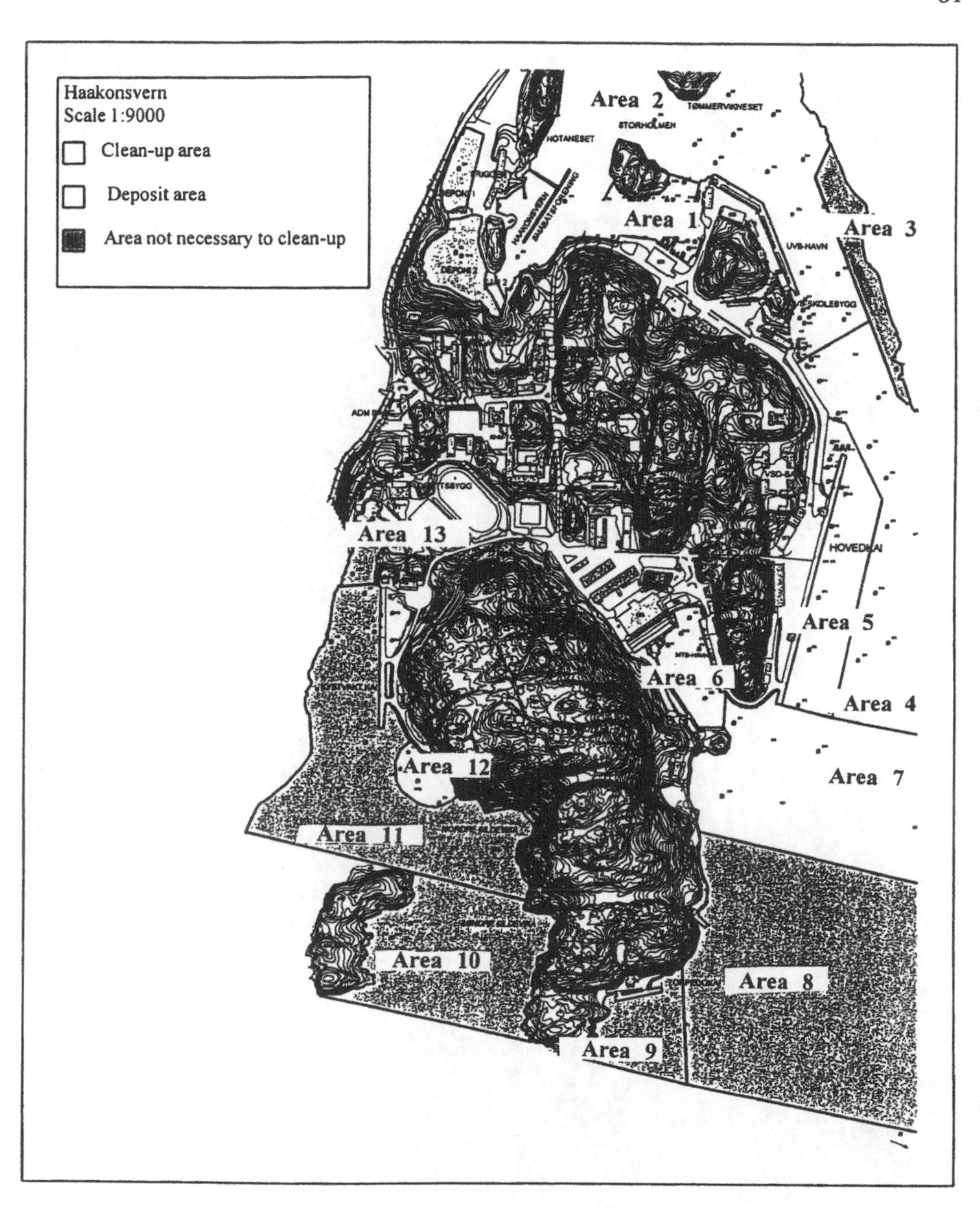

Figure 6. Overview of the clean-up area.

CHEMICAL ASPECTS OF ENVIRONMENTAL CONTAMINATION AT MILITARY SITES

S. NEFFE

Institute of Chemistry and Protection against Chemical Warfare Agents
Military University of Technology
Kaliskiego St. 2 01-489 Warsaw. Poland

Abstract

Military forces are producer of large quantity of toxic and harmful wastes. Many of these wastes are similar to those of large civilian industrial organizations, but some are peculiar to the Army mission. The paper presents the chemical aspects of environmental impact of former and present military activities. Up-to-date information and experience show that on military sites a majority of the suspected contamination is aliphatic, as well as aromatic hydrocarbons including polyaromatic hydrocarbons, chlorinated aliphatic hydrocarbons, heavy metals (Cr Cu, Pb, Hg, Zn and Cd), organic amines, plastic, inorganic salts, acids and bases. Relatively few suspected contamination caused by accidents have been registered in the sites evaluated so far.

1. Introduction

The Army is a huge and diverse organization with heavy demands for equipment, chemicals and other materials. Military forces are also a producer of large quantity of toxic and harmful wastes. Many of these wastes are similar to those of large civilian industrial organizations, but some are peculiar to the Army mission. For example. the Army has major responsibility for energetic materials - propellants and explosives, and for NBC defence. These tasks and activities have strong influence on the environment throughout generating toxic and harmful waste. The world wide disarmament process offers the opportunity to integrate ecological interests in the defense mandate. As a logical consequence, an increasing number of people are devoting particular attention to the complex problems relating to "environmental protection and defense mandate".

The peace time activities have more or less severe influence on the environment. In most countries, the armed forces by their size, contributes substantially to the countries environmental condition.

After the end of the "cold war" large former military areas in Europe have become available for conversion to civil use. However serious environmental problems have to

F. Fonnum et al. (eds.),
***Environmental Contamination and Remediation Practices at Former and Present Military Bases*, 83–92.**

be solved on abandoned and present military sites. It is typical feature of military areas that relatively small but highly polluted sites are neighbors to vast areas which are free from pollution and thus could be re-used for civil purposes [1-6]. The chemical aspects of environmenta~ impact of former and present military activities is the subject of this paper.

2. Types of military sites and their characteristic pollution

Military bases vary considerably in size, in the width of activities which were conducted on site, and in the duration of use for military operations. As a result, the potential suite of environmental problems which may be encountered is site-specific. Typical military base includes installations that conducts various industrial operations, conducts training, houses troops, maintains combat equipment, and does all the other things a modern army must to do to stay operationally ready. The most complex and potentially most harmful contaminated sites occur on military installations involved in manufacturing and industrial operations. Activities at military bases causing pollution can be enumerated as follow:

- Production of energy for military installations,
- Accommodation of soldiers and their families, administrative buildings, headquarters.educational buildings ,
- Development. production and testing weapons and weapons system,
- Stores of military materials, including fuels, munitions,
- Construction of defense related infrastructure, .
- Transportation activities (road, rail and air),
- Troops exercising,
- Telecommunication activities.

Functionally military bases may be viewed as small cities or settlements with regard to the municipal. commercial and technical mix of potential wastes. Military bases or complexes usually consist of typical installations and infrastructure (Table 1).

The military activities allow contaminants to enter the environment. While some of these contaminants are normally present at low concentrations. other contaminants are at very high concentrations.

The causes for pollution at military bases that ultimately lead to contamination are listed below. They can be classified into 3 groups: storage of substances, operating pollution - such as sense. which include leaks in pipes and tanks, leaky sewer systems. as well as actual accidents and fires[6].

TABLE 1. Types of military sites and their typical infrastructure.

Type of military site	Military installations and infrastructure
Garrisons	Housing area (troop accommodation and residential areas), Storage area, Heating plants with coal or heating oil storage areas. Solid waste disposal, Facilities for water supply, Cookhouses and mess facilities. wet and dry washing Animal husbandry,
Maintenance facilities and motor vehicle garages	Workshops and repair facilities Maintenance hangars. Parking area Deposit areas. Refueiing points. W'ashing and repair ramps.
Air bases and airfields	Runway Taxiway
	Fuel depots, Refueling points, Shelters and hangars. Munitions bunkers Housing area Telecommunication facilities
Exercise and training grounds	Shooting ranges Exercise grounds. Maneuver areas Roads and paths NBC training areas Special terrain areas, Driving school tracks Target areas Logistic facilities Camps
Storage and handling of hazardous substances -short term storage -long term storage Harbors and navy facilities	Fuel depots Refuelling stations Munitions magazines Chemical substances stockpiles Supply stores Fuel depots Refuelling points Munitions magaziness Telecommunication facilities Workshops Housing area
Others	Hospitals Shops Social and recreational facilities

Causes for Pollution that may Result in Contamination :

- Storage of environmentally hazardous substances,
- Leaks in pipes and storage tanks
- Overfilling of storage tanks
- Drip and spill lossesLeaky sewer systems
- Accidents
- Contaminated fire-fighting water.

This list is certainly not complete, but it shows that, in contrast to many industrial sites. all these potential causes are often found at military sites. Almost all of the contamination in military bases originate from spills or leakage of fuel or solvents such as kerosene and chlorinated hydrocarbons.

Potential waste types that may be expected at big former military facilities include:

- Aircraft fuel - kerosene,
- Motor fuel- petrol and diesel fuel,
- Lubricating and cutting oils,
- Cleaning solvents - chlorinated hydrocarbons,
- Rocket fuel and oxidant - amines and nitric acid,
- Insulation materials, asbestos,
- Explosives and munitions,
- Laboratory reagents and photochemicals. .
- Paints
- De-icing substances,
- Municipal-type solid wastes,
- Medical wastes and drugs,
- Agricultural materials, fertilizers,
- Pesticides.

In military firing ranges several tons of copper, zinc, lead. cadmium and antimony are deposited annually. Studies of heavy metal pollution from these activities have been carried out in many countries. The general conclusion is that there are minor local pollution in the target and impact areas, but this has contributed significantly to the aquatic ecosystem.

3. Chemical substances as pollutants at military sites

A wide variety of pollutants are commonly found at military sites. In addition to the products commonly used for industrial activities the pollutants found at many military sites have a distinct military application.

TABLE 2. Shows the potential environmental risk caused by substances, materials and equipment used in military sites.

Kind of substance. material or equipment	Kind of chemical substance	Environmental risk
Decontaminants chlorine-based oxidants	- chlorinated lime	noxious
	- chloramine B	toxic
	- calcium hypoclorite	toxic
	- dichloramine	toxic
	- hexachloromelamine	toxic
Decontaminants strong bases	- sodium hydroxide	corrosive
	- sodium sulfide	corrosive, toxic
	- sodium carbonate	corrosive
	- sodium alcoholate	toxic
	- sodium phenolate	toxic
	- ammonia water	noxious
Decontaminants weak bases	- sodium bicarbonate	neutral
	- monoethanoloamine	noxious
	- diethyienetriami	noxious
Salts	- calcium chloride	noxious
	- magnesium chloride	noxious
	- zinc chloride	noxious
	- potassium chlorate	toxic
Decontaminants anticorrosive additives	- chromates	toxic
	- phosphates	noxious
Solvents	- dichloroethane	toxic
	- ethanol	noxious
	- ethylene glycol	toxic
	- petrol	noxious
	- diesel fuel	noxious
Disinfecting agents	- phenol	toxic
	- cresol	toxic
	- formaldehyde	toxic
Flame agents	- napalm (petrol + Al salts of fatty acids)	noxious
	- triethyl aluminium	toxic
Smokes	- low viscosity petroleum oil	noxious
	- parafi-free kerosene	noxious
	- silicon tetrachloride	toxic
	- titanium tetrachloride	toxic
	- carbon tetrachloride	toxic
	- white (red) phosphorous	toxic
	- sulfur trioxide	corrosive
	- chlorosulfonic acid	corrosive
	- zinc oxide	neutral
	- ammonium chloride	noxious
	- anthracene	noxious
	- naphthalene	noxious
	- grained aluminium	neutral

From chemical view point the following substances or substance groups are usually anticipate on the military sites:

- Mineral oil hydrocarbons (MHC),
- Monoaromatic hydrocarbons - benzene. toluene. xylenes(BETEX),
- Highly volatile halogenated hydrocarbons (HVHC).
- Polycyclic aromatic hydrocarbons (PCAH),
- Heavy metals (e.g. arsenic, nickel, mercury, selenium, thallium, Lead, cooper,
- cadmium, zinc).
- Nitrate, nitrite.
- Aromatic and aliphatic amines
- Nitric and suifuric acid.

3.1.1. *Halogenated hydrocarbons*

hexachlorocyclohexene (mix), 1,2,3-trichlorobenzene. hexachlorobenzene, DDT, PCB, dioxins/furans.

3.1.2. *Phenols*

phenol. cresols, monochlorophenol, 2,4-dichlorophenol, 2,4,5trichlorophenol. 2,4.6-trichlorophenol. tetrachlorophenol, pentachlorophenol

3.1.3. *Explosives*

Table 3 shows a representative list of explosives derived substances that may be of interest in the investigation of abandoned military bases [5]:

TABLE 3. Explosives derived substances that may be found at abandoned military sites.

No.	Chemical name of explosive	Abbreviation
1	2-mononitrotoluene	2-MNT
2	3-mononitrotoluene	3-MNT
3	4-mononitrotoluene	4-MNT
4	2,4-dinitrotoluene	2,4-DNT
5	2,6-dinitrotoluene	2,6-DNT
6	3,4-dinitrotoluene	3,4-DNT
7	3,5-dinitrotoluene	3,5-DNT
8	2,4,6-trinitrotoluene	2,4,6-TNT
9	4-amino-2,6-dinitrotoluene	4-amino-2,6-DNT
10	2-amino-4,6-dinitrotoluene	2-amino-4,6-DNT
11	Hexahydro-1 ,3,5-trinitro-1 ,3,5-triazine	RDX
12	Hexanitrodinhenvlamine	Hexvl
13	1 ,3,5-trinitrobenzene	1,3,5-TNB
14	1 ,3-dinitrobenzene	1,3-DNB
15	Octohydro-1, 3, 5, 7-tetranitro-1, 3, 5, 7-tetrazine	HMX

De-icing agents: at airfields several tones of ethylene or propylene glycol is used annually as runway de-icing agent. Several tons of inorganic salts or urea are also used annually for melting ice on runways.

Liquid rocket fuel - samine: triethylamine. diethylamine. dimethylaniline.

4. Remediation of military sites

The basic question in the clean-up decision process is "how clean is clean". The issue of "how clean is clean" is often framed as a choice between using generic numerical criteria, such as the "Dutch list", or a site-specific risk assessment approach. In practice, the generic criteria approach has often been coupled with a general clean-up level to achieve soil "multifunctionality". The site-specific risk assessment, has usually been coupled to a clean-up level matched to the intended future use of a specific site [7].

As pollutants move through the geological environment they can be:

- Eliminated due to the natural degradation processes (biodegradation. oxidation.
- photodegradation, evaporation),

- Dispersed. resulting in a decrease in pollutant concentration.
- Accumulated without chemical changes (e.g heavy metals. DDT),
- Transformed into new chemical compounds which may have different chemical and toxicological properties.

The model, called the Prioritization Model, is intended to provide a rational approach to assigning relative ranking among military contaminated sites [4, 7]. The Prioritization Model is divided into three components:

- Potential to release - identifying whether the waste has released and penetrated the groundwater or whether the site conditions are such that the pollutant can easily leach into the groundwater. Potential to release considers measures of net precipitation, depth to aquifer, permeability, and degree of containment.
- Pollution characteristic - identifying how toxic and mobile constituents are in the waste and the quantity of pollution that may be potentially released into the groundwater. Pollution characteristic portion of the Model groups contaminants into 14 classes, based on toxicity/mobility considerations.
- Potentially affected population -identifying the nearest groundwater well and the size of the potentially affected population served by the well or group of wells.

To examine an area the typical approach is a combination of chemical analyses and risk assessment which comprise the most significant proportion of the potential risks in each environment medium (e.g.! groundwater, soils, surface water, sediments). Based on the detected soil, sediment. surface water or groundwater concentrations and, to the extent possible, the established toxicity benchmarks or Carcinogenic Slope Factor, indicator analyses are identified which individually represent a significant proportion to the potential risk. This selection step ensures that the risk assessment will emphasize the compounds which are the principal contributors to potential site hazards [7].

For the decontamination of polluted sites there are the following methods accessible [8-1 3]:

- encapsulation or insulation
- off-site removal to special waste deposits
- pumping off water or air out of the ground
- physico-chemical or washing treatments
- microbiological treatment
- thermal treatment by incineration or
- pyrolysis
- other physicochemical methods

The use of these techniques depends on:

- the structure of the contaminated soil

- the type of contamination
- the proposed use of the decontaminated site

The thermal techniques are the most effective ones, they can be used for all soil structures and organic contamination. The washing techniques are able to clear the soil from organic and inorganic contamination and is very useful for soil polluted with military decontaminants. Depending on the structure of the contaminated soil there is a residue of about 10 to 30 % of the decontaminated soil. which must be deposited or otherwise treated.

The microbiological treatment uses the natural capacities of microorganisms to convert organic contamination into water and carbon dioxide. The method is able to clean mineral oil. Depending on the contamination the treatment takes about 12 to 24 months.

5. Conclusions

Many different environmental problems exist in the military. Different harmful and toxic materials, both dilute and concentrated. as well as mixed with a variety of other substances are generated during the Army activity. It is unlikely that one method will be effective in dealing with all these problems. The military forces should support research and development programs to advance fundamental knowledge in a variety of areas, all with strong potential to provide useful clean-up technology in the near future. The clean-up technologies developed will have importance beyond their specific military applications. They can be useful in destroying a broad range of toxic wastes - herbicides, pesticides and other troublesome organic compounds from civilian industry.

Up-to-date information and experience show that on military training areas a majority of the suspected contamination is aliphatic, as well as aromatic hydrocarbons including polyaromatic hydrocarbons, chlorinated aliphatic hydrocarbons, heavy metals (Cr, Cu, Pb, Hg, Zn and Cd), organic amines, plastic, inorganic salts, acids and bases. Relatively few accidents causing contamination have been registered at the sites evaluated so far. But this is due to the fact that files on accidents, fires, explosions e.t.c. are rarely available.

In summary, it require close examination to determine extent of a possible contamination of the subsoil (e.g. whether the groundwater is affected). In the future one can assume that former military training areas have to be thoroughly examined and rehabilitated whereas many other sites can be available after environmental risk assessment without rehabilitation.

References

1. Schafer R., Results of the contaminated-site survey, *Proceedings of the International Workshop on Military and Armament Contaminated Sites*. 19-22 June 1995, Berl in-Schoneberg, pp. 309 - 322.
2. Franzius V., Winde Ch., Armament-Contaminated sites in the Federal Republic of Germany, *Proceedings of the International Workshop on Military and Armament Contaminated Sites,* 19-22 June 1995, Berlin-Schoneberg, pp.146 - 173.
3. Borner S.,Evaluation of groundwater quality in the meadows along the river Elbe near Torgau with special attention to the Armament-Contaminated sites. *Proceedings of the International Workshop on Military and Armament Contaminated Sites*. 19-22 June 1995. Berlin-Schoneberg, pp.186 - 213.
4. LePage A. C.. (1995), *Cleanup strategies for U.S. Army. Europe. military bases*. NATO ASI Series.
5. Aulenbacher R., Kiefhaber K.P., Registration of contamination potential at military installations subject to conversion in Rheinland-Pfalz, *Proceedings of Environmental Restoration Conference,* Munich 25-27 October 1994, pp.434 - 447
6. Kramer J., Registration of contamination potential at military installations subject to conversion in Rheinland-Pfalz, *Proceedings of Environmental Restoration Conference.* Munich 25-27 October 1994. pp.434 - 447.
7. Hinsenveld M., (1995), Remediation strategies for contaminated (former) military sites, NATO ASI Series, Environment-Vol.1 *Clean-up of Former Soviet Military Installation,* R. C. Hendron et al. (eds.), Springer-Verlag, pp. 57 - 78.
8. Spyra W., Mobile disposal of conventional munitions with special risk potential. *Proceedings of the International Workshop on Military and Armament Contaminated Sites,* 19-22 June 1995, Berlin-Schoneberg, pp. 214 - 223.
9. Jenkins R.O., Leach C.K.,(1996), Priorities and progress in the application of micro-organisms to the management of toxic metals/radionuclides pollution, *Environmental Engineering and Pollution Prevention*, J. Wotte et al. (eds.), pp. 195 - 216.
10. Teaf Ch. M., (1995), Human health and environmental risks associated with contaminated military sites, NATO ASI Series, Environment-Vol.1 *Clean-up of Former Soviet Military Installation,* R.C. Hendron et al. (eds.),Springer-Verlag, pp 31 - 44.
11. Ondruschka, B., Hirth, Th. (1996), Advanced demiiitarization technologies: disposal of hazardous military waste by supercritical. ultrasonic or microwave irradiation, *Environmental Engineering and Pollution Prevention*, J. Wotte et al. (eds.), pp. 111 - 121.
12. Hirth Th., Bunte G., Eisenreich N., Krause H., Decontamination of soil and groundwater by supercritical fluids, *Proceedings of Environmental Restoration Conference*. Munich 25-27 October 1994. pp.320 - 333.
13. Shaw R. W., Research on the destruction of chemical warfare agents. *Proceedings of 4th Int. Symp. Protection Against Chemical Warfare Agents*. Stockholm. Sweden 8-12 June 1992, pp. 131 - 137.

US ENVIRONMENTAL PROTECTION AGENCY'S BALTIC TECHNICAL ASSISTANCE PROGRAM: FROM CAPACITY BUILDING TO ENVIRONMENTAL SECURITY

VACYS J. SAULYS
Office of International Activities
US EPA Region 5
77 W. Jackson Blvd. R-19J
Chicago, IL 60604, USA

1 Introduction

The United States Environmental Protection Agency's (US EPA) Baltic assistance program had its genesis under the United States/ Union of Soviet Socialist Republics (US/USSR) Environmental Exchange Program in the mid-1980's when, with significant resistance from the Soviet side, a Lithuanian component was recognized in the formal agreement. Attempts to develop Latvian or Estonian components were not successful. The Lithuanian project focused on environmental degradation rates of toxic substances, water quality modeling and biological effects of environmental contamination. The program introduced studies of neoplasms in natural systems, screened the Nemunas River mainstem for toxic hot spots, trained Lithuanian scientists in Qual IIe and other models and produced an analysis of Hydromet's historical water quality data for Latvia. It became dominant during the 1992-93 period as the Agency's priorities switched from scientific exchange programs in Eastern Europe to government capacity building focus driven by US Agency for International Development funding.

The US AID funded environmental programs, initiated in 1992 as part of US government's assistance to Central Europe's reestablished democratic governments, required focusing US EPA assistance program to Estonia, Latvia and Lithuania on:

1. Establishing democratic principles in government
2. Building capacity in the new environmental ministries

The US EPA used 6 methods to achieve these objectives:

1. Quick Response Standardized Environmental Training Programs;
2. Targeted Management Training;
3. Demonstration Projects, focused on "Hotspot" issues of major national and local importance;

Disclaimer: This document has not completed review by US EPA and reflects only the opinion of the author.

F. Fonnum et al. (eds.),
***Environmental Contamination and Remediation Practices at Former and Present Military Bases*, 93–102.**

4. Targeted roadblocks to timely production and analysis of environmental information for management decisions;
5 Identified key low cost technology and information to encourage desired change in existing practices;
6. Direct technical consulting.

2. Standardized Training

Since the funding for US EPA environmental programs was extremely limited ($ 100-200,000 per country per year), the funding was stretched by EPA and its cooperators by contributing not only salary expenses, covering administrative expenses but also at times paying for international travel. Even so the programs had to be very narrow to avoid dilution. As a result the standardized training approach provided trained government personnel and developed local training capacity(where noted by *) for:

A. Principles of Environmental Impact Assessment*
B. Chemical Emergency Health & Safety (in cooperation with US Air Force)*
C. Forest Health Monitoring Field (with US Forest Service and Bowling Green Ohio University)
   1. Field data collection
   2. Computerized data base management and statistical analysis
D. Computer Operations
   1. Local Area Network Administration
   2. Unix GIS Workstation Operations
   3. Introduction to INTERNET (through CIESIN)

3. Management Training

The management training was targeted to provide specialized technical familiarization in key project areas to both senior specialists and management. It ranged from participation in US conferences, formal EPA training courses (both in country and the United States), to tailored briefings at US EPA, State and municipal agencies; and, to actual participation in field surveys and laboratory studies with government personnel in the United States. This was particularly useful in the areas of GIS and INTERNET capacity development, in pesticide management, water quality monitoring, and hazardous waste management and evaluation. This approach was primarily used in Latvia and Lithuania due to the broader nature of their assistance programs.

4. Demonstration Program

The US AID, US EPA and respective Baltic environmental ministries negotiated the selection of demonstration projects and sites. They were generally targeted to assist the countries address commitments in their respective Environmental Action Plans. Estonia requested assistance to conduct an Environmental Impact Assessment on the expansion of their existing oil shale mining operation. In Latvia the focus was narrowed to issues dealing with the City of Daugavpils drinking water supply. In Lithuania, assistance was requested to help evaluate the threats to the Siauliai City's potable groundwater supplies by Zokniai Airbase and to identify potential pollution prevention sources at the Mazeikiai Oil Refinery. Three of the four demonstrations had a community participation component that required citizen participation, education and crossgovernment involvement. The projects were designed to deliberately force working relatiorsh ps across agencies to accomplish project goals and involved not only government officials at the agencies or delegate responsibilities and initiatives to local govemments countercurrent to centralization tendencies of the new national institutions at the given moment.

4.1. ESTONIA

The Estonian Kurtna Mines EIA demonstration project was complex in that not only the local, regional and national government agency staff had be trained in the EIA process and the public persuaded to participate, but also mitigation alternatives to current and potential future impacts had to be identified. As part of the project, training was conducted in Principles of Environmental hnpact Assessment, on public partiåpation program design, and on modeling and assessing contaminant movement in groundwater. The demonstration program successfully completed the assessment, producing a decision document that reflected both local government and public input for the national government to act upon. The decision document was accepted. Some of the recommended changes to practices and infrastructure investments have been financed by local government, others have been presented to the donor community for potential grant funding. A monitoring program for the lakes has been initiated by the local Sillamae University, which had been designated as an EPA literature and software repository at the request of the US Embassy. The lake levels have been reported to be returning to their fommer levels. A new enviror~mental impact assessment law is under review. A certification program was created for EIA practitioners based on US EPA's course.

4.2 LATVIA

In 1993, the City of Daugavpils, Latvia, attempt to switch its drinking water supply from the Dauguva River to nearby groundwater supply had stalled due to lack of funding. The current system delivered water often with an unacceptable chlorine content ir~ addition to other problems associated with an aging distribution system. The intake was sanding in and the water supply was subject to contamination by upstream sources. The city wæ regarded by donors to be of high financial risk, a hopeless backwater with a low economic development potential and a management team destined to be stuck in the former system. US EPA and WDNR project design mission met with the City leaders explained that there was little funding. However, the US would work with the city to design some activities that would provide not only physical benefits but also raise their profile; but the city leadership would have to work jointly with us. Wisconsin Department of Natural Resources (WDNR) and Rust Engineering designed and ordered a chlorine metering system for the drinking water plant, associated chlorine gas warning and emergency response equipment, and provided basic low cost improvements to the analytical control laboratory. Training to operate the equipment was provided respectively by the vendor and the WDNR.

As the next step, the US EPA and WDNR assistance team persuaded the City to develop an American style Wellhead Protection Plan to protect it's future drinking water supply and to demonstrate to outsiders the initiative and the self discipline needed to undertake more complex projects. A community based stakeholder steering committee was established in 24 hours, representing municipal and regional government interests, education, industry and environmental interests. The World Bank was advised by EPA of the project and kept informed of progress. An environmental literature and environmental software repository was established at the local university by US EPA. The information was used to design local school environmental curri';uia, develop public information fact sheets and deploy a limited public information program. By the time the plan was presented at the First Groundwater Protection Conference to Baltic municipal and environmental agency specialists, the World Bank was negotiating with the Latvian Government to fund Daugavpils water and wastewater infrastructure modernization. The WB also participated in the Conference. This past Spring the general contractor was tasked by Municipality to revise the Wellhead Protection Plan as part of the required overall facility management plan.

The planning process identified a number of regulatory deficiencies for Latvian groundwater protection as well as a number of potential contamination sources of the local drinking water aquifers. The Latvian Environmental Protection and Regional Development Ministry (LEPRDM) requested that, rather replicating the planning process at other municipalities, the program be redefined to assist the Ministry to improve the national groundwater protection program and to advise Daugavpils municipal government on interim remedial actions. WDNR presented its program and legislation as a guide to the Ministry. The American examples were analyzed and compared to the Latvian requirements and laws. A workshop involving specialists from all government levels and affected agencies was held this June. An Action Plan to improve the groundwater program was drafted, reviewed and is being finalized for submittal to the LEPRDM. A Groundwater Modeling Workshop to build modeling expertise for wellhead protection area delineation was held the previous year. Participants included specialists from all 3 project countries.

Most of groundwater threats in Daugavpils were due to leaky oil tanks at both military and industrial fuel oil depots or historical industrial waste oil dumping. A US EPA review indicated that situations were well described by Latvian consultant studies. Information on available environmental criteria, remediation technology, treatment processes and options was provided for the time when either public or private sector funding was available. Two of the sites were an inactive municipal waste landfill and existing landfill operation. Both were known to be leaching into the groundwater. The WDNR sent an inspection team which reviewed the municipal waste management collection and disposal practices. The team on exiting made recommendation for changes in operations at the current landfill and flagged the historical landfill as posing a greater threat due destruction of the cover by scrap metal scavengers. WDNR proposed an outline for an interim management manual of the landfills until funding could be identified to reconstruct the current operations. The draft of the manual was recently completed and is under review. The City is seeking funding for an engineering and design study as part of the first step in securing funds to eliminate further contamination of its groundwater aquifers and divert leachate to a treatment system.

4.3. LITHUANIA

The Lithuanian Environmental Protection Ministrv (LEPM) requested two demonstration projects: I. A pollution prevention overview of Mazeikiai Refinery to help the Ministry's predecessor Department define environmental control restrictions in the event the facility received funding for reconstruction; and, 2.) Assess the environmental and public health risks posed by the Zokniai Air Base, particularly to the City of Siauliai drinking water aquifers.The refinery evaluation went the traditional route of using an EPA contractor that had just finished some similar refinery evaluations in Trinidad. Suitcases of reports in Russian, Lithuanian and English were reviewed prior conducting a screening survey in Lithuania. Based on these reviews it was decided to combine the next field sampling mission with a training exercise for regional inspectors.

The survey course presented an overview of US regulatory program, provided copies of models and training in field emissions modeling and emphasized sampling and analytical quality assurance. The field survey flagged that the area's dominant source of potential ambient noncompliance with S02 criteria as being the local heating plant and not the refinery; recommended a leak management program and investment in new product loading system as the prevention payback areas for volatile organics. Its survey raised issues about the technical adequacy of the existing groundwater monitoring system and questioned the need for a diversion system out of the yenta River watershed, given the use of secondary biological treatment, followed by a six month or greater holding and polishing pond capacity. The major differences in the 3 way split samples were reflective of analytical method differences but underscored the need to improve the analytical technical base to analyze at criteria levels. The product samples, taken at LEPM's request, showed it generally comparable to US gasoline for measured parameters and demonstrated the refinery's ability to produce lead-free gasoline. The study also demonstrated the difficulty to work with a facility whose feedstock and product delivery were at the whim of international and national politics.

The Zokniai Air Base project was a more complicated program. At the onset of the program the Danish EPA committed to fund an investigation of the base leading to remediation of the petroleum product contamination. An agreement was made with the Lithuanian Environmental Protection Ministry (LEPM); the Danish EPA; its consultants, Kruger Consult and the Baltic Consulting Group to have the US EPA focus on offsite monitoring issues, provide consulting reviews to LEPM of technical proposals and interim work products and use the base territory as necessary for hazardous waste site assessment training programs. This resulted in a multilevel capacity building program centered on the City of Siauliai:

1: Siauliai municipal groundwater baseline study: to provide a snapshot Siauliai groundwater quality. At the City's request. this was expanded to replicate the Daugavpils Wellhead Protection Planning Program.
2. Development of Standard Operating Procedures for Hazardous Waste Site Assessment.
3. Hazardous Waste Site Assessment Workshop to demonstrate the US approach and key field technologies.
4. Chemical Emergency Response Health and Safety for first responders and hazardous waste site investigators.

The demonstration program has yielded a number of outputs. Most important of these is the forging of both of formal and informal working relationships between the LEPM Central Laboratory, its field staff and the Lithuanian Geological Survey.This is followed by the multi level participation by environmental health and other agencies in Siauliai program. The Lithuanian Geological Survey has initiated a new groundwater monitoring program tor petroleum products and chlorinated solvents. The Hazardous Waste Workshop surfaced for senior environmental officials the potential for chlorinated solvent contamination of groundwater in Lithuania and demonstrated more efficient and sensitive field technologies. The Siauliu Groundwater Baseline Study flagged chlorinated solvent presence in the Zokniai area and the importance of modern sampling and analytical technology to gather relevant data for environmental management decisions.

The basic US EPA Chemical Emergency Health and Safety course filled a vacuum for training on preplanned response to chemical emergencies at the first responder level to minimize risks to the environment, the public and the responder. At the Latvian and the Estonian governments request the training was repeated for their nationals, training nearly eighty fire fighters and other specialists. Nine fully qualified trainers were prepared to continue the course in their respective countries. By the end of the first course the Lithuanian fire fighters applied the training to a rocket fuel spill at Zokniai; subsequent waste pesticide repository fires in Lithuania; and an oil spill in Latvia. A follow-up training video, based on the NATO BERE '96 Exercise, is a available in English and will be shortly released in Estonian, Latvian and Lithuanian languages. This training program was feasible only through the joint efforts of US EPA and the US Air Force's European Command. The Air Force in part under the Military to Military program assisted with an airlift of 3.5 tons of EPA equipment and training materials, provide co-trainers and made available its Fire Fighters Academy in Ramstein Air Force Base for trainer familiarization. This program served as model for development of the current US Environmental Security pilots in the Baltic Countries.

The City of Siauliai was a key energy source in the demonstration program. It fully participated in defining the extent of the demonstration project activities and the development of its products A deliberate effort was made to cross link with other projects, using the City as a display of municipal level monitoring and information management. This resulted in establishing cooperative relationships with Illinois EPA technical staff who provided free consultations on wastewater treatment system design and technologies, pretreatment issues even some used analytical equipment. Siauliai University also has been designated as an EPA literature and software repository.

The program is notable also in the barriers it encountered. The US EPA Region 5 Standard Operating Procedures for Hazardous Assessment was translated into Lithuanian and prox ided to the Ministry as a baseline from which to develop Lithuanian SOPs for use in field investigations in Lithuania. Reputedly, due to donor requirements that their field proceeds be followed this initiative lost its impetus, particularly since it became a inter-Ministry issue. This appears to be a symptomatic situation, reflected in the apparent lack of ownership for the European Commission Directive on "Good Laboratory Practices".

5. Regional Monitoring and Information Management

The primary objective of the US EPA program in the Baltics was to improve environmental data collection, analysis and delivery to decision makers. It was quickly evident that Ministries needed to be quickly computerized. steered in the direction special data base development, the extensive monitoring networks (compared to the US systems) reassessed for current relevance and the data collection and analytical technology base modernized. It was clear that the US program would not be able to accomplish this by itself and that it would have to be structured in such a way to encourage participation by other donor organizations and the recipient government organizations. The project was planned in three phases:

Phase 1: Monitoring Network Analysis
Phase 2: Monitoring System Needs
Phase 3: Implementation.

Due to the importance of computerization of the Baltic Ministries to improve their overall management capacity, it was decided to proceed directly to implementation without waiting overall system assessment by focusing on administrative systems and ensuring Geographic Information System capacity at each Ministry.

5.1 NETWORK ANALVSIS

Information on existing networks was inventoried, digitized maps were created and locational data, monitoring parameters and objectives entered from the questionnaires. The analysis, data base and associated software were presented at a general meeting to the client country monitoring management, visually identifying the network data base shortcomings. Estonia and Lithuania elected to proceed with water quality network review utilizing internal resources. Latvia received US technical assistance. Illinois EPA, US EPA and ECAT Latvia conducted a surface water quality workshop for Latvian national, regional and local monitoring specialists. The workshop developed an action plan and schedule for implementing a systems approach to redesign of the Surface Water Quality Network. Related standard operating procedures are being developed to ensure compatibility with European Commission Directives. These activities were further funded by the US program and are should be finalized this month.

5.2 NEEDS ASSESSMENT

All major donor organizations recognized that the archaic analytical and monitoring technology was a major barrier to improved monitoring results. However, neither of the 3 Ministries were able to present analysis of their current situation and needs to support a major equipment funding commitment. Agreement was reached with EU PHARE program on the overall scope of the monitoring needs assessment. Three US teams, staffed primarily by Illinois EPA and supported by staff from US EPA Wisconsin Department of Natural Resources, EU PHARE contractors and Bowling Green University, were launched simultaneously to conduct the assessments. The Reports were the basis of 3 PHARE grant applications which secured a total commitments of nearly

85 million USD. The Reports also recommend strengthened Quality Assurance Management programs and organizational changes.

All 3 Central Laboratories have improved their national quality control programs. The Latvian Central laboratory has been certified and Lithuaniane Certification package has been prepared. The Latvian Government simplified their monitoring program management by merging Hydromet into the Latvian Environmental Protection and Regional Development Ministry (LEPRD). The framework approach provided coherent rationale for upgrading the laboratory support system by the Baltic governments or other donors. The need to coordinate donor funding led the 3 Baltic Ministries and US EPA to establish annual, rotating donor and Baltic Country planning meetings which now appear to have been institutionalized.

6. Information Management

The framework approach was also used in computerization of the Ministries. The US EPA specialist worked in tandem with Ministry specialists to determine the overall objectives of the computerization effort, identify optimal Local Area Network design, identify INTERNET connectivity issues, and help establish some of the early Wide Area Network priorities and locate the Unix based GIS workstation. The project cooperators included CIECIN, University of Bowling Green, and the German State of Nordheim-Westfalen. A closeout international conference was held in Jurmala, Latvia, which brought together donor and Baltic specialists? to compare current achievements to future regional needs. The presentation of the Great Lakes Information Network catalyzed the formation of BALLERINA, a Baltic Sea regional network. The remaining US EPA assistance will provide improved linkages for Latvian and Lithuanian monitoring programs to INTERNET and BALLERINA.

7. Technology Transfer

While the US EPA program early on recognized that it could not provide significant funding for equipment, certain technologies were key to having changes in practices. It is apparent that, if the objective of the assistance program is efficient computerization of administrative and scientific information management, the recommended design has to be wired and equipped in at least a skeleton fashion to allow its stepwise completion by the client Ministry or other donors. Similarly, it is obvious that to ensure that training in modeling is used it is important that not only the model is made available and the computers to run the models are available. What is sometime forgotten. that the regulatory infrastructure must be open to use of models (or even new monitoring technology); and. the academic and developing consulting community communities needs to be co-trained to both carry on the training and to provide support to both the regulatory agencies and the regulated community.

With small project budgets, it is becomes harder to identify the key reinforcing technology. In Latvian and Lithuanian groundwater monitoring programs it became apparent that basic groundwater sampling equipment, sample transport coolers and basic accurate field analytical equipment were key to eliminate the fatalist complacency with high detection and variable sample quality. It also has the side benefit of improving the productivity of the field work and reducing some of the personal technical drudgery.

In the information management area, INTERNET connectivity was introduced in part as a way for the Ministries to communicate more efficiently by e-mail and transfer information internally. Secondarily, it was viewed by the new countries as a way to advertise themselves. It is now increasingly being used as a means of public outreach and is facilitating an increase in regional information exchange on environmental programs and environmental monitoring results.

The classical technology transfer medium: the book was not neglected in the technology transfer program. Initially, this program targeted the Ministries, and was credited with early the establishment of at least the Lithuanian Ministry's library. The repository system was expanded to Universities and Collages in project areas and eventual to 16 sites consisting primarily of university collections in the Baltic Sea watershed. It was supported by consisted primarily of culled holdings from the EPA Region 5 library, supplemented intermittently with new EPA publications. The repositories were expanded by a suite of environmental tutorial software. EPA environmental models and specially produced Technical Information Packages. The future updates are planned primarily through CDROM and INTERNET. The updates, like the previous material, will cover the entire spectrum of US EPA policies, methods, regulations and US technologies.

8. Consultation Support

It is apparent from the above program description that the Military site remediation, was not a strong focus of the capacity building program. This was reflective of early Baltic concentration on the classic water and wastewater treatment infrastructure. Despite the prominence of military base contamination. it initially was a distant third if not fourth on the Baltic governments environmental agendas, behind air, solid waste management and competed with orphan pesticide disposal.

The requests for direct consultation ranged from options and process for restructuring the environmental agencies (Latvia and Lithuania); reviews of environmental impacts of new projects (Butinge Terminal, Latvian Greenfield Papermill proposal and Siauliai Wastewater plant); harbor dredging (Liepaja and Klaipeda); pesticide disposal options (Lithuania and Latvia); analytical sample confirmation and methodology support; to EIA and other regulation reviews (all 3). However, the Lithuanians have led with regular consultation requests on proposals for contaminated site studies. subsequent interim contractor work products and final proposals for remediation. Probably, the most notable are the Klaipeda Oil Terminal (and its bioremediation facility), Zokniai Air Field and the "inventory of Damage and Cost Estimate of Remediation of Former Military Sites in Lithuania." More limited projects were reviewed in Estonia and Latvia. The reviews conducted by EPA identified several common elements in the proposals and work products:

1. Lack of local risk based criteria or criteria based on designated uses for aquifers or contaminated territories. This is still the major handicap to deriving consistent and reproducible management decision.
2. Contamination screening surveys limited to a narrow suite of pollutants by funding limitation or arbitrary donor agency assumptions.

3. Willingness to make decisions made on a limited suite of data, often on poorly documented analytical data quality from samples collected with technology unacceptable for the needed data quality objectives.
4. The remediation proposals have grown increasingly sophisticated, reflecting the private sector consultants increased familiarity with remediation technology options and hopefully, that of the government specialists.
5. Poor to non-existent public information much less public participation procedures.

9. Post US AID Technical Assistance

The US EPA technical assistance program, cofunded by US AID has been closed in Estonia, will end this year in Latvia, and will be completed in Lithuania by October, 1998. The current projects focus on Regional Monitoring issues.US EPA, the Department of Defense and Department of Energy have agreed with the Department of State to cooperate in the area of Environmental Security. Environmental military base management and orderly transfer of military bases to civilian use are recognized environmental security issues in the Baltic Countries. In fact, the lack of trained military environmental personnel makes it difficult for the environmental and defense agencies find common ground and priorities in some instances.

Discussions have been initiated with the three countries to select pilot projects which could serve as models for all three countries. As mentioned earlier, the projects are expected to draw upon the respective expertise of not only the military but civilian experts. They are expected to define a framework which will define an approach and will be completed by either the client governments themselves or with the assistance of other donors. The participating Ministries are expected to work closely together with their US counterparts. Based on the experiences described above, interested European partners can expect to be welcomed to help assure continuity. Agreement on the projects is expected by the end of 1997 and the start of implementation early next year.

RESULTS OF MASS TRANSPORT MODELLING AT SOME WELLFIELDS AND FORMER SOVIET MILITARY BASES IN LITHUANIA

M. GREGORAUSKAS
Vilnius Hydrogeology Ltd.
Basanaviciaus 37-1
2009 Vilnius
Lithuania

1. Introduction

Changes in groundwater quality are effected by a whole complex of natural and anthropogenic factors. Geological-hydrogeological conditions of the area, groundwater abstraction and pollution are the most significant among them. Groundwater pollution affects groundwater quality (its physical, chemical, and biological properties) making it partially or completely unusable. The extent of groundwater pollution is expressed by comparing concentrations of various dissolved components with the values of their highest permissible concentrations.

There are many sources and reasons for groundwater pollution. Former military bases, especially if they are situated near the wellfields for centralised water supplies, are a significant source of pollution.

In Lithuania only groundwater is used for centralised water supplies. Recently with the groundwater quality in main cities becoming worse, the prognosis of a deterioration in the quality of the groundwater is being borne out. The main methods for water quality predictions are analytical calculations and mathematical modelling.

Analytical methods for the evaluation of pollution migration have many limitations. Analytical solutions of mass transport differential equations are used mostly for simple hydrogeological conditions and filtration schemes. As a rule in complicated hydrogeological systems filtration properties of aquifers, boundary conditions, vertical leakage, processes of advection, dispersion, sorbtion and biodegradation all vary in time or space. It is necessary to take these factors into account. The usefulness of analytical methods under these conditions is very limited and for some complicated hydrogeological conditions, particularly when regional evaluations are carried out, analytical solutions do not exist. Recently the ability of digital methods to solve differential mass transport equations has improved and the increased power of modern computers has made mathematical modelling the preferred method for mass transport simulations.

F. Fonnum et al. (eds.),
Environmental Contamination and Remediation Practices at Former and Present Military Bases, **103–122.**

A mass transport mathematical model is a representation of a complicated natural system by a set of algorithms derived from simplified relations between various parameters, properties and stresses existing in the aquifer system. Mass transport modelling is carried out in several steps. Firstly, a groundwater flow model is constructed. Particle tracking analysis is then used for the delineation of recharge and discharge areas followed by the determination of the flow path for contaminant migration analysis as well as the delineation of capture zones of the water supply wells. Finally mass transport modelling is carried out.

In Lithuania the software developed in the United States Geological Survey (USGS) and other companies is used for mass transport modelling: MODFLOW (groundwater flow model), MODPATH and WHPA (particle tracking), MT3D96 (mass transport modelling), WATERQ4F and PHREEQC (thermodynamic calculations), ARMOS (movement of water and free phase hydrocarbon), etc. [1,2,3,4,5,6,7].

All the models of pollution migration created during the last few years in Lithuania for various hydrogeological systems (wellfields, urban areas, landfills, military bases) can be divided into three groups:

1. Models where migration of contaminants dissolved in groundwater is simulated directly by taking into account processes of advection, dispersion, sorbtion and degradation.
2. Models which do not directly simulate the migration of contaminants dissolved in groundwater. This model is a combination of a groundwater flow model (MODFLOW) and a particle tracking model (MODPATH). This enables to determine the water budget, mixing proportions and the development of the catchment area with time. At a later stage a complicated mathematical background for the calculation of thermodynamic equilibrium conditions (WATERQ4F, PHREEQC) is used to evaluate the concentrations of contaminants dissolved in the groundwater.
3. Models which simulate the simultaneous movement of shallow groundwater and free phase pollutants, for example free hydrocarbon (ARMOS).

Below is a short description of the results obtained using the above-mentioned groups of models.

2. Description of the obtained results

An example of the migration of dissolved material in groundwater is the model of the hydrogeological system of Panevezys city near which the former Soviet military airbase - Pajuoste airport - is situated. The main aim of the model was to evaluate the influence of the urban area of the city on the quality of wellfield groundwater; the secondary aim was to evaluate the influence of the military base on the groundwater quality.

There are three main sources of drinking groundwater in Panevezys:

1. Shallow groundwater which is abstracted by dug wells in the northern and central part of the city.
2. The Suosa-Kupiskis aquifer in the dolomites of the Devonian age. This lies at a depth of 30-50 m and is abstracted by some factories and wells in the city.
3. The Sventoji-Upninkai aquifer in the sandstones of the Devonian age. This is a very thick aquifer (250 m) lying at the depth of 50-65 m and abstracted in the city's water supply wellfield.

The results of our investigations show that the shallow groundwater is heavily polluted over the whole area of the city. Even the average values of water hardness, nitrates, unoxidized organic matter exceed the highest allowable concentrations (Fig. 1).

The groundwater of the Suosa-Kupiskis aquifer in the city is also somewhat polluted. There were high concentrations of unoxidized organic matter, sulphates and chlorides. Signs of pollution in the upper part of Sventoji-Upninkai aquifer in the city have been observed since 1968.

In the Panevezys wellfield the groundwater is much less mineralised and cleaner than in the city. However in the upper part of Sventoji-Upninkai aquifer the amount of unoxidized organic matter and water hardness is high (up to 6.6-6.7 meq/l). It is clear that the polluted shallow groundwater in the city penetrates into the deeper aquifers and reaches the wellfield.

To predict groundwater quality, a mathematical model of the hydrogeological system of Panevezys city and its environs was constructed. Modelling was carried out in several stages. In the first stage, the regional groundwater model was constructed where the abstraction regime of all the wellfields in the region was reconstructed. Later, part of this model which included Panevezys city and its environs was modelled in detail.

Modelling results show that at present and in the future about 30% of groundwater resources are formed from the vertical leakage from the shallow groundwater aquifer. The rest of the resources come from the strong lateral groundwater flow in the abstraction aquifer.

The catchment area of the wellfield formed over the entire period that water has been abstracted from it (1960-1997) is shown in Fig. 2. It covers almost the half area of the city and the Pajuoste military base. It means that the shallow groundwater from this area reached the wellfield and took part in forming groundwater resources and its quality.

The three-dimensional mass transport model was used to predict the water hardness related to the mixing of groundwater flows from the shallow, Suosa-Kupiskis and Sventoji-Upninkai aquifers.

The model results show that in the future, assuming abstraction from the wellfield at the maximum discharge rate over a period of 50 years, there will be an increase in water hardness in those parts of the city where shallow groundwater is strongly polluted (Fig. 3, 4). For example, in the northern and central parts of the city, where hardness of shallow groundwater is up to 12-18 meq/l, an increase in

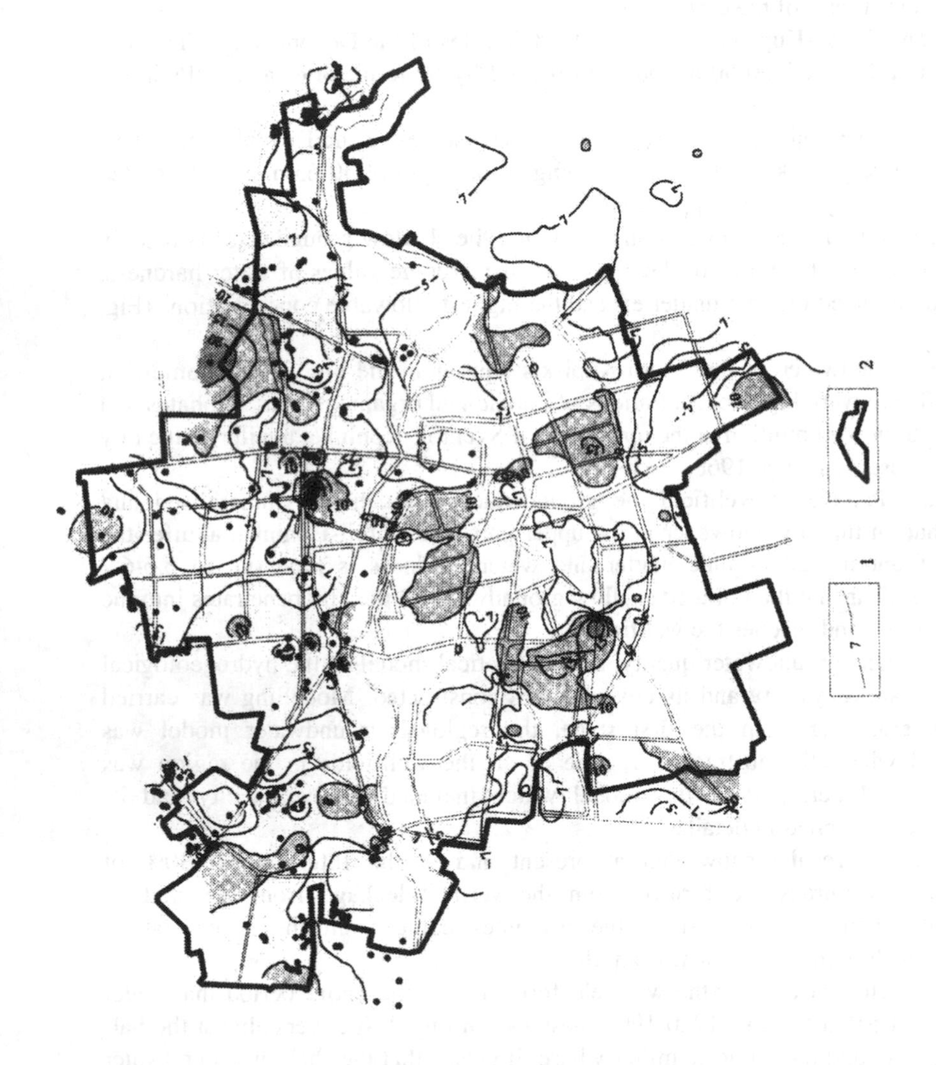

Figure 1. Total hardness of shallow groundwater in Panevezys city

1 - isoline of total hardness, mcq/l; 2 - territory of the city

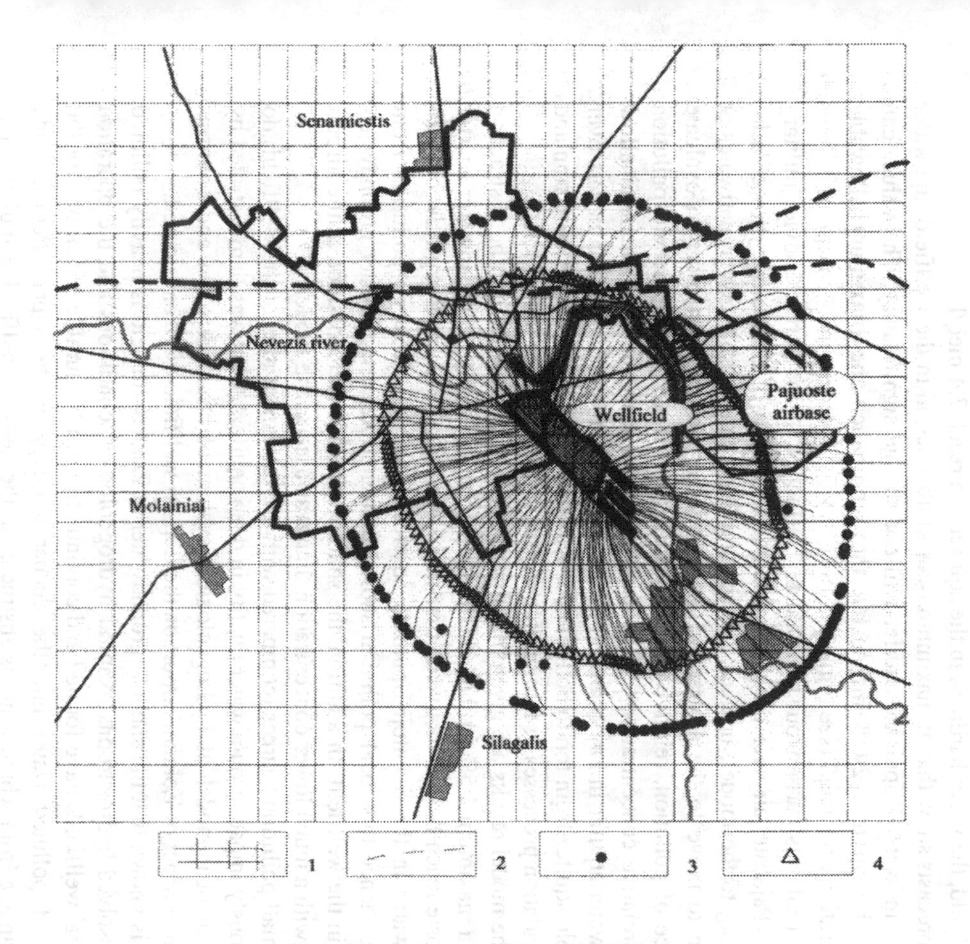

Figure 2. Catchment area of Panevezys wellfield

1 - model grid; 2 - pathline; 3- catchment area in productive aquifer; 4 - catchment area in shallow groundwater aquifer

hardness in the Suosa-Kupiskis aquifer up to 7.5-8.5 meq/l is possible (see Fig. 3). In the wellfield, the water hardness in the aquifer can reach 7.4 meq/l.

Forecasts show that at maximum rates of abstraction in the wellfield, the water hardness in sventoji-upninkai aquifer will not change significantly. The values can increase by up to 0.2-0.3 meq/l, but they can reach the maximum allowable concentration of 7 meq/l (see Fig. 4).

A trial was carried out to evaluate influence of the former Soviet military airbase (Pajuoste airport) on the groundwater quality in Panevezys wellfield. According to the model, it takes 17 years for a particle in shallow groundwater at Pajuoste to move to the abstraction aquifer in the wellfield. In order to evaluate influence of pollution, relative concentrations were used. A hypothetical pollution with a relative concentration 1 was used as a boundary condition in the shallow groundwater aquifer in Pajuoste, and it was taken that this concentration occurs over the whole aquifer from top to bottom over the whole area of the airport. Sorbtion and biodegradation processes in this aquifer were not taken into account.

The model results are given in Fig. 5. They indicate that during the abstraction period, a maximum 1.5% of the contaminants from the shallow groundwater aquifer in Pajuoste reach the productive aquifer in the wellfield. A very strong flow of fresh groundwater in the Sventoji-Upninkai aquifer dilutes the front of pollution from Pajuoste; hence this local pollution source does not seriously impact groundwater quality in the wellfield. In addition, the shallow groundwater pollution at the airport is local with a much lower concentration than was used in the model.

Small pollution sources (compared with the catchment area of the wellfield) do not seriously impact groundwater quality in deep aquifers. This was demonstrated by a mathematical model of hydrogeological system of Siauliai town and environs where the impact of Zokniai airport on the Lepsiai wellfield were considered.

It is known that the shallow groundwater at Zokniai airfield is heavily polluted with dissolved hydrocarbons. Natural hydrogeological conditions and the operation of Siauliai wellfields have formed and continue to form favourable conditions for the leakage of polluted water into the deeper aquifers of the Upper Permian and Devonian age from which water is abstracted at the Lepsiai wellfield. Simulation has shown that the catchment area of the wellfield formed during the period 1950-1997 in the Devonian, Permian and water table aquifers covers part of the Zokniai airport (Fig. 6.). This means that shallow groundwater leaks into the abstracted Devonian aquifer and that the mixture of the shallow groundwater with the water of Permian and Devonian aquifers has already reached the wellfield abstraction wells.

The same hypothetical situation of contaminant migration as described for Pajuoste airport has been simulated at the Lepsiai wellfield and Zokniai airport. The results of the simulation are shown on Fig. 7. It may be seen that over the whole period its of operation of Lepsiai wellfield, 1.5% of contaminants from the polluted shallow groundwater at Zokniai airfield have leaked into the Permian aquifer and only 0.5% has leaked into the abstracted Devonian aquifer. Therefore the impact of Zokniai airfield on abstracted water quality at the Lepsiai wellfield is not significant. It confirms the results of groundwater monitoring carried out from the start of

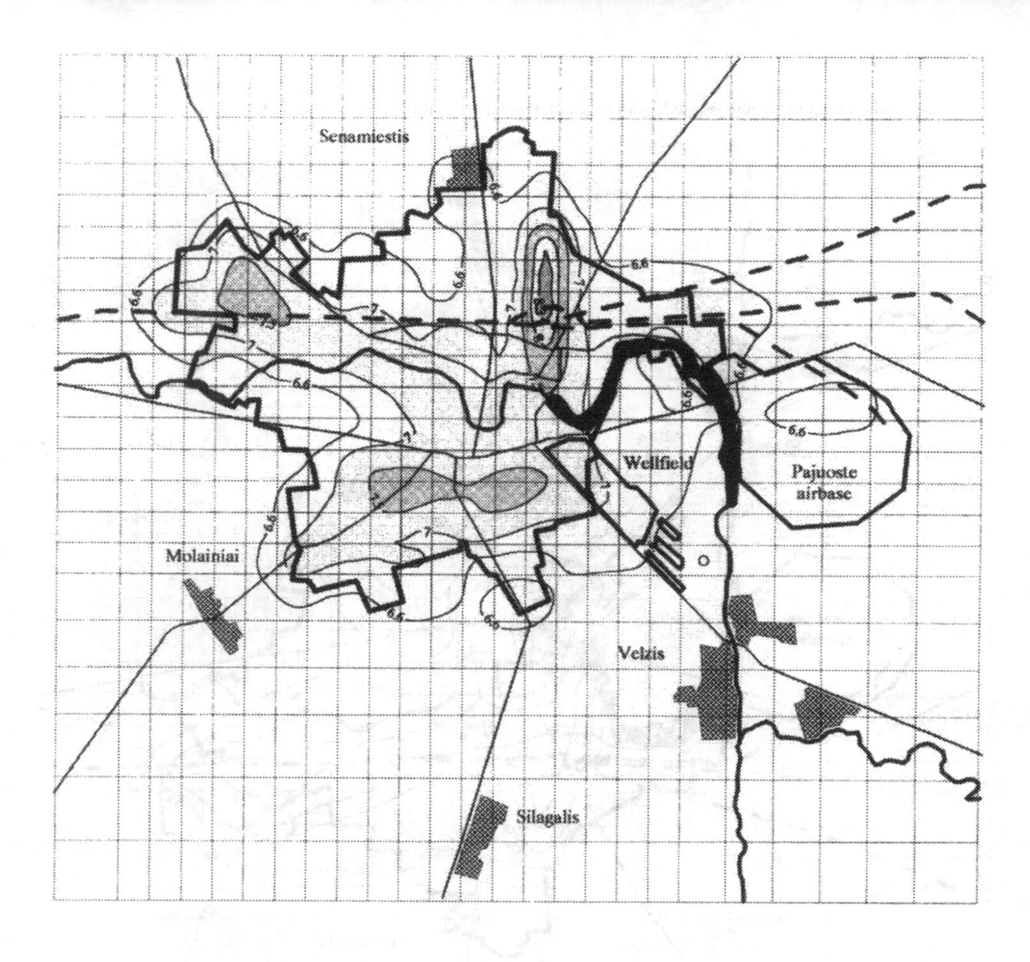

Figure 3. Predicted total hardness in the Suosa-Kupiskis aquifer in Panevezys

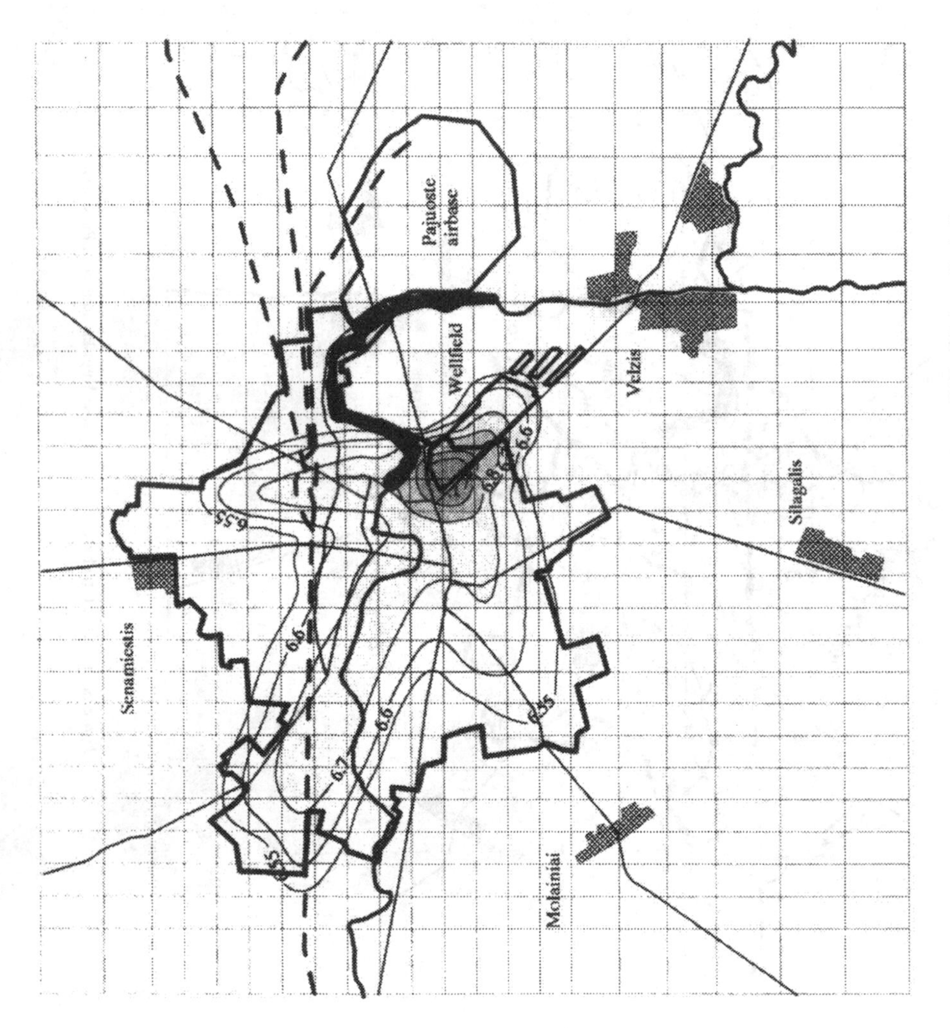

Figure 4. Predicted total hardness in the Sventoji-Upninkai aquifer in Panevezys

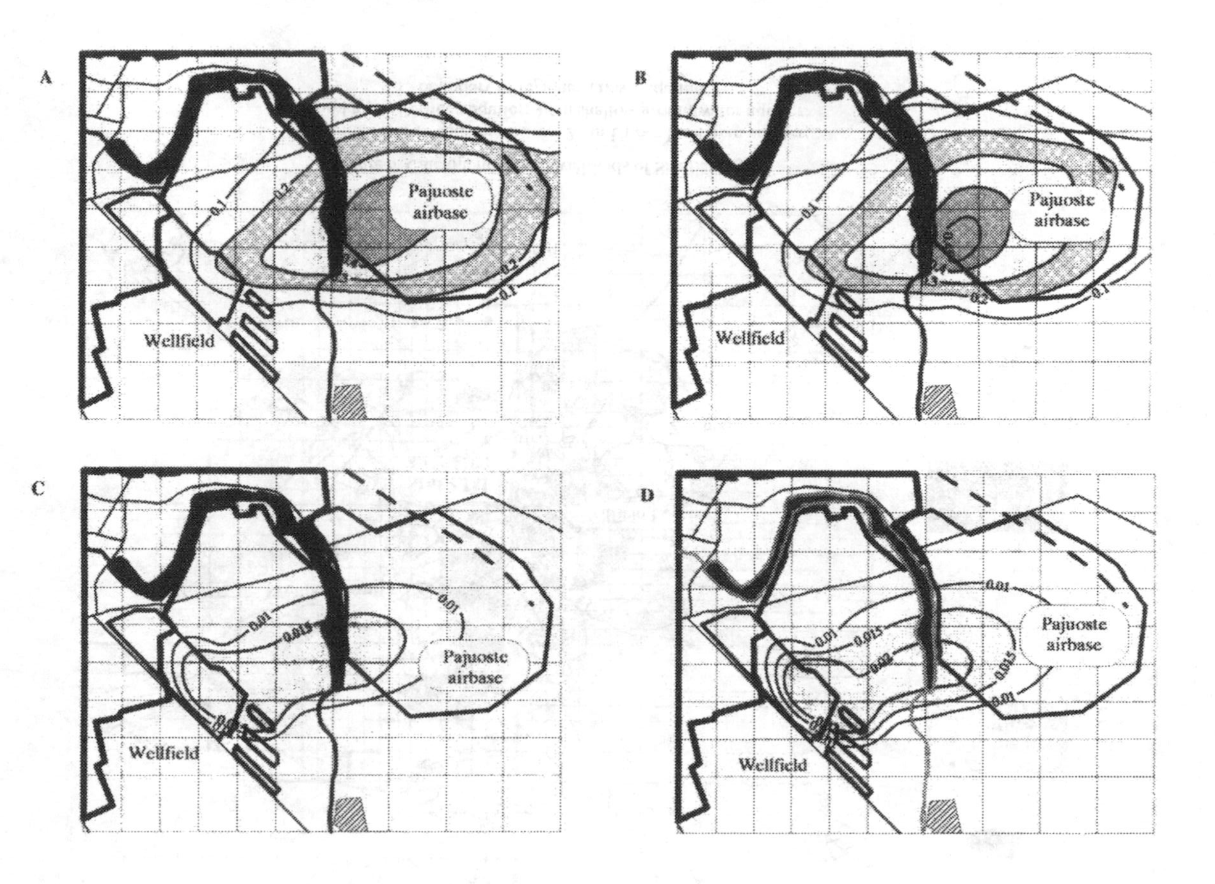

Figure 5. Simulation results of pollution transport from Pajuoste military airbase (relative concentrations)

A - present front of pollution in the Suosa-Kupiskis aquifer, B - predicted
C - present front of pollution in the Sventoji-Upninkai aquifer, D - predicted

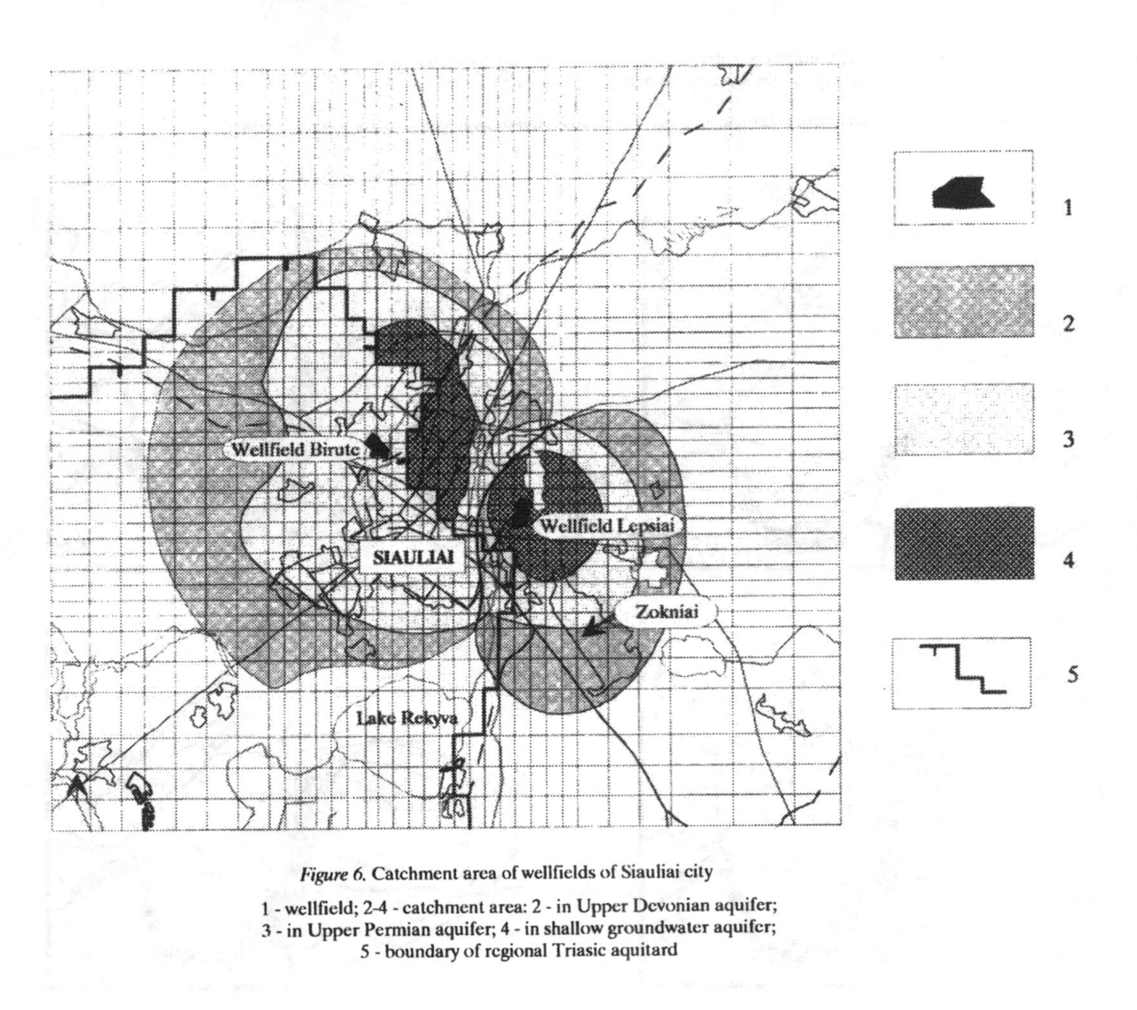

Figure 6. Catchment area of wellfields of Siauliai city

1 - wellfield; 2-4 - catchment area: 2 - in Upper Devonian aquifer;
3 - in Upper Permian aquifer; 4 - in shallow groundwater aquifer;
5 - boundary of regional Triasic aquitard

operation of the wellfield - an increase in the concentration of hydrocarbons or the products of their destruction has not been observed.

An example of the second type of model is the model for the Raudondvaris wellfield, city of Kaunas.

Raudondvaris wellfield, at the confluence of the rivers Nemunas and Nevezis where there is an alluvial aquifer, was surveyed in 1987 and recommended for water supply. Once developed, it will be a typical riverbank wellfield where the two sources of surface water mentioned above will control the volume and the quality of the abstracted groundwater. Due to the specific hydrochemical environment formed in the alluvial aquifer, the groundwater contains a high concentration of iron (up to 10.5 mg/l) and manganese (up to 3.6 mg/l). The average concentrations of those metals in the groundwater are 4 and 1 mg/l respectively.

A particular hydrochemical environment in the Raudondvaris wellfield has been formed by the original geological-hydrogeological conditions of the Nemunas and Nevezis rivers confluence: the upper part of the geological section is presented by a layer of mud, aleurite and clay. It covers a thick, permeable (with a co-efficient of transmissivity of up to 3000-7000 m^2/d) but stagnant aquifer. There is effectively no oxygen in this aquifer. Therefore mobile forms of iron and manganese produce the low negative values of redox potential Eh. On the other hand, an environment that is slightly alkaline, poor in oxygen but rich in carbonates and bicarbonates is very favourable for the accumulation of the mobile forms of bivalent Fe and Mn ions and weakly associated with bicarbonate ions [6]. Furthermore, this type of hydrochemical environment is typical of the former military areas polluted by organic substances (hydrocarbons).

It is clear that operation of the wellfield will change the quality of abstracted water and the concentrations of Fe and Mn in it. The main effect will be the flow of slightly alkaline, rich in oxygen surface water from the Nemunas river into the abstraction wells. Therefore an appreciable amount of iron and manganese will be transformed into the weakly mobile hydrolysed and carbonate forms.

In this case, a mathematical simulation was used to forecast the changes in the Fe and Mn concentrations during the course of water abstraction and to answer the question - what concentrations of these metals must be removed from the water.

Three options were simulated. First and foremost, the dynamics of water levels, of the catchment area and the groundwater budget formation at an abstraction rate of 35 000 m^3/d (the approved safe yield of the wellfield) was simulated. The other two options simulated were for abstraction rates of 5000 and 10 000 m^3/d.

The dynamics of catchment area formation at a pumping rate 35 000 m^3/d are shown in Fig. 8. Simulation data show that the first portion of Nemunas river water will reach the pumped wells after 15 days and the last portion after 75 days. From Fig. 8 it may be seen that after 50 days, the catchment area did not include the Nevezis river, but after 100 days water from both rivers contributes to the safe yield of the wellfield. After the 3 years the catchment area covers almost the entire area modelled.

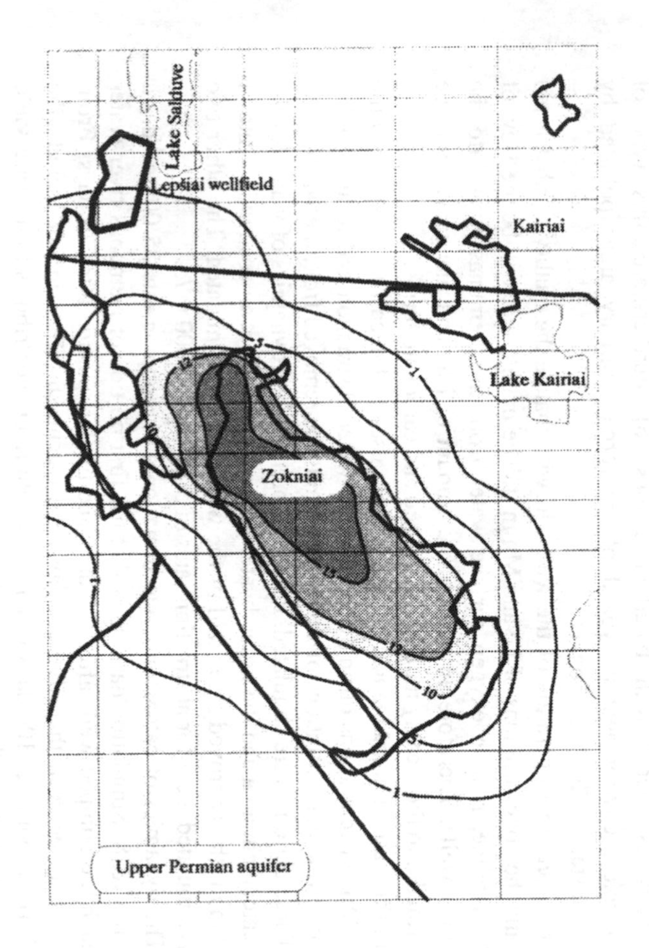

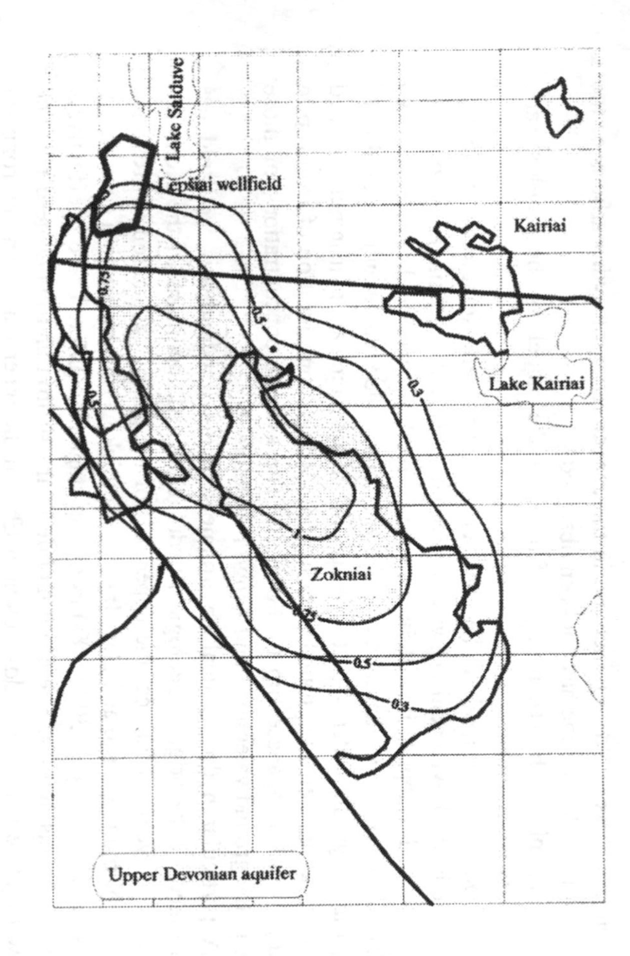

Figure 7. Simulation results of pollution transport from Zokniai military airbase (relative concentrations, %)

The simulated dynamics of the water budget formation show that at the maximum pumping rate of 35 000 m^3/d, after 50 days, flow from the rivers account for 14 100 m^3/d; after 100 days - 21 800 m^3/d and after one year - 26 400 m^3/d, or 40, 62 and 75% of the abstraction rate respectively. After approximately 70 days, the input of groundwater flow into catchment area through the lower part of the abstracted aquifer decreases from 50 to 14%. Meanwhile, the total input from both rivers increases consistently up to 82% (2 years after the start). There is an insignificant increase to 85% after 25 years of water abstraction. This means that after 2 years at maximum output of the wellfield, the catchment area of the wellfield will effectively cover both rivers and will reach the contours of the recharge.

It is clear that the amount of Fe and Mn to be removed depends upon the extent of the changes in the hydrogeochemical environment in the abstraction aquifer. On the other hand, the extent will depend on the rate of mixing of groundwater and surface water representing different hydrochemical environments. Essential parameters of the hydrochemical environment are Eh and pH. Both of them are seasonally variable. Subsequently the following procedure for forecasting the hydrochemistry was used:

1. the mixing rate of surface and groundwater at different capacities of the wellfield were determined from the flow model MODFLOW
2. the change in the mixing rate with time and its value for a period of constant flow was determined from the transport model MODPATH
3. corresponding values of pH and Eh for different mixing rates were determined considering seasonal variations of these parameters
4. corresponding forms and concentrations of Fe and Mn in the mixture were calculated using the WATEQ 4F computer program.

Calculations are shown in Fig. 9. The comments "winter " and "summer" on the figure mean that river water with the winter or summer parameters will reach the wells at the time shown. From the figure it follows that only the concentration of Fe will decrease significantly and only at a high enough pumping rate and after sufficient time. Even in the best case, its concentration will exceed the permitted concentration of 0.3 mg/l. In addition, the relatively low concentrations of Fe in the groundwater will occur only if the summer water reaches the wells. When winter water reaches them, the Fe concentration will increase about three fold.

An example of the third type of model is the model for the migration of hydrocarbons at the Klaipeda Oil Terminal. Here the situation resembles groundwater pollution on large military bases. The American company Environmental Systems and Technologies Inc.'s program ARMOS was used to build a model to solve this kind of problem.

At the oil products distribution ramp at the Klaipeda Oil Terminal, a 60 cm thick layer of hydrocarbons has been found above the water table (in general a mixture of fuel oil, diesel fuel and kerosene). Additionally, the water table aquifer itself is polluted by the dissolved hydrocarbons. Groundwater flows from the ramp and discharges into the Kurish lagoon (see Fig. 10). This discharge is not small - simulation shows that 70% of the total flow from the Terminal area flows into the

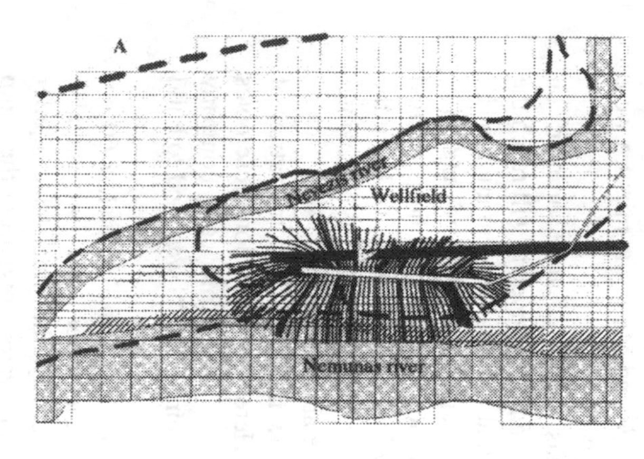

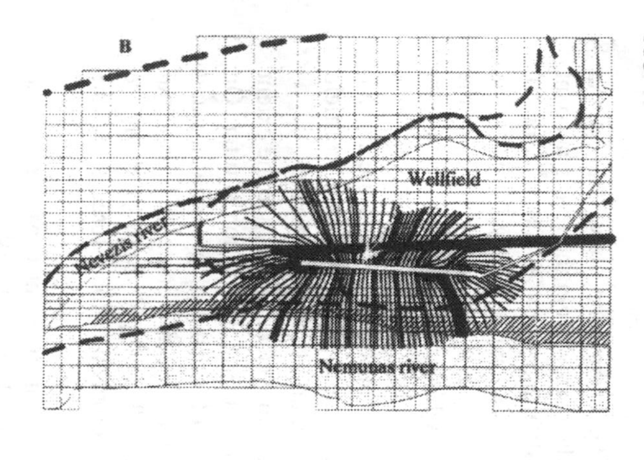

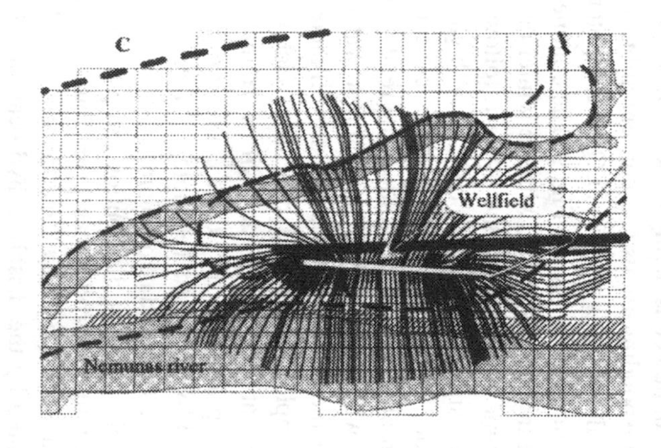

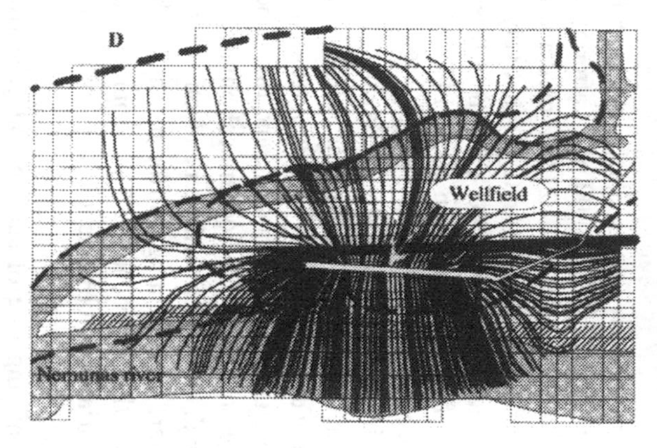

Figure 8. Dynamics of catchment area formation at a pumping rate of 35000 m^3/d at Raudondvaris wellfield

A - after 50 days; B - after 100 days; C - after 1 year; D - after 2 years

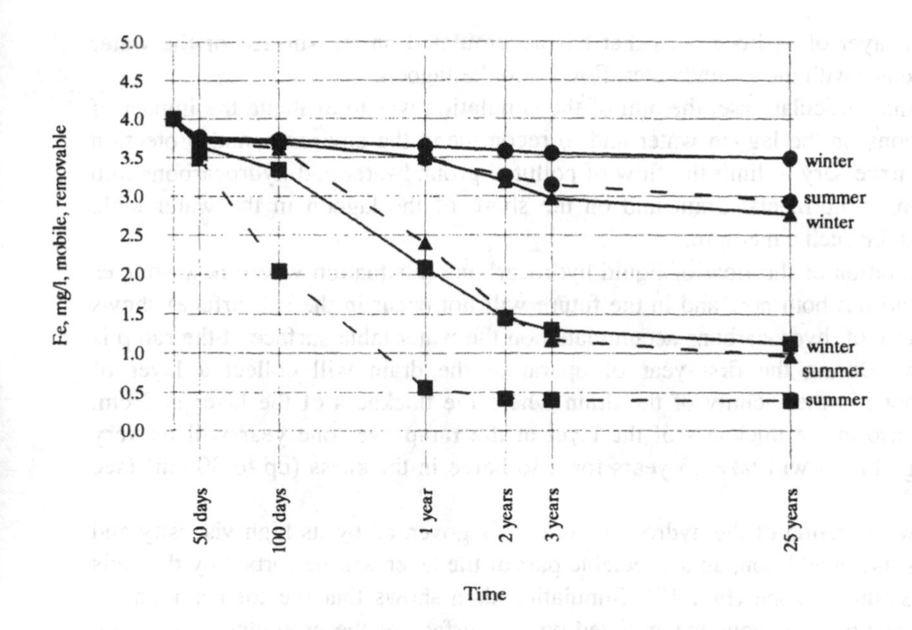

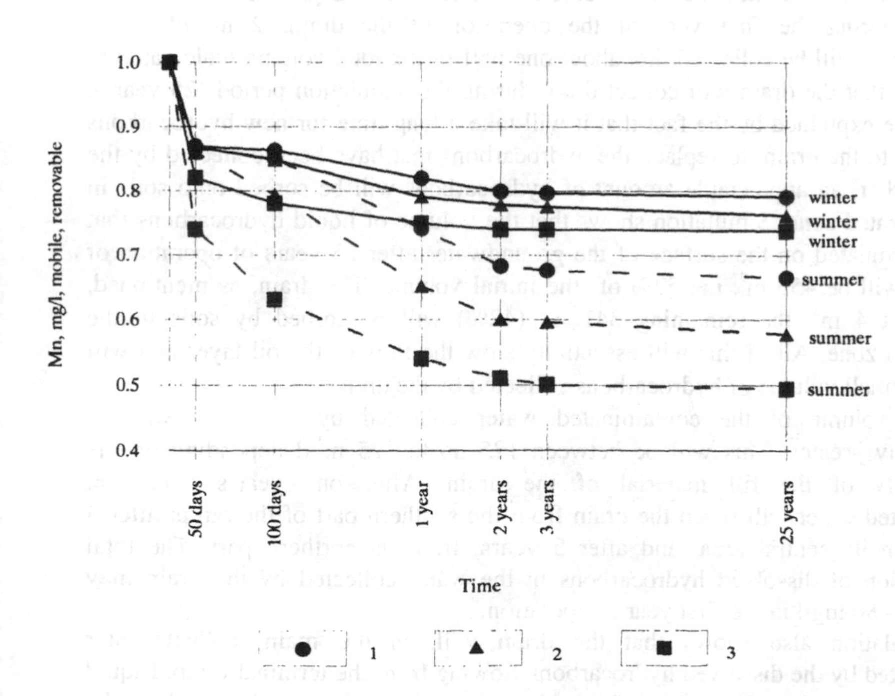

Figure 9. Removable concentrations of iron and manganese at different pumping rates of wellfield

1-3 - pumping rate: 1 - 5000 m^3/d; 2 - 10000 m^3/d; 3 - 35000 m^3/d lagoon. A layer of hydrocarbons that has accumulated on the surface of the water table, together with the groundwater, flows into the lagoon.

In this particular case, the aim of the simulation was to evaluate the impact of hydrocarbons on the lagoon water and to recommend the environmental protection measures necessary to limit the flow of polluted groundwater and hydrocarbons into the lagoon. A horizontal drain laid on the shore of the lagoon in the water table aquifer can be such a measure.

Simulation of the flow of liquid hydrocarbons (a situation where new sources of hydrocarbons both now and in the future will not occur in the subsurface) shows that the flow of hydrocarbons accumulated on the water table surface at the ramp is very slow. During the first year of operation the drain will collect a layer of hydrocarbons in the vicinity of the drain where the thickness of the layer is 7 cm. The decrease in the thickness of the layer at the ramp over one year will be very small (Fig. 11). It will take 25 years for it to halve in thickness (up to 30 cm) (see Fig. 11).

Slow migration of the hydrocarbon layer is governed by its high viscosity and by its density. In addition, an appreciable part of the layer will be sorbed by the soils in the unsaturated zone (Fig. 12). Simulation data shows that the total amount of hydrocarbons that has now accumulated on the surface of the groundwater is about 800 m^3. During the first year of the operation of the drain, 2 m^3 of liquid hydrocarbons will be collected, i.e. about one half of the total volume (calculated to be 4.2 m^3) that the drain will collect drain during the simulation period (25 years). This can be explained by the fact that it will take a long time for new hydrocarbons to migrate to the drain, to replace the hydrocarbons that have been collected by the drain. Further, an appreciable amount of hydrocarbons will be sorbed onto soils in the unsaturated zone. Simulation shows that the volume of liquid hydrocarbons that have accumulated on the surface of the groundwater after 25 years of operation of the drain will be 455 m^3, i.e. 57% of the initial volume. The drain, as mentioned, will collect 4 m^3, the remaining 341 m^3 (42%) will be sorbed by soils in the unsaturated zone. All of this will essentially slow the flow of the oil layer and will affect the small volume of hydrocarbons collected by the drain.

The volume of the contaminated water collected by the drain will be significantly greater. This will be between 125 up to 225 m^3/d depending on the permeability of the fill material of the drain. After one year's operation, contaminated water will reach the drain from the southern part of the ramp; after 3 years, from its central area, and after 5 years, from its northern part. The total concentration of dissolved hydrocarbons in the water collected by the drain may exceed 30 - 80 mg/l in the first year of operation.

Simulation also shows that the drain will, in the main, collect water contaminated by the dissolved hydrocarbons flowing from the terminal ramp. Liquid hydrocarbons will be collected only in small volumes. Therefore the drain will not be an effective method of cleaning up an area contaminated by liquid hydrocarbons flowing on the water table surface. However, it will be an effective method of

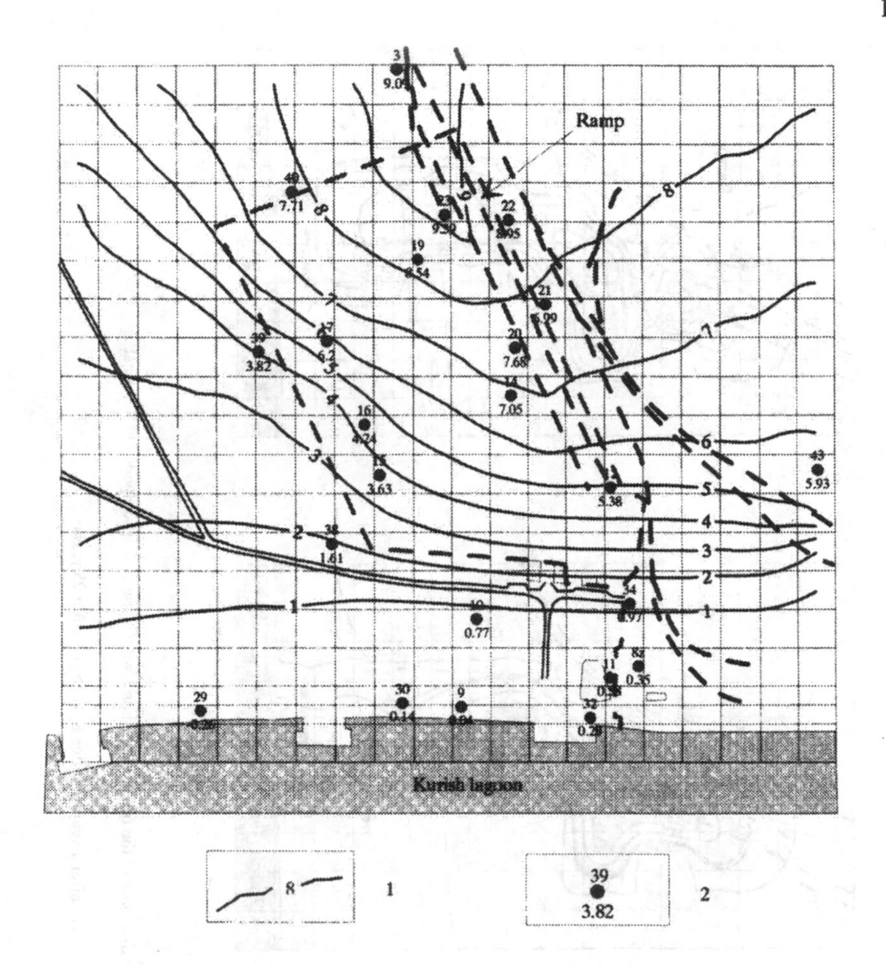

Figure 10. Simulated water table countours at Klaipeda Oil Terminal

1 - simulated water tabel countour, m a.s.l.; 2 - monitoring well: above - well number; below - water table, m a.b.s.l.

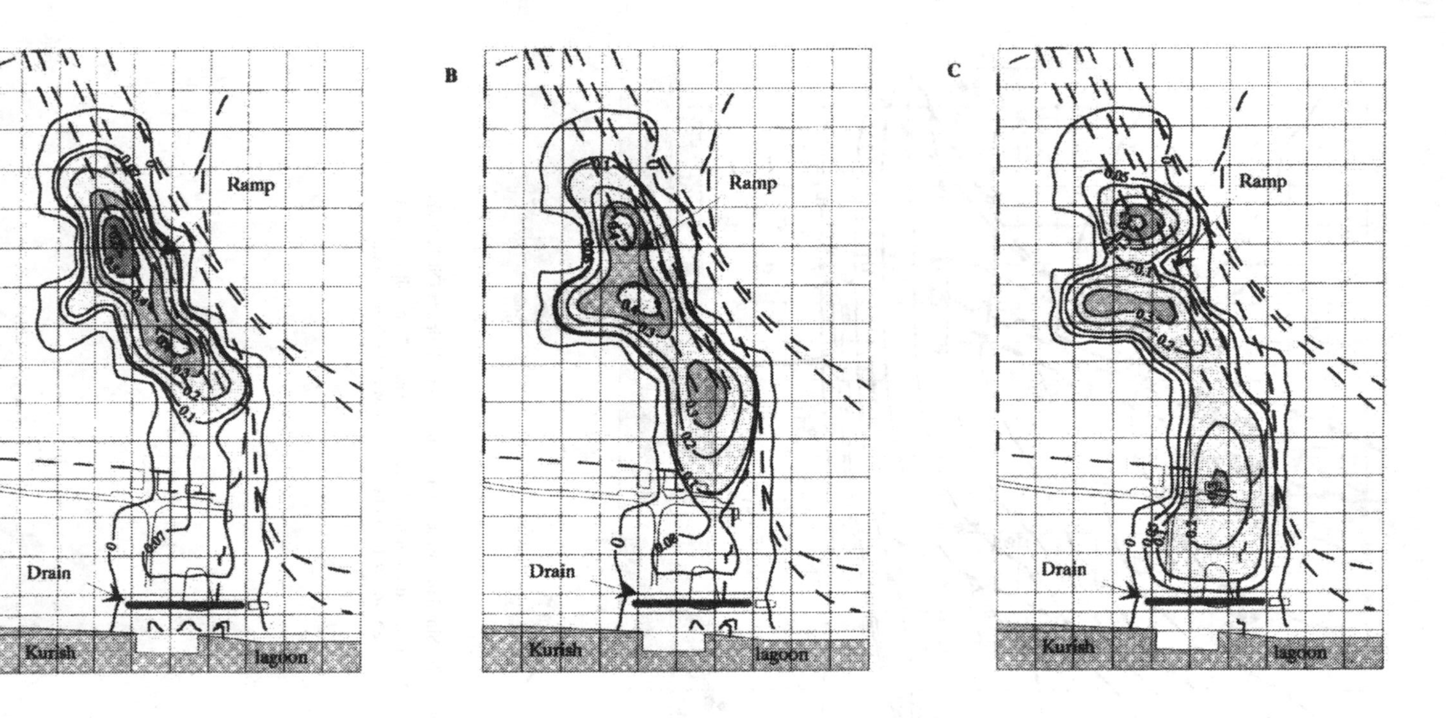

Figure 11. Simulated dynamics of the free hydrocarbon layer (in meters) at Klaipeda Oil Terminal

A - after 1 year; B - after 10 years; C - after 25 years

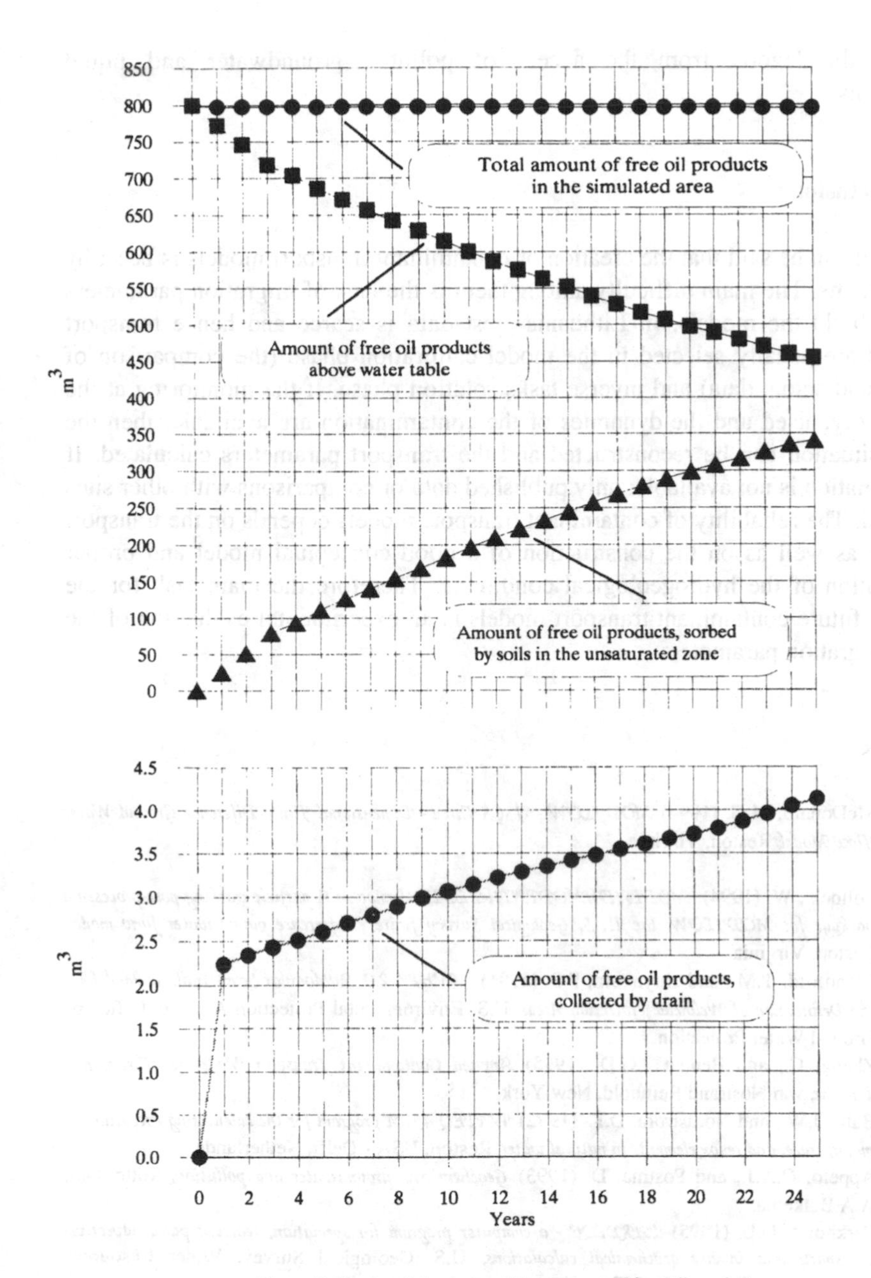

Figure 12. Simulated dynamics of the amount of free oil products

protecting the lagoon from the access of polluted groundwater and liquid hydrocarbons.

3. Conclusions

In general it can be said that the creation of contaminant transport models is beset by many problems. The main difficulty among them is the lack of migration parameters needed to build the models. In Lithuania test data is scarce and hence transport parameters are usually selected in the model calibration phase (the comparison of simulated and actual data) and inverse tasks solution phase. If the monitoring at the test site is organised and the dynamics of the contamination are available, then the observed situation can be reconstructed and the transport parameters calculated. If such information is not available, only published data or comparisons with other sites can be used. The reliability of contaminant transport models depends on the transport parameters as well as on the construction of a good conceptual model and proper schematization of the hydrogeological conditions. Therefore the main task for the creation of future contaminant transport models is an experimental evaluation of the required migration parameters.

References

1. McDonald, M.G. (1994) *MODFLOW - USGS Three - Dimensional Finite-Diference Ground-Water Flow Model*, Reston, Virginia.

2. Pollock., W. (1994) *MODPATH/MODPATH-PLOT, Version 3: A particle tracking post-processing package for MODFLOW, the U. S. Geological Survey finite - difference gound-water flow model.* Reston, Virginia.
3. Blandford, T.M., and Huyakorn, P.S. (1991) *WHPA 2.0: A Modular Semi-Analytic Model for the Delineation of Wellhead Protection Areas*, U.S. Environmental Protection Agency, Office of Ground Water Protection.
4. Zheng, C., and Bennet, G.D. (1995) *Applied Contaminant Transport Modeling: Theory and Practise*, van Nostrand Reinhold, New York.
5. Ball, J.W., and Nordstrom, D.K. (1992) *WATEQ 4F. A program for the calculating speciation of major, trace, and redox elements in natural water*, Reston, USA - Delft, Netherlands.
6. Appelo, C.A.J., and Postma, D. (1993) *Geochemistry, groundwater and pollution*, Rotterdam, A.A.Balkema.
7. Parkhurst, D.L. (1995) *PHREEQC - a computer program for speciation, reaction path, advective-transport, and inverse geochemical calculations*, U.S. Geological Survey, Water Resources Investigations Report 95-4227.

MASS TRANSPORT MODELLING IN GROUNDWATER STUDIES. ACHIEVEMENTS OF LATVIAN SCIENTISTS

A. SPALVINS
Environment Modelling Centre, Riga Technical University
9 Ausekla Str., Riga, LV - 1010, Latvia

Abstract.

Migration processes of immiscible light petroleum hydrocarbons (called oil in this paper) in groundwater differ considerably from the mass transport of dissolved contaminants. For this reason, a specialised software must be applied for simulating problems of oil movement. Two rather different cases are presented concerning oil contamination problems at the former Rumbula airbase and at the Ilukste oil storage terminal where the ARMOS code has been applied for modelling transient oil transport processes. Advances and drawbacks of ARMOS are discussed and some important simulation results reported.

1. Introduction

At present, only two teams of groundwater modellers are active in Latvia: the group from the State Geological Survey of Latvia (SGSL) and the Environment Modelling Centre (EMC) of the Riga Technical University. Just recently, both teams were involved in projects regarding building of hydrogeological models for the investigation of various problems of a drinking water supply. For example, the EMC team together with specialists from SGSL have developed the Regional Hydrogeological Model (REMO) "Large Riga" [1]. REMO includes the central part of Latvia, the northern part of Lithuania and a large fragment of the Gulf of Riga. By applying REMO, the EMC and SGSL teams modelled new well fields for the water supply of Riga [2]. These projects dealt only indirectly with the mass transportation processes when catchment areas of well fields were investigated regarding possible contaminant sources of groundwater.

Since 1996, the EMC team under contracts with the Baltec Associates, Inc. (BA) has been involved in projects on problems of oil migration in groundwater. These differ strongly from the well-investigated transportation processes of dissolved contaminants and for that reason, specialised software tools should be used. The ARMOS code developed by the Environmental Systems and Technologies, Inc. (ES & T) has been applied. Annotations of two other ES & T codes (SpillCAD and BioTRANS) clarify why they have not been used by the EMC team.

The mathematics used in ARMOS are explained in brief, in order to clarify the basic ideas running this code and to discuss the origins of certain ARMOS limitations and

F. Fonnum et al. (eds.),
***Environmental Contamination and Remediation Practices at Former and Present Military Bases*, 123–142.**

errors. Close attention is paid to Van Genuchten's (VG) capillary model and the problem of coupling water and oil flow equations.

The advantages and drawbacks of ARMOS are demonstrated in this paper by a case analysis of two different oil contaminated objects: 1) the former Rumbula airbase; 2) the Ilukste oil storage terminal.

In order to overcome some of the ARMOS limitations, it was necessary to apply additional software tools. Some of them are being developed by the EMC team and may be of interest to specialists engaged in simulation of contamination processes in groundwater and with interpolating of various hydrogeological data for digital mapmaking.

2. A brief description of the ARMOS code

The ARMOS code has two versions: standard (750 nodes on an xy -grid) and professional (3,000 nodes). ARMOS simulates an oil plume movement and its thickness distribution in an unconfined aquifer. Free oil recovery measures can be simulated as well. Volumes of free and trapped oil can be calculated by accounting for the influence of groundwater table fluctuations.

ARMOS integrates a database, the KRIGING type interpolator for 2D-mapmaking, the module for the simulation of the oil plume migration, the module for computing volumes of free and trapped oil and some graphics for visualisation of various results. Theoretically, only ARMOS is needed to obtain the necessary results, but as is shown later, in practice this is not the case.

For basic initial data, ARMOS applies apparent oil thicknesses H_o observed in screened monitoring wells (see *Fig. 1a*). The parameter H_o roughly corresponds with the thickness of the oil contaminated soil. One should be aware that $H_o >> V_o$ where V_o is the total specific oil volume (see *Fig. 1e, 1f*) corresponding with the real thickness of the oil fluid.

The complex non-linear relationship H_o versus V_o in ARMOS is provided by the VG soil capillary model. Here this relationship is examined more closely because the correct choice of VG model parameters largely determines all of the results of ARMOS. As is explained later, it is possible that some of the components of this model are incorrectly used in ARMOS.

3. Van Genuchten's capillary model

The VG model is applied under the assumption that air, oil and water fluids in the soil quickly reach a hydrostatic equilibrium and, for that reason, no vertical flow of the fluids exists. It is clear from *Fig. 1a* that, in the screened well, two fluid interface elevations z measured above mean sea level (a msl) can be observed $z = Z_{ao}$ (air-oil) and $z = Z_{ow}$ (oil-water). The elevation $z = Z_a$ can be obtained analytically as explained later. The space between elevations Z_{ao} and Z_a is the contaminated aeration zone. The difference is $Z_{ao} - Z_{ow} = H_o$ and the oil-water saturated zone is located within this interval. The auxiliary variable $Z = z - Z_{ow}$ is also applied in this paper.

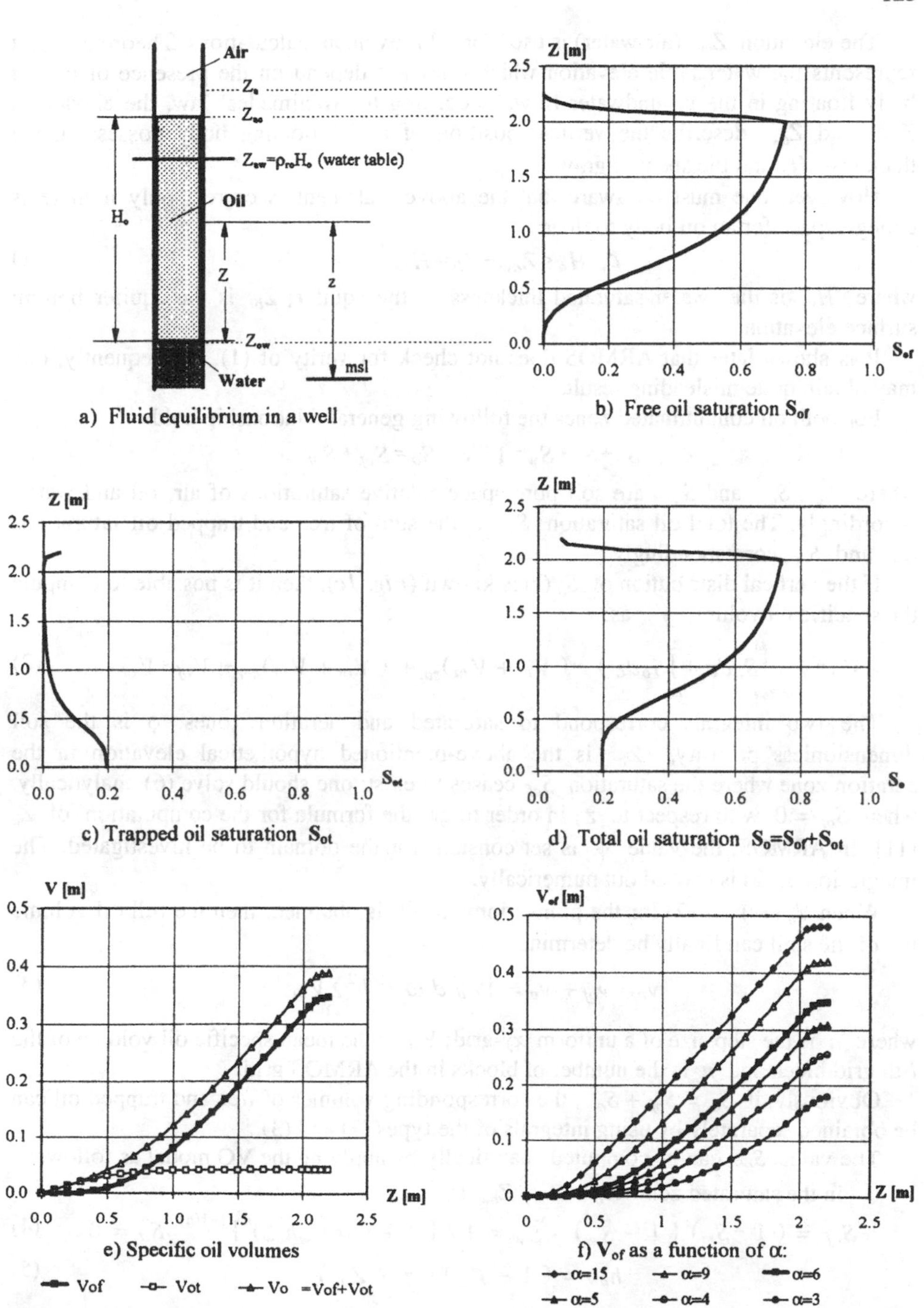

Figure 1. **Components of the VG soil capillarity model. Graphs b) - e) represent the Ilukste oil terminal case with the parameters: $n = 3.4$, $\alpha = 6.0$, $\rho_{ro} = 0.82$, $S_m = 0.2$, $S_{or} = 0.2$, $S_{og} = 0.05$, $\phi = 0.35$, $\beta_{ao} = 2.8$, $\beta_{ow} = 1.4$, $H_o = 2.0$**

The elevation Z_{aw} (air-water) is used for oil movement calculations. Theoretically, it represents the water table elevation which does not depend on the presence of the oil body floating in the groundwater flow. According to Archimedes' law, the elevations Z_{ao} and Z_{ow} describe the vertical position of a flat floating body possessing the thickness H_o and the specific gravity ρ_{ro}.

However, one must be aware that the above statement is correct only if there is enough space for an oil body to float:

$$\rho_{ro} H_o < Z_{aw} - Z_b = H_w \tag{1}$$

where H_w is the water saturated thickness of the aquifer; Z_b is the aquifer bottom surface elevation.

It is shown later that ARMOS does not check for verity of (1). Consequently, one may obtain quite misleading results.

For both oil contaminated zones the following general relationship holds:

$$S_a + S_o + S_w = 1 \quad , \quad S_o = S_{of} + S_{ot}$$

where S_a , S_o and S_w are soil pore space relative saturations of air, oil and water, accordingly. The total oil saturation S_o is the sum of free and trapped oil saturations S_{of} and S_{ot}, correspondingly.

If the vertical distribution of $S_o(z)$ is known (*Fig. 1c*), then it is possible to compute the specific oil volume V_o as:

$$V_o = \phi\left(\int_{Z_{ow}}^{Z_{ao}} S_o dz + \int_{Z_{ao}}^{Z_a} S_o dz \right) = (V_{of} + V_{ot})_{sat} + (V_{of} + V_{ot})_{aer} = V_{of} + V_{ot} \quad . \tag{2}$$

The two integrals correspond to saturated and aeration zones: ϕ is the soil dimensionless porosity, Z_a is the above-mentioned hypothetical elevation in the aeration zone where the saturation S_{of} ceases to exist: one should solve (6) analytically, when $S_{of} = 0$ with respect to z, in order to get the formula for the computation of Z_a [11]. In ARMOS, the value ϕ is set constant for the domain to be investigated. The integration of (2) is carried out numerically.

When $V_o(xy)$ of (2) for the plume domain Ω is obtained, then the full oil volume v_o of the spill can finally be determined:

$$v_o = v_{of} + v_{ot} = \int_{\Omega} V_o \, d\omega \approx h^2 \sum_{i=1}^{n} V_{oi} \tag{3}$$

where h is the step size of a uniform xy-grid; V_{oi} is the mean specific oil volume of the i-th grid block and n is the number of blocks in the ARMOS grid.

Obviously, if $S_o = S_{of} + S_{ot}$, the corresponding volumes of free and trapped oil can be obtained separately by using integrals of the types (2) and (3).

The value S_{of} can be computed analytically by applying the VG model as follows:

in the saturated zone $H_o = Z_{ao} - Z_{ow}$:

$$S_{of} = (1 - S_m)(1 - \tilde{S}_w) \, , \, \tilde{S}_w = 1 / [1 + (\alpha\beta_{ow} h_{ow})^n]^{(1-1/n)} \, , S_a = 0 \, , \tag{4}$$

$$h_{ow} = (1 - \rho_{ro})(z - Z_{ow}) \; ; \tag{5}$$

in the aeration zone Z_a - Z_{ao} :

$$S_{of} = (1 - S_m) (\hat{S} - \tilde{S}_w) \quad , \hat{S} = 1 / [1 + (\alpha\beta_{ao} h_{ao})^n]^{(1-1/n)}, \tag{6}$$

$$h_{ao} = \rho_{ro}(z - Z_{ao}) \tag{7}$$

where S_m is the residual water saturation; $\tilde{S}_w$ is the apparent water saturation presented by the notable "S" - shape function introduced by M.Th. van Genuchten [5] and adapted for the light oil case by J.C.Parker and R.J.Lenhard; [9]; $\tilde{S}_w$ holds in both zones; $\hat{S}$ is the VG type function used in the aeration zone; n , α are the variable VG parameters; β_{ow} , β_{ao} are the dimensionless oil-water and air-oil surface tension scaling factors, accordingly; h_{ow} , h_{ao} are the current oil-water and air-oil capillary heads, correspondingly; ρ_{ro} is the dimensionless relative oil gravity and ρ_{ro} = 1.0 , for water, but for light oil ρ_{ro} < 1.0 . The graphs of (4) and (6) are shown in *Fig. 1b*. The VG model of S_{of} integrates the properties of soil and oil into a united system which can be applied for any combination of oil and soil types due to an enormous flexibility of the above-mentioned VG functions.

The shape of S_{of} is extremely responsive to the choice of the VG parameters α and n. The value of n mostly determines the general form of S_{of} , but α serves as an additional scaling factor of the current variable capillary heads of (5) and (7). *Fig. 1f* illustrates the extreme sensitivity of the value of V_{of} versus changes of α .

Hence, free oil migration simulation is the principal task of ARMOS, and one can imagine that the above described VG model of S_{of} covers all possible needs and that no calculations of the component S_{ot} are needed. Unfortunately, this is incorrect, for no movement of free oil is physically possible without the presence in soil pores of trapped (immobile) oil [6]. Moreover, computing volumes of trapped oil may be important for biological and other clean-up measures or for the investigation of oil leakage sources.

In ARMOS, for computing S_{ot} (see *Fig.1c*), the empirical relationship, suggested by C.S.Land [7], is used. It was transformed by us in the following general form:

in the saturated zone:

$$S_{ot} = \bar{S}_{or} - S_{of} S_{or} / (S_{of} + S_{or} \tilde{S}_w) \quad ,$$

$$\bar{S}_{or} = S_{of}^{\max} S_{or} / (S_{of}^{\max} + S_{or} \tilde{S}_w^{\min}) \quad , \tag{8}$$

in the aeration zone:

$$S_{ot} = \bar{S}_{og} - S_{of} \; S_{og} / (S_{of} + S_{og} \tilde{S}_w) \quad ,$$

$$\bar{S}_{og} = S_{of}^{\max} \; S_{og} / (S_{of}^{\max} + S_{og} \tilde{S}_w^{\min}) \quad . \tag{9}$$

The constants S_{or} (oil residual saturation), S_{og} (oil gravitational saturation) and $\bar{S}_{or}$, $\bar{S}_{og}$ are maximal asymptotic and current historical trapped oil saturations S_{ot} in saturated and aeration zones if $z = Z_{ow}$, $z = Z_a$, accordingly; S_{ot} = 0 when $z = Z_{ao}$. The constants $S_{of}^{\max}$, $\tilde{S}_w^{\min}$ are the corresponding values of (4) if $z = Z_{ao}$ ($Z = H_o$).

When z rises, then asymptotically $S_{of}^{\max} \rightarrow (1 - S_m)$, $\tilde{S}_w^{\min} \rightarrow 0$. These constants are included in (8), (9) in such a way, that $\bar{S}_{or}, \bar{S}_{og}$ are only slightly smaller than S_{or}, S_{og}, respectively, even if H_o is relatively small. For example, in *Fig. 1b*. $S_{of}^{\max} = 0.74 > 0.8 = 1 - S_m$. Nevertheless, in *Fig. 1c* $\bar{S}_{or} = 0.196 < 0.2 = S_{or}$.

The expressions (8), (9) are not an explicit part of the classic VG model. Nevertheless, they enable us to obtain the total oil saturation S_o (see *Fig.1d*). For this reason, we conditionally consider them as components of the VG model.

It is rather strange than in the ARMOS manual [11] the formula (8) is given but it is not used as an obligatory counterpart of the expression (4). No oil movement in soil is possible if trapped oil has not been accumulated. This fact causes some inconsistent ARMOS results discussed later.

It has been observed that apparent thicknesses H_o in monitoring wells can change considerably during time. It means that some transformation $V_{of} \leftrightarrow V_{ot}$ takes place even if V_o is constant. The phenomenon can be partly explained by the VG model as a result of water table elevation Z_{aw} fluctuations [8]. If Z_{aw} rises, H_o decreases because some part of the free oil becomes trapped ($V_{of} \rightarrow V_{ot}$). Just the opposite happens when Z_{aw} falls. In this case some amount of trapped oil is released ($V_{ot} \rightarrow V_{of}$), and H_o increases.

ARMOS can provide a response to the fluctuation history of Z_{aw} with a constant fluctuation magnitude in the domain. We have not investigated this ARMOS option because fluctuations do not change V_o and they may only partly explain the behaviour of H_o in monitoring wells.

In reality, there are other important factors: no vertical equilibrium of oil and water, the influence of a well screen, a vertically non-homogeneous aquifer, a locally confined aquifer, etc.

4. Oil and water flow equations

In ARMOS, *xy*-flows are used for describing oil and water movement under the assumption that fluids in the vertical *z*-direction of the soil are in a local hydrostatic equilibrium. In this light, it is possible to compute the areal distribution of H_o without involving conventional time-consuming 3D-modelling.

The governing equations for the *xy*-flows of water and oil are [11]:

$$\frac{\partial V_w}{\partial t} = \frac{\partial}{\partial x}\left(T_{wxx}\frac{\partial Z_{aw}}{\partial x} + T_{wxy}\frac{\partial Z_{aw}}{\partial y}\right) + \frac{\partial}{\partial y}\left(T_{wyy}\frac{\partial Z_{aw}}{\partial y} + T_{wyx}\frac{\partial Z_{aw}}{\partial x}\right) + I_w, \quad (10)$$

$$\frac{\partial V_o}{\partial t} = \frac{\partial}{\partial x}\left(T_{oxx}\frac{\partial Z_{ao}}{\partial x} + T_{oxy}\frac{\partial Z_{ao}}{\partial y}\right) + \frac{\partial}{\partial y}\left(T_{oyy}\frac{\partial Z_{ao}}{\partial y} + T_{oyx}\frac{\partial Z_{ao}}{\partial x}\right) + I_o, \quad (11)$$

$$T_w = \begin{vmatrix} T_{wxx} & T_{wxy} \\ T_{wyx} & T_{wyy} \end{vmatrix}, \quad T_o = \begin{vmatrix} T_{oxx} & T_{oxy} \\ T_{oyx} & T_{oyy} \end{vmatrix} \quad (12)$$

where Z_{aw}, Z_{ao} are fluid elevations from *Fig. 1a*; V_w, V_o are water and oil specific volumes (for V_o see (2)); I_w, I_o are vertically integrated source - sink terms of water and oil; T_w and T_o are water and oil transmissivity tensors obtained by integrating

$$T_w = \int_{Z_b}^{Z_{aw}} K_w dz \ , \ T_o = \int_{Z_{ow}}^{Z_a} K_o dz \tag{13}$$

of the soil conductivity tensors K_w , K_o for water and oil phases, correspondingly. In ARMOS, elements of K_w are computed as follows:

$$K_{wxx} = 2\overline{K}_w(\cos^2\theta + R\sin^2\theta) \ , \ K_{wyy} = 2\overline{K}_w(R\cos^2\theta + \sin^2\theta) \ ,$$
$$K_{wxy} = K_{wyx} = \overline{K}_w(R-1) \sin 2\theta \ ,$$
$$\overline{K}_w = 05 \ (K_{wx} + K_{wy})/ \ (R+1) \ , \ 1 \le R = K_{wx}/ \ K_{wy} \ \text{or} \ K_{wy}/ \ K_{wx} \tag{14}$$

where θ is the angle between the y-axis of the domain and the principal axis of K_w ; K_{wx} , K_{wy} are diagonal elements of K_w when $\theta = 0$; R is the anisotropy ratio ≥ 1 by the definition. In ARMOS, a constant K_w value can be used for a domain. In our experiments, $R = 1$.

The oil phase conductivity is assumed to be of the form

$$K_o = k_{ro}\mu_{ro}^{-1} K_{sw} \ , \ k_{ro} = (1 - \tilde{S}_w)^{1/2} (1 - \tilde{S}_w^{1/m})^{2m} \ , \ m = 1 - 1/n \tag{15}$$

where μ_{ro} is the ratio of oil to water viscosity; K_{sw} is the saturated conductivity of water ($S_w = 1.0$); k_{ro} is the relative oil conductivity expressed by the VG model [5].

In general, V_w and V_o are functions of both Z_{aw} and Z_{ao}, and the time-derivative terms of (10) , (11) may be expanded as the following full differentials:

$$\frac{\partial V_w}{\partial t} = \frac{\partial V_w}{\partial Z_{aw}} \frac{\partial Z_{aw}}{\partial t} + \frac{\partial V_w}{\partial Z_{ao}} \frac{\partial Z_{ao}}{\partial t} \ , \tag{16}$$

$$\frac{\partial V_o}{\partial t} = \frac{\partial V_o}{\partial Z_{aw}} \frac{\partial Z_{aw}}{\partial t} + \frac{\partial V_o}{\partial Z_{ao}} \frac{\partial Z_{ao}}{\partial t} \ . \tag{17}$$

Now, we observe that (10) and (11) are coupled through the time-derivative terms of (16) and (17). Some coupling also occurs via the transmissivity terms (12) because, according to (15), T_o can be obtained as a function of T_w . However, the coefficient $\partial V_w / \partial Z_{ao}$ is rather small. Therefore, (16) can be simplified:

$$\frac{\partial V_w}{\partial t} \approx \phi \ (1 - S_m) \ \frac{\partial Z_{aw}}{\partial t} \tag{18}$$

and the water flow of (10) becomes independent from the oil flow of (11).

In our case studies, we applied stationary water flows ($\partial Z_{aw} / \partial t = 0$) of (10); then term (17) of the oil flow also takes a simpler form.

A Galerkin finite element method is used in ARMOS to approximate the flow equations (10) , (11). Normally, linear rectangular elements with sides parallel to the axes x, y are applied.

In this paper details of numerical fluid flow modelling are not discussed due to the lack of critical comments about this part of ARMOS.

5. First experiments with ARMOS at the former Rumbula airbase

We became familiar with ARMOS through the international project "Cooperation between Latvia and Denmark on transfer and know-how concerning investigation and remedy of oil pollution in groundwater" when BA asked the EMC team to perform the following simulation for the former Rumbula airbase which had been selected as the object of this investigation:

- to systematise initial information from various sources and to create databases;
- to build up the digital hydrogeological model (HM) of the contaminated site;
- to evaluate volumes of aviation fuel spill plumes and to compute velocities and directions of their movement;
- to simulate measures taken for the free oil phase clean-up of the site.

The airbase is located in the southern part of Riga on the right bank of the Daugava river about 15 kilometres from the centre of the city. The base was operational over a 24 year period and was shut down almost twenty years ago (1978). Since then, no oil spills have been recorded here. Consequently, this investigation dealt with a historical contamination when no reliable information about sources of pollution was available. However, the BA team was successful in contouring six spill plumes by using as a guide existing and removed objects of the base fuelling system. The plume identifiers used in *Fig.2* (b6, b9, b12, b15, b24, b34) coincide with the codes of the corresponding BA monitoring wells where, within the plume area, the maximal free oil thickness H_o has been observed.

It may be concluded from the groundwater table elevation distribution of *Fig. 2* that, in general, in the HM centre, the groundwater flow is directed towards the Daugava river. In particular, even this central region is under a considerable influence of local groundwater table maximums and minimums which are caused by hillocks and drainages (brooks, ditches). HM does not include the vertical leakage flow passing to the confined middle Devonian layer through the 0.5 - 4.0 metre thick moraine because ARMOS cannot account for this process.

ARMOS professional, SpillCAD and BioTRANS standard codes developed by ES & T were recommended and provided by the Danish Hedeselskabet Company as the fuel spill simulation tools. Only ARMOS was used during the reported preliminary investigation stage when it was necessary to forecast the free oil movement and to evaluate the extent of a free oil volume.

SpillCAD provides the same services as ARMOS, but it also permits a simulation of a dissolved oil product migration. Unfortunately, SpillCAD does not permit the application of imported HM maps prepared by other software tools. This fact results in a catastrophic limitation of SpillCAD because this code applies an ideal uniform groundwater flow for HM. One can change only a slope and a direction of this flow. Nothing like the HM of *Fig. 2* or *Fig. 3* is admissible for SpillCAD !

BioTRANS deals with dissolved oil products only and it is used mostly when a clean-up of trapped oil is considered. This code allows the use of prepared HM.

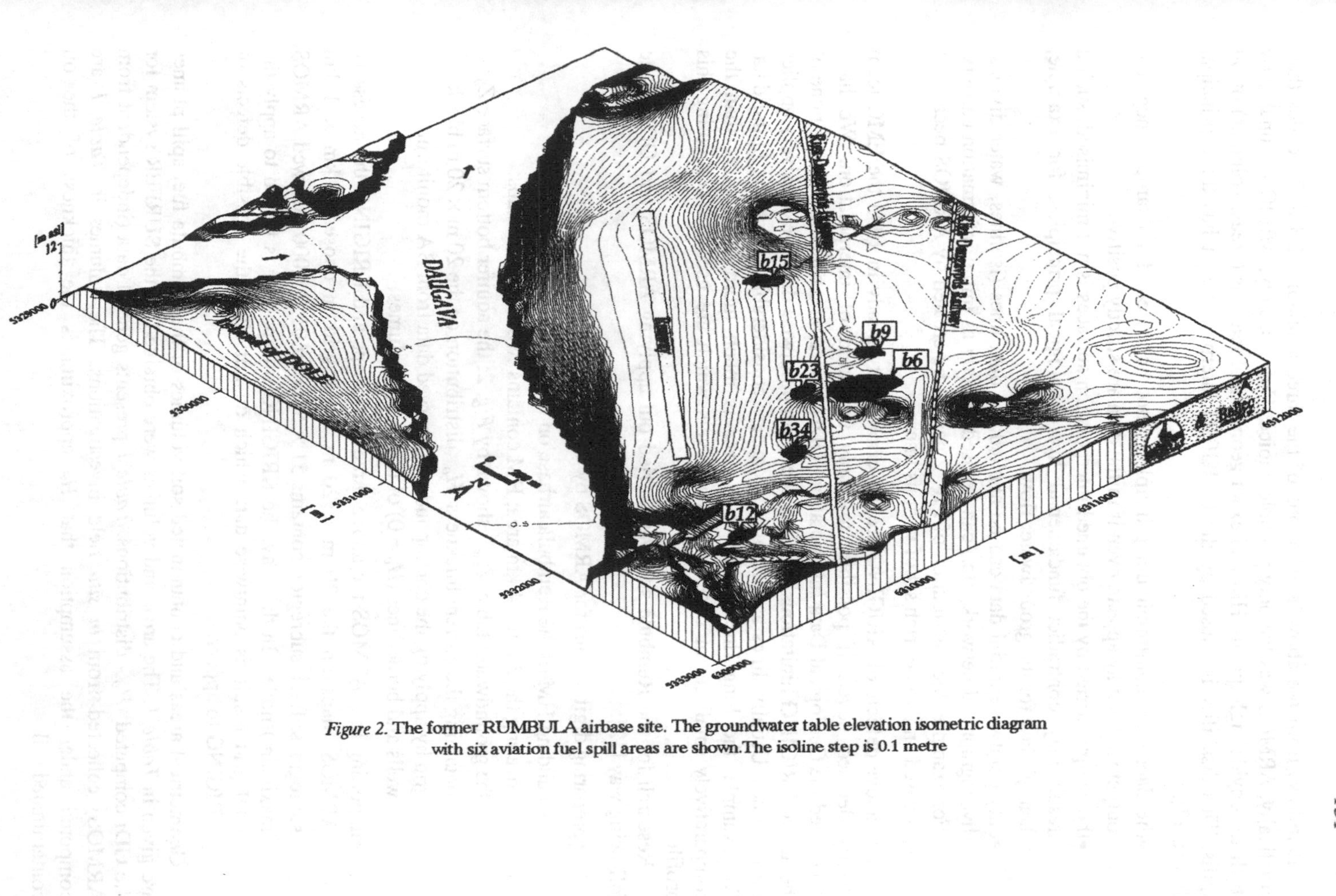

Figure 2. The former RUMBULA airbase site. The groundwater table elevation isometric diagram with six aviation fuel spill areas are shown. The isoline step is 0.1 metre

It follows from the above annotations of the three available ES & T codes that, practically, ARMOS was the only possible choice. Unfortunately, ARMOS is unable by itself to provide high quality HM and exact geometrical features (areas, volumes) of oil spills. This disability is caused by the KRIGING interpolation. It has three principal defects [3]:

- the data maximum/minimum principle is not honoured; for example, negative output values may appear, even if initial data are strictly positive;
- the energy carried by the obtained interpolation surfaces is not minimised; such a result may contradict fundamental hydrogeological principles, for example, Darcy's law for the groundwater flow;
- only pointwise initial data can be applied; no data lines (isolines, water divides, hydrographical network, etc.) can be used directly as initial information carriers. For example, the distribution of *Fig. 2* cannot be created by ARMOS because it is based on a rather rich set of various lines.

The above defects of KRIGING are so serious that already in 1993 the EMC team strictly decided never, if possible, to apply this kind of interpolation. We have developed the Geological Data Interpolation (GDI) programme which possesses none of the above KRIGING limitations [4]. Together with pointwise data it extensively applies various lines. Usually, lines as information carriers considerably enrich the initial data. For example, the long line profile of a river is much more informative than the comparatively small set of observed water levels which are used for computing this profile.

Research for the Rumbula site was carried out applying ARMOS and GDI in the following way [12]:

- preparing digital maps for ARMOS by GDI:
  - building HM of the Rumbula airbase on the 40 m × 40 m block grid covering the 3 km × 4 km area; HM contains two elevation maps: the groundwater table Z_{aw} shown by *Fig. 2*; the aquifer bottom surface Z_b;
  - obtaining the current thickness H_o distribution on the 20 m × 20 m block grid by applying the corresponding observed data from BA monitoring wells and borderlines ($H_o = 0$) of the six spill plumes;
- importing into ARMOS the above three GDI maps and KRIGING the necessary ARMOS maps on the 20 m × 20 m block grid in chosen 1 km × 1 km subregions. Each subregion contains 51 × 51 = 2,601 < 3,000 (allowed ARMOS maximum) nodes. In this way, the KRIGING interpolator is forced to apply the GDI grid result as pointwise data which do not allow the possible defects of KRIGING to appear.

Geometrical areas and contaminated soil volumes of the Rumbula fuel spill plumes are given in *Table 1*. The areas and volumes were obtained by the SURFER system for the GDI computed H_o - distributions. *Table 1* presents general data (independent from ARMOS) collected from *in situ* field investigations. The volumes in *Table 1* are computed under the assumption that H_o approximates the thickness of the oil contaminated soil.

Table 1. **Geometrical areas and total contaminated soil volumes of the Rumbula airbase spill plumes**

Plume ID [1]	Max. plume thickness (7 Oct 96) [m]	Geometrical area [m²]	Number of grid blocks in plume	Contaminated soil volume [m³]	Mean contaminated thickness [2] [m]
b6	0.95	43200	108	18621	0.43
b9	0.93	7600	19	3652	0.48
b12	0.24	5200	13	789	0.15
b15	0.62	9600	24	1097	0.11
b23	0.15	6400	16	332	0.05
b34	0.32	5200	13	1017	0.20
	Total:	**77220**	**193**	**25509**	**0.33**

[1] *the identifier (ID) code denotes the monitoring well with the max. oil thickness observed;*
[2] *mean thickness = volume / area*

Table 2. **ARMOS estimated parameters of the Rumbula airbase spill plumes for a one year period**

General Conditions: Steady state water flow. Z_{aw}, H_o and Z_b generated by GDI.
Constant total oil volumes. Time = 0(initial); 360 [days] (final).
Oil Properties: Kerosene, $\rho_{ro} = 0.84$, $\mu_{ro} = 2.30$, $\beta_{ao} = 2.6$, $\beta_{ow} = 1.5$
Soil Properties: $K_{sw} = 4.0$[m/day], $\phi = 0.35$, $S_m = 0.25$, $\alpha = 10$[1/m], $n = 3.98$,
$S_{or} = 0.2$, $S_{og} = 0.05$, ratio = 1.0, angle = 0.0 [degree]
Kriging Parameters: area around plume 1000m • 1000m, grid step $h_x = h_y = 20$m, number of nodes 2601

Plume ID	Total spill volume [m³]	Total residual volume [m³]	Unsaturated residual volume [m³]	Total oil area [m²]		Mass balance error [%]
	initial	final	final	initial	final	final
b6	1427.6	39.4	2.1	35930	37550	0.013
b9	201.7	8.2	0.4	5650	7260	0.008
b12	2.0	0.0	0.0	2496	2496	5.136
b15	17.8	0.8	0.1	3215	4010	0.033
b23	0.1	0.0	0.0	405	405	0.000
b34	6.8	1.0	0.0	4040	4040	0.103
Total:	**1656.0**	**49.4**	**2.6**	**51736**	**55761**	

According to *Table 1*, the largest plume is the b6 fuel plume (area = 43,200 m², volume = 18,621 m³, mean thickness = 0.43 m); but b9 has the highest mean thickness (0.48 m), its soil volume being the second largest one (3,652 m³).

To run ARMOS, it is necessary to enter the oil and soil properties of the VG model. We applied the parameters of kerosene and the sandy Rumbula soil. *Table 2* presents estimated initial and final parameters of the Rumbula migrating plumes for a one year (360 days) period if the steady state distribution Z_{aw} (water table elevations of *Fig.* 2) is used and oil volumes of plumes are constant.

The following comments can be made on these results:

- the initial areas of all spill plumes are smaller than their geometrical areas from *Table 1*. For example, the area of the b23 plume (405 m^2) is much smaller than 6,400 m^2 from *Table 1*. This phenomenon represents a principal disagreement between the input data and the ARMOS results because the computed initial total oil areas should be equal to the geometrical areas;
- total oil areas of the three largest plumes b6, b9, b15 have increased during one year as they should be due to a movement of the fuel spill;
- the mass balance error is acceptable for five plumes but is poor for the b12 plume (5.13%);
- volumes of trapped (residual) oil in the aeration (unsaturated) zone are insignificant for all the plumes.

More careful studies of the causes of the above "areal paradox" of ARMOS lead to the inevitable conclusion that ARMOS does not account for the initial trapped (residual) oil phase specific volume V_{ot} (see *Fig. 1c, d, e*). The residual oil volume of *Table 2* is due only to some plume surface changes during its movement. Some minor errors in the plume area calculations may also have originated with the KRIGING interpolator, especially when H_o is small. These errors can be explained by the influence of some control constants which are applied to prevent obtaining negative H_o.. Unfortunately, no information is available from [11] about these and how to change these constants.

In order to explore ARMOS responses more carefully with respect to area and volume calculations, the most important b6 plume was investigated and the following conclusions were drawn [12]:

- the total spill volume is very sensitive with respect to the chosen oil and soil properties (from 2,578 m^3 till 0.83 m^2 in sand and silty clay, correspondingly). It should be so due to VG model properties;
- the total initial spill area is smaller than the geometrical one and strongly depends on the chosen oil and soil properties (from 42,370 m^2 till to 405 m^2 in sand and silty clay, correspondingly). This should not be the case (paradox of areas !);
- the total final spill area increases if the saturated water conductivity K_{sw} increases as it should be due to the increase of the plume migration velocity.

In situ field investigations and ARMOS simulation results showed that no significant movement of spill plumes may be expected at present. Initially, it was hard to believe that such a forecast may be right. However, during twenty years, five plumes have moved only 50 - 100 metres with respect to their possible spill sources. Only the b12 plume has moved from the fuel storage as the probable spill source about 200 metres down along the water divide of two brooks running there (see *Fig. 2*). If one considers the graph S_o of *Fig. 1d* carefully, then in the zone of small H_o values, an entry threshold can be observed. No oil can replace water until some critical oil pressure is reached. This fact explains the existence of the dynamic equilibrium of the Rumbula spill plumes with slowly drifting borderlines. However, the relative oil conductivity k_{ro} of (15) behaves such that the central maximum of a plume moves faster than its frontline. The observed b6 plume thickness peak at the plume front area can be partly explained by such a phenomenon. The following maximal H_o values have been predicted by ARMOS for b6 during the next 20 years: initial - 0.95 m; after 2, 10, 15 and 20 years - 1.19 m, 1.38 m and 1.46 m, respectively.

The above facts show in the case of an oil migration how misleading a simple straightforward judgement of a hydrogeologist who has dealt only with a dissolved contaminant transportation in groundwater may be. In reality, at the location of the b6 plume, the mean hydraulic gradient is about 0.003 and, if $K_{sw} = 4.0$ m/day, $\phi = 0.35$ then a water particle will travel about 12.5 metres = $365 \times 4.0 \times 0.003/0.35$ during one year. One naturally expects that the oil velocity will have about the same value. The Rumbula case proves that an oil body may move much slower than a water flow. It may even stop moving altogether.

The ARMOS simulation showed that, in the case of the Rumbula site, the change $V_{of} \rightarrow V_{ot}$ from free to trapped oil phase appears to be more alarming than the slow drift of the plumes in the direction of the Daugava river. This transformation proceeds rather quickly for small plumes [12].

Only preliminary free oil remediation tests have been performed for the b6 plume under the following assumptions:

- no water pumping; the b6 well is used for oil skimming;
- the oil pumping rate is controlled by the condition $H_o \sim 0$ in b6.

ARMOS forecasts that the average oil skimming rate of the b6 well should be about 0.05 m^3/day.

Our experiments with ARMOS in the Rumbula case have provided valuable knowledge about oil transportation in groundwater. An improper ARMOS response has been discovered regarding the calculations of trapped oil volumes. To compensate this limitation, the EMC team has developed the VOF code [10]. It performs oil volume calculations by applying of VG model described in this paper.

6. Analysis of the Ilukste oil storage terminal case

The Ilukste oil storage terminal (called base in this section) is located in the southern part of Latvia near the town Ilukste. The base is situated on a steep bank of the Ilukste river which is a tributary of the Daugava river (see *Fig. 3*).

Over twenty years, diesel fuel has been leaking through the lower part of the Ilukste river bank. A special pond has been constructed to prevent straight dripping of oil into the river. The contents of the pond are periodically removed.

Numerous teams of researchers have been unsuccessful in discovering the nature of and real reasons for this leakage phenomenon. However, the EMC team accepted the BA proposal to simulate the Ilukste site situation by using the ARMOS code. The main goals of this project were as follows:

- to create numerical HM and an oil transport model;
- by using these models
  - to determine the area and the volume of the contaminated domain;
  - to find the distribution and intensity of the oil leakage sources;
  - to evaluate the effectiveness of various proposed free oil clean-up measures.

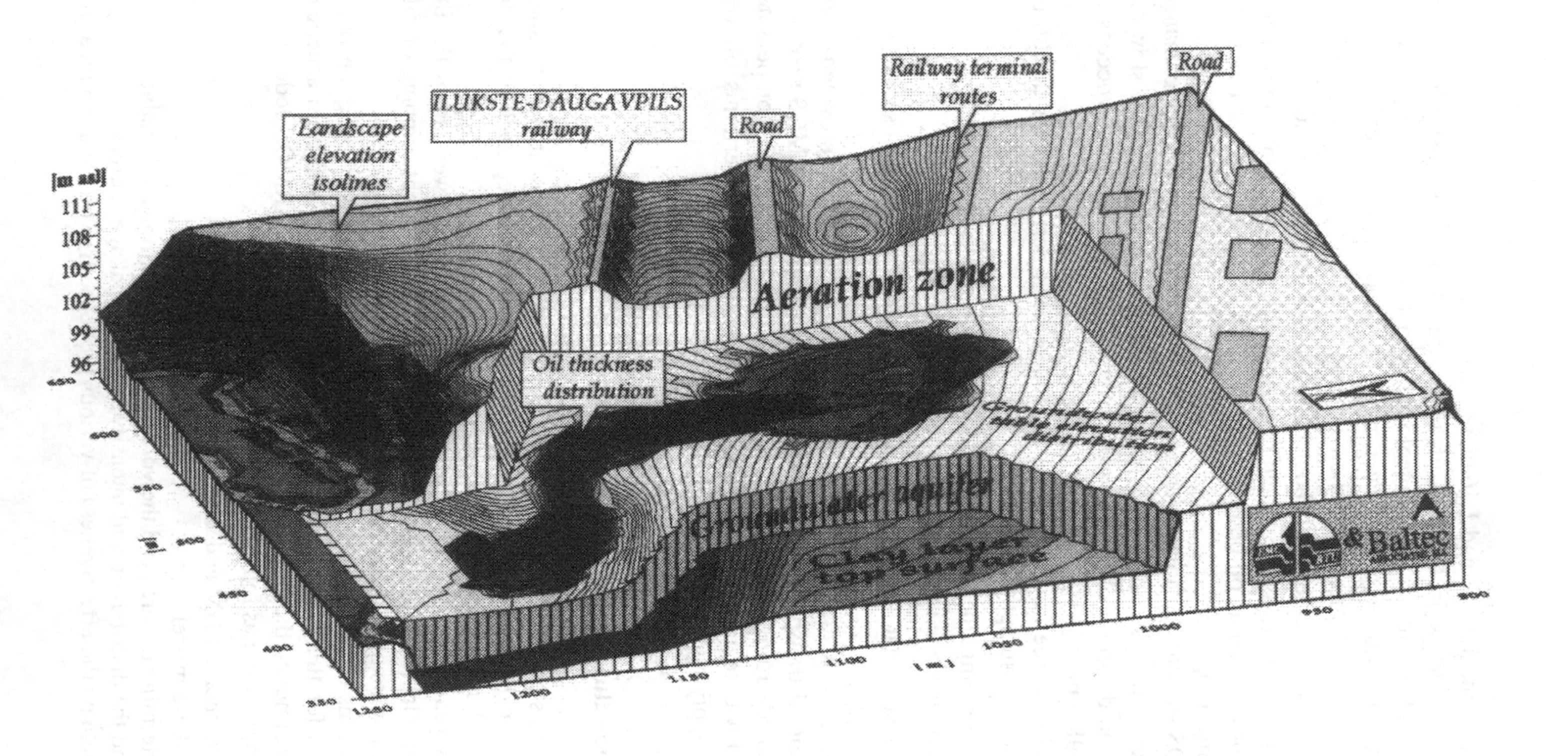

Figure 3. The ILUKSTE oil storage terminal site. The complex isometric diagram with elevation distributions of : the landscape relief, the groundwater table, the clay layer top surface and the simulated oil spill thickness. The z-step for all kinds of isolines is 0.1 metre

The hydrogeological situation at the Ilukste base site is much more complex than that at the Rumbula airbase for the following reasons:

- •an uneven landscape relief causes local water divides and other roughnesses of the groundwater table;
- •high hydraulic gradients (about 0.04) of the groundwater flow;
- •perched groundwater bodies exist there due to the presence of numerous clay interlayers.

Generally, the Ilukste oil base site is a part of the groundwater recharge zone of Latvia. Its minimal and maximal relief elevation marks are 98.2 and 118.0 metres a msl. These correspond to the mean level of the Ilukste river and a hillock near the base, accordingly. The thickness of the aeration zone is 4 - 8 metres with respect to the principal oil-contaminated aquifer.

In order to overcome some of the ARMOS limitations already detected, the following additional software tools were applied in the Ilukste base case:

- •the VOF oil volume calculation program, since, ARMOS accounts for the trapped oil volume incorrectly;
- •the SURFER for Windows program by Golden Software, Inc. for replacing ARMOS poor graphics when final results are to be displayed and documented;
- •the Data Base Management System ACCESS by Microsoft for creating a common data base for all software tools involved in the modelling.

The 5×5 m block grid was used for approximating of the base place domain.

First, it was necessary to understand by which route diesel fuel can reach the oil pond and what the shape of a hypothetical spill plume would be? *In situ* investigations of BA yielded the following hypothesis: "The oil flow originates in whereabouts of the railway terminal of the base. While going down to the North, the flow is gradually focused into a narrow stream. Before the Daugavpils - Ilukste railway, this stream takes a sharp left turn, becomes more centred and runs down along or beneath the railway for a distance of about 50 metres. Finally, the oil stream turns right and, gradually broadening out into a fan-shape reaches the pond". The toned form in *Fig. 4* represents the hypothetical spill plume shape based on *in situ* observations of H_o and the above hypothesis.

We assumed that the oil plume is caused by some constant leakage near the base railway terminal routes used for oil transportation. Consequently, the plume must reach a steady state form after some time period which is longer than the free oil travel duration from the base to the pond. Chemical analysis of fuel taken from the pond stated that the oil there is about 12 years old. So, there was enough time to reach such a steady plume condition, hence, the base had been operational for more than 20 years.

The next task was to obtain the distribution of H_o for the steady oil plume contoured under the already explained assumptions. There were plenty of *in situ* data of H_o for the "broad areas" of the plume. Unfortunately, no direct observations of H_o were available for the narrow "imaginary" oil stream running close or beneath the railway because, normally, no well drilling should be allowed in this zone. To forecast the expected values of H_o here, we applied the basic principle of Bernoulli's rule for moving liquid bodies: "The liquid velocity $\times$ the body cross section area = constant

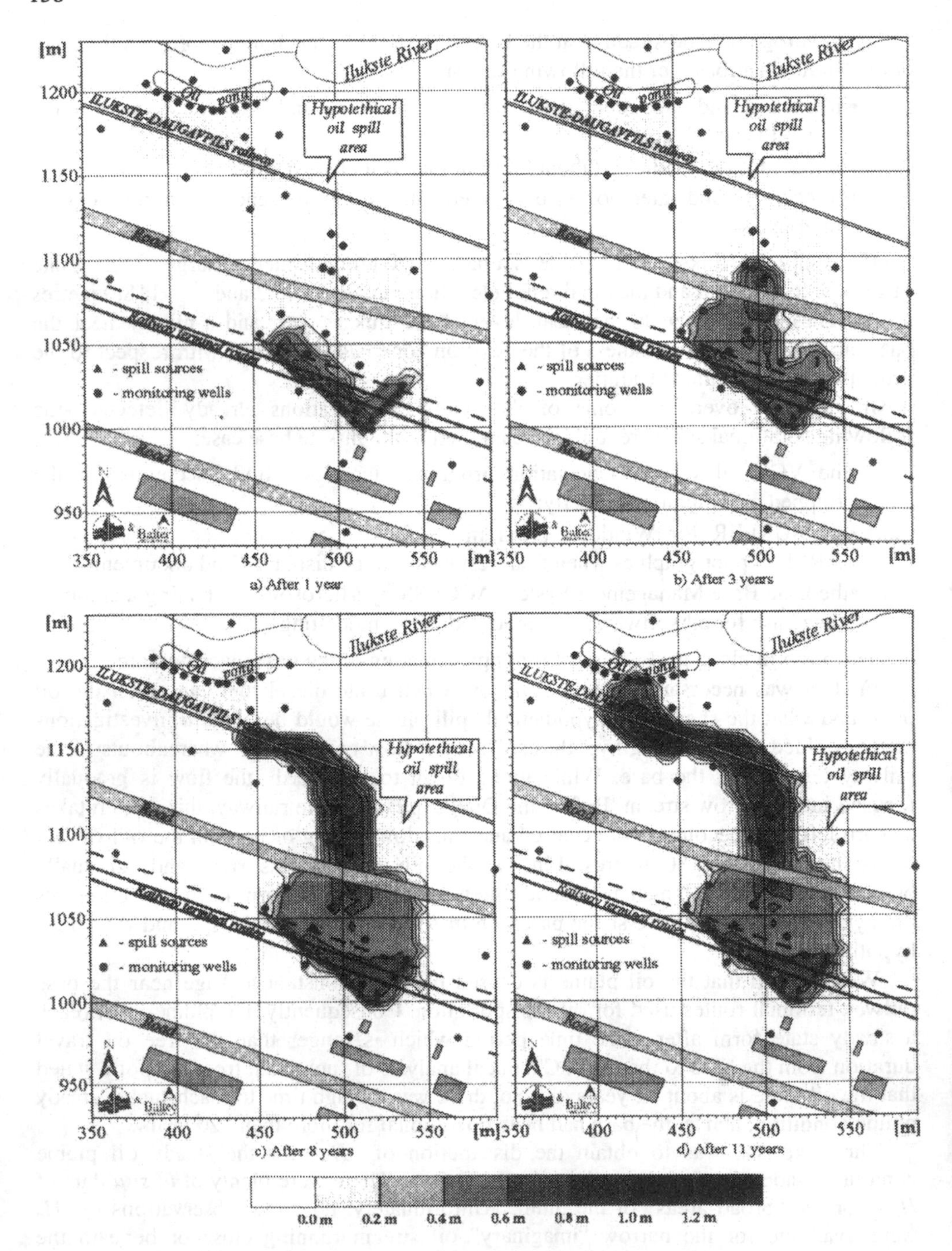

Figure 4. **The ARMOS simulated oil spill shape and the thickness development at the ILUKSTE oil storage terminal site.The thickness H_0 isoline step is 0.2 metre**

in all cross sections, if there are no liquid losses". Therefore, one may expect the H_o maximum in the narrow part this oil flow because an increase of H_o naturally compensates a cross section width decrease there. As a preliminary we used, H_o ~ 1.3 m in this measurement inadmissible zone. The GDI program prepared the hypothetical spill plume distribution of H_o. It was used as a standard during the next experiments with ARMOS.

No one knows, for certain, how the above plume is formed in nature. Obviously, the main element in this process is the groundwater table Z_{aw} of (10). The influence of a probable highly conductive soil lens beneath the railway and flow blocking clay interlayers may be extremely important in the reported case as the objects of the transmissivity tensors T_w , T_o of (13). Unfortunately, ARMOS does not permit the area variable conductivities K_w , K_o which are needed there.

Moreover, an improper ARMOS response was detected when we tried to stop the free oil flow movement locally by setting $Z_{aw} - Z_b = 0$ when, according to (13), $T_w \to 0$. Instead of stopping, the flow actually accelerated within the chosen local domain because of the corresponding increase of the groundwater hydraulic gradient there. This finding discovers a principal drawback the ARMOS oil and water flow simulation because the condition (1) is ignored completely.

So, in ARMOS, Z_{aw} must be used as the only tool for direct controlling the spill plume movement. However, VG parameters may also indirectly change the movability of oil. An oil leakage intensity and its areal distribution likewise can modify the plume.

We approached and solved the above problem as follows:

- all unknown elements should be searched iteratively by using GDI and ARMOS;
- the total oil leakage intensity should be determined when Z_{aw} and VG parameters are chosen;
- the distribution of oil leakage sources should be iteratively determined; Z_{aw} and VG parameters adjusted by simulating the steady state plume creation.

The main element of applied Z_{aw} was the waterdivide located along the narrow part of the oil plume and acting as a barrier for water and oil flows coming down from the contaminated base site. The second important element was the hollow of Z_{aw} operating as the drain of the focused oil stream there.

Thanks to GDI, it was relatively easy to obtain the first approximation of Z_{aw} to be used in determining the value of the total oil leakage intensity. It was done by an ARMOS simulation of a spill plume vanishing if the oil leakage ended. The H_o standard was applied as the initial oil distribution. In the oil pond, a set of ARMOS fuel skimming wells were installed to measure the volume of free oil flowing into this pond.

During the first phase of intensive oil run-off, the relationship of volume versus time is a practically linear one and its slope volume / time is the total oil leakage intensity searched for. It supports the steady state form of the chosen standard plume and, for the Ilukste base, the leakage intensity is about 70 - 90 litres/day.

In order to investigate various possibilities for removing free oil, the described spill plume run-off experiment was repeated after the final adjusting of Z_{aw} , the VG parameters and after obtaining the ARMOS-simulated steady state plume. These results showed that clean-up facilities should be constructed near the oil pool to exploit only the natural oil run off process. No oil skimming wells located at the oil leakage sources are expected to be effective for they will cease to function soon after ending the leakage [13].

It took about ten iterations to calibrate Z_{aw} , VG parameters and the oil leakage distribution to satisfy the 10 - 20% accuracy when the following conditions had to be complied with simultaneously :

- the steady state ARMOS-simulated spill plume parameters were close to those of the standard one (shape, formation time, contaminated soil volume);
- the simulated parameters of operational oil skimming wells had to be within close proximity of the real ones.

The main software tool for making of Z_{aw} was GDI. Practically no help was provided by classic particle tracking techniques, for they fail to properly describe the movement of large liquid bodies.

The VG parameter α served as the variable to be calibrated. The final set of the VG parameters is represented by *Fig. 1* where *Fig. 1f* illustrates the great responsiveness of V_{of} versus α changes. Moreover, α also greatly affects the oil moveability. This phenomenon may be observed as changes of the oil plume shape and formation time.

We discovered that the "L"-shaped configuration of constant uniform oil sources (see *Fig. 4*) provided a good approximation of the real oil leakage. It was proven that the observed H_o distribution near the base did not originate in any single source or point [13]. A recommendation was made to check over the diesel fuel sewage system constructed to protect the base from oil spills. This system, if not built properly, may become transformed into a space-distributed oil leakage source because it contains numerous pipes joined by seams which are potential local sources of oil leakage. They may become active when the slopes of the sewage pipes are incorrect, thus allowing some amount of oil to become stagnant and to accumulate within the pipes. This theory proved to be correct because stagnant oil with thickness about 20 cm had been discovered in one part of the sewage system! So, the old mystery on oil leakages in the Ilukste oil storage terminal was explained due to a successful interpretation of ARMOS simulation results.

Fig. 4 gives some dynamics of the simulated spill plume over the eleven year period when the moving plume reached the pond. It takes about a further three years to form the full shape and the volume of the steady state plume (see *Fig. 3*). One may observe that the simulated plume indeed has the predicted maximum of H_o in its narrow part.

Due to the VOF code, it was possible to verify the ARMOS oil volume calculations. The results of this comparison are given in *Table 3*.

Table 3. A comparison of Ilukste oil spill plume volumes obtained by ARMOS and VOF codes

Code	v_{of} [m^3]	v_{ot} [m^3]	$v_o = v_{of} + v_{ot}$ [m^3]
ARMOS	302	31	333
VOF	307	198	505

The oil volume v_{of} computed by VOF is only slightly larger (307 m^3 > 302 m^3) than that obtained by ARMOS. Hence, both codes apply the formulae (4) , (6) in order to obtain S_{of}, this small difference may be explained by the already mentioned influence of some ARMOS control constants.

The volumes of trapped oil v_{ot} differ considerably (198 m^3 >> 31 m^3) for reasons already explained .

The reported Ilukste oil terminal case was difficult but it provided a considerable amount of new knowledge. We understood that in cases dealing with an intensively moving oil flow, it would be necessary to account for a variability in aquifer parameters. The ARMOS code does not allow it.

7. Conclusions

The ARMOS code has been used for simulating oil mass transportation processes in groundwater at two rather different oil contaminated places.

The former Rumbula airbase case was a rather simple one. It dealt with stagnant slowly drifting aviation fuel plumes. However, due to this fact, an apparently fundamental error of ARMOS in the form of trapped (immobile) oil volume calculations has been discovered. The EMC team developed the VOF code for obtaining all kinds of oil volumes. VOF was applied later in the Ilukste case. Hence, since ARMOS cannot provide good initial data interpolation, the GDI code was used for the digital mapmaking.

The Ilukste oil storage terminal case was a really difficult one. It was necessary to simulate an active oil plume of a very complex shape. Oil leakage sources of the plume have not yet been discovered despite numerous efforts. Through this project it was learned that the inability of ARMOS to account for variable aquifer parameters was a serious limitation of this code.

Utilising the ARMOS module to perform oil flow simulations has proved to be a good quality tool for the regime of a steady water flow. However, an improper response of the module was detected: it did not check for the existence of a water bearing aquifer. This slip may result in serious mistakes in the cases dealing with ample oil bodies or small saturated aquifer thicknesses.

The Ilukste case was a great success for the EMC team because, due to mathematical simulations, old mysteries of oil contamination origins were partially solved and practical recommendations for a free oil clean-up were provided.

8. References

1. Spalvins, A., Janbickis, R., Slangens, J., Gosk, E., Lace, I., Viksne, Z., Atruskievics, J., Levina, N., Tolstovs, J. (1996) *Hydrogeological Model "Large Riga". Atlas of Maps,* Riga-Copenhagen, 102 p. (Boundary Field Problems and Computers; 37-th issue; bilingual: Latvian and English).
2. Spalvins, A., Gosk, E., Grikevich, E., Tolstov, J. (1996) *Modelling new well fields for providing Riga with drinking water,* Riga-Copenhagen, 40 p. (Boundary Field Problems and Computers; 38-th issue).
3. Spalvins , A., Slangens, J. (1994) *Numerical interpolation of geological environment data, Proc. of Latvian-Danish on "Groundwater and Geothermal Energy",* Vol. 2, Riga-Copenhagen, pp. 181 - 196. (Boundary Field Problems and Computers; 35-th issue).
4. Spalvins, A., Slangens, J. (1995) *Updating of geological data interpolation programme, Proc. of International Seminar on "Environment Modelling",* Vol. 1, Riga-Copenhagen, pp. 175 - 192. (Boundary Field Problems and Computers; 36-th issue).
5. Van Genuchten, M.Th. (1980) A closed-form equation for predicting the hydraulic conductivity of unsaturated soils, *Soil Sci. Soc. Amer. J.,* 44, pp. 892 - 898.
6. Stallman, R.W. (1964) *Multiphase fluids in porous media - a review of theories pertinent to hydrologic studies, Fluid Movement in Earth Materials, Geological Survey Professional Paper* 411-E. US Government Printing Office, Washington, 50.p.
7. Land, C.S. (1968) Calculation of imbibition relative permeability for two and three-phase flow from rock properties, *Transactions of American Institute Mining Metallurgical and Petroleum Engineering* 243, pp. 149 - 156.
8. Kemblowsky, M.W., Chiang, C.Y.(1990) Hydrocarbon thickness fluctuations in monitoring wells. *Ground Water,* 28, pp. 244 - 252.
9. Parker, J.C., Lenhard, R.J. (1989) Vertical integration of three phase flow equations for analysis of light hydrocarbon plume movement. *Transport in Porous Media,* 5, pp. 187 - 206.
10. Spalvins, A., Lace, I. (in press) *Estimating of free and trapped oil volumes for light hydrocarbon plumes in groundwater,* Riga (Boundary Field Problems and Computers, 40-th issue).
11. ARMOS, 1988 - 96, *User and Technical Guide,* Environmental Systems and Technologies, Inc.
12. Contract No 6153/96 (BC-001) report (1996) *Modelling of the groundwater flow dynamics and the contaminant movement for the Rumbula airbase place by SpillCAD, ARMOS and BioTRANS,* Riga Technical University, Environment Modelling Centre, 113.p.
13. Contract No 6190/97 report (1997) *Modelling of oil flow movement and recovery for the Ilukste oil storage terminal,* Riga Technical University, Environment Modelling Centre, 67 p.(in Latvian).

GIS AND MATHEMATICAL MODELLING FOR THE ASSESSMENT OF VULNERABILITY AND GEOGRAPHICAL ZONING FOR GROUNDWATER MANAGEMENT AND PROTECTION

JOÃO PAULO LOBO-FERREIRA
Dr.-Ing., Head of the Groundwater Research Group
Laboratório Nacional de Engenharia Civil, Av. do Brasil 101
P-1799 LISBOA Codex, Portugal,
Tel. (351 1) 8482131, Fax: (351 1) 8473845
E-mail: LFerreira@LNEC.PT ; http://www-dh.lnec.pt/gias/gias.html

1. Concepts of Groundwater Vulnerability to Pollution

1.1. INTRODUCTION

The groundwater group of the European Community Commission (EEC) held a meeting on February 91, in the European Water Institute, in Brussels, with the purpose of establishing an international agreement on common methodologies for the elaboration of a groundwater resource inventory for all Member-States. Such an inventory has been made in the past, between 1979 and 1981, for all countries that were Community Members-States at the time. The existing inventory needed to be updated to new Members-States, e.g. Spain, Portugal and the new Länder of Germany.

The new inventory is to include the aspects of groundwater resource quantity, demand, and vulnerability to pollution. The limitations of the existing inventory were pointed out at the meeting, and it is desired to improve the methodological practices of its elaboration. An important limitation that was emphasised is that the different vulnerability maps, independently of their individual merit, are not comparable, as the key parameters and criteria underlying the elaboration of each are not the same examples of "classic" vulnerability maps include that for France [1], for Germany [2], for the Southern region of Portugal [3], for Denmark [4], and the study (not yet published) for some EEC Member-States [5]. All these maps consider more and less vulnerable categories (e.g. high, medium and low vulnerability categories) but the key parameters selected and criteria used in the ranking are different.

It was decided in the meeting that it is necessary to uniformize the criteria and procedures used by each Member-State to evaluate, rank, and map groundwater pollution vulnerability. The advantages of this uniformization are that:

a) maps generated in one country are readily understandable by all;
b) maps from different countries are comparable and can be put together to cover cross-boundary and multinational regions;

F. Fonnum et al. (eds.),
***Environmental Contamination and Remediation Practices at Former and Present Military Bases*, 143–170.**

c) the dialogue among professionals from various countries is facilitated and is desirable as each country may benefit from the experience acquired by the rest; and

d) it becomes possible to create a compatible and coherent data base that includes data for all EEC Member-States, whenever possible based on the experience gathered in the EEC CORINE Project.

The establishment of standardised criteria must of course be made by international agreement before the groundwater inventory update can be initiated. In the EEC meeting it was requested that each participant submit a proposal for an "operational definition of vulnerability".

It was suggested at the EEC meeting that, once standardised criteria have been established, a pilot study be conducted by the new Member-States, e.g. Spain, Portugal and the new Länder of Germany. The methodological experience acquired in this pilot study may then be extended to the remaining Member-States of the EU. In our opinion such pilot studies, as the one presented in Section 1.6, are advantageous as they may serve as examples of standardised procedures and as guidance to the other countries. The adopted scale is 1:500 000. This seems to us to be the appropriate scale, as a larger scale would require much more detailed information than is generally available.

In Section 1.2 below, we review the criteria of vulnerability ranking used in the elaboration of a few existing EEC vulnerability maps. What we consider to be some of their limitations are pointed out. In Section 1.3, we propose a definition of the term vulnerability in the context of groundwater pollution. In Section 1.4, we suggest a system of evaluation and ranking of groundwater pollution vulnerability as defined in 1.3, to be adopted by all EU Member-States in the elaboration of their new vulnerability maps. Conclusions of this section of the paper and the Portuguese DRASTIC groundwater vulnerability map are presented in Sections 1.5 and 1.6.

1.2. REVIEW OF "CLASSIC" CRITERIA UTILISED FOR VULNERABILITY CLASSIFICATION IN THE 70-80'S EEC GROUNDWATER RESOURCE INVENTORY

As referred previously, the key factors and criteria underlying the elaboration of the existing vulnerability maps were different for different countries. Here we review the classification criteria utilised in a few EEC groundwater vulnerability maps.

1.2.1. *The vulnerability map of Valence, France, by BRGM [1]*

The vulnerability map of Valence, France, by BRGM [1] consider four vulnerability categories: very high, high, variable, and low. The key factors considered in the evaluations are:

a) present value of the aquifer for water supply
b) protection capacity relative to pollutants loaded at the soil surface
c) travel time of the pollutant from the surface to the aquifer
d) persistence of the pollutant in the aquifer

Thus, the category of very high vulnerability is characterised as follows:

- Regions with extensive aquifers that are intensely used by large urban and industrial centres
- Reduced protection to pollutants loaded at the soil surface
- High permeability due to fissures or fractures, with a velocity of pollutant propagation approx. a few meters per day
- High persistence of pollutants in the aquifer

The concept of vulnerability implied by these factors has a clearly composite nature. Factor a) is dynamic i.e., it changes in time by human influence; while factors b), c) and d) are static i.e., they depend on soil and aquifer characteristics, which change over much longer periods of time. The travel time of the pollutant (factor c)) is an arguably valid parameter to include per se, as is explained in 1.2.2, below. Factors a) and d) would not be included in a more restrictive definition of vulnerability.Factor a) considers the value of the aquifer as a water supply resource. Only a broad definition of vulnerability can include value as a factor. Factor d) could perhaps be better described as aquifer resilience rather than included in the concept of vulnerability. In summary, we think that the implicit concept of vulnerability is too broad and in fact mixes together distinct concepts. Vulnerability in the strict sense is not distinguished from resilience (see Section 1.3), and both static and dynamic factors are considered together.

1.2.2. *The vulnerability maps for the U.K. by the Institute of Geological Sciences*

The vulnerability maps for the U.K. by the Institute of Geological Sciences (IGS) considered, in 1983, four vulnerability categories (and three categories that represent areas with variable vulnerability, areas with insufficient information and areas with no aquifers present). The categorization is based on a single parameter, the residence time in the unsaturated zone of a conservative, non-absorbable and non-adsorbable pollutant with the physical properties of water. Thus, the four vulnerability categories correspond to a residence time greater than 20 years; between 1 and 20 years; between 1 week and 1 year; and less than a week.

The IGS authors of such maps point out themselves some limitations of this classification system:

a) The system leads to evaluation errors in:

- aquifer discharge areas. According to this system, these areas are classified as highly vulnerable because travel time to the aquifer is zero. However, pollutants cannot penetrate the aquifer as it is discharging.
- areas of shallow confined aquifers. This system classifies such areas as highly vulnerable given the short travel time. However, these aquifers may be well protected by a thin near-impermeable overlying layer.

b) Therefore the IGS authors mentioned that "most bodies in the United Kingdom who are responsible for the management of aquifers and groundwater resources do not accept this definition of aquifer vulnerability."

1.2.3. *The "classic" vulnerability map of Portuguese southern regions of Alentejo and Algarve, by Laboratório Nacional de Engenharia Civil (LNEC)*

The "classic" vulnerability map of Portuguese southern regions of Alentejo and

Algarve, by LNEC [3] also considers the four vulnerability categories, very high, high, variable, and low. However the factors considered in the evaluation and categorisation are not the same as for the BRGM map [1]. These factors are mostly hydrogeologic in nature and were estimated from the "Lithological Map of Portugal for Hydrogeologic Use" [6].

The key factors considered were:

a) Permeability of the unsaturated zone and of the aquifer
b) Aquifer recharge rate
c) Attenuation capacity of the unsaturated zone
d) Velocity of pollutant propagation in the unsaturated zone

According to these factors, the vulnerability categories were made correspond to the following hydrogeological groups:

a) Very high vulnerability:

- Karst formations of medium to high permeability. Velocity of pollutant propagation in the order of 10 m/d.
- Fluvial and dune alluvial formations of medium to high permeability.Velocity of pollutant propagation in the order of 1 to 10 m/d.

b) High vulnerability:

- Gabbro and diorite rock formations. Velocity of pollutant propagation in the order of 1 m/d.
- Detritic cover formations of medium to low permeability.
- Mesozoic and Cenozoic porous and fractured sedimentary formations of medium to low permeability.
- Clay, sand and marl formations (Grés de Silves).

c) Highly variable vulnerability:

- Fractured igneous formations, mainly granites.Variable pollutant propagation velocity.
- Fractured metamorphic formations, mainly schists and grauvaques.

Figure 1 shows the "classic" vulnerability assessment of Portugal, developed by LNEC [3], for Alentejo and Algarve, and extended to the rest of the country in [6] following this procedure. In this assessment the solely consideration of the above mentioned factors, heavily dependent on hydrogeological factors, was not considered to be the optimal one, but a first assessment of the hydrogeological data available in Portugal. The study co-financed by the Portuguese General-Directorate for the Quality of the Environment and by LNEC, was a step to extend the 70-80's vulnerability mapping of the EEC, to include Portugal.

1.3. A NEW CONCEPT FOR THE DEFINITION OF GROUNDWATER VULNERABILITY TO POLLUTION

Before we can consider the evaluation of groundwater vulnerability to pollution, it is necessary to define the term vulnerability. The term vulnerability has been defined and used before in the area of water resources, but within the context of system performance evaluation, e.g. the definition given by Hashimoto *et al.* [7].

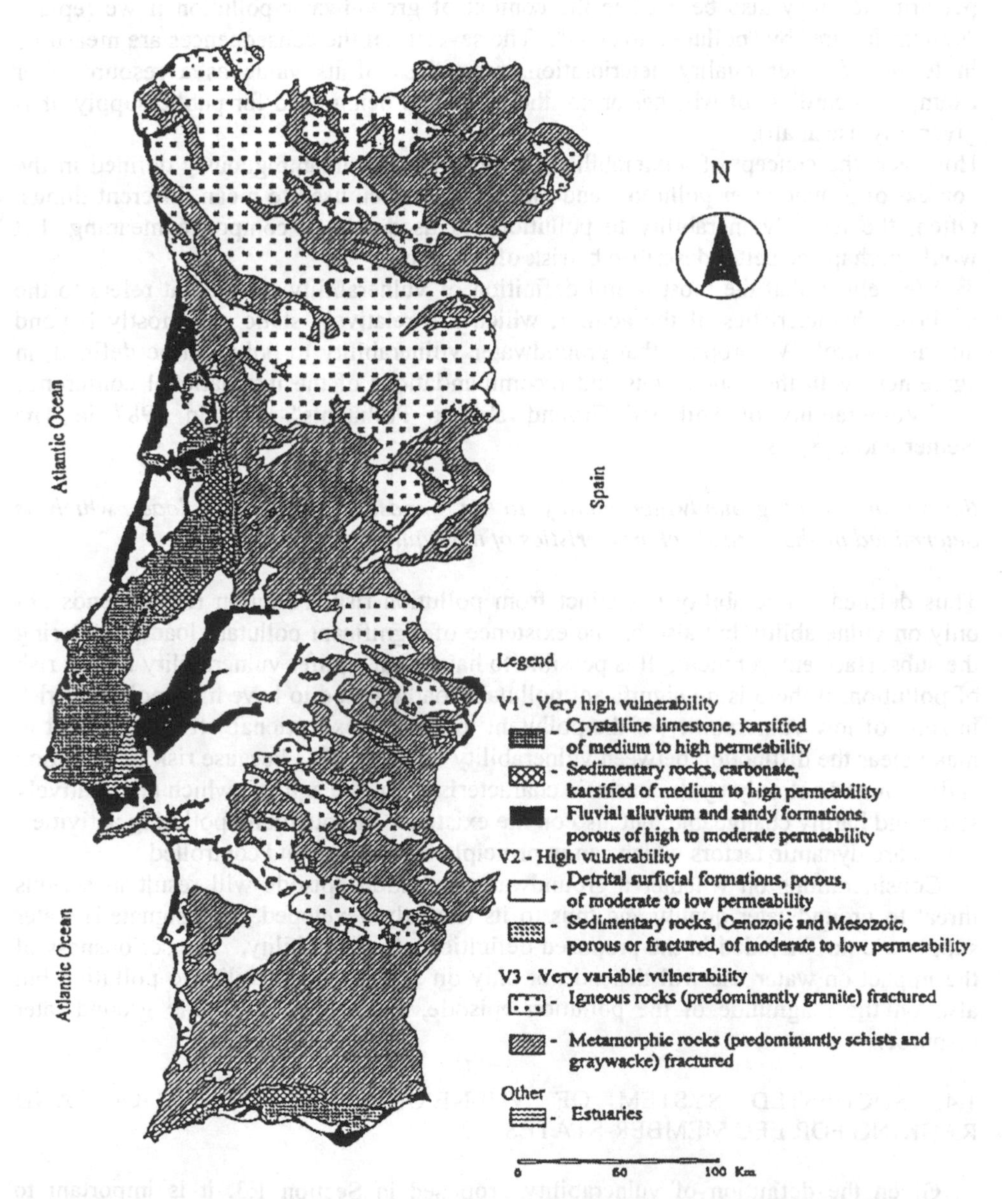

Figure 1. "Classic" vulnerability assessment of Portugal developed by LNEC [3 and 6]

These authors present an analysis of system performance which focuses on system failure. They define three concepts that provide useful measures of system performance: (1) how likely the system is to fail is measured by its *reliability*, (2) how quickly the

system returns to a satisfactory state once a failure has occurred is expressed by its *resiliency*, and (3) how severe the likely consequences of failure may be is measured by its *vulnerability*. This concept of vulnerability defined in the context of system performance may also be used in the context of groundwater pollution if we replace "system failure" by "pollutant loading". The severity of the consequences are measured in terms of water quality deterioration, regardless of its value as a resource (for example, regardless of whether or not the aquifer is being used for public supply or is given any use at all).
However, the concept of vulnerability has not yet been unambiguously defined in the context of groundwater pollution, and the term has been used to mean different things. Often, the term "vulnerability to pollution" is used with a composite meaning that would perhaps be better described by risk of pollution.

We believe that the most useful definition of vulnerability is one that refers to the intrinsic characteristics of the aquifer, which are relatively static and mostly beyond human control. We propose that groundwater vulnerability to pollution be defined, in agreement with the conclusions and recommendations of the international conference on "Vulnerability of Soil and Groundwater to Pollutants", held in 1987 in The Netherlands, [8], as

the sensitivity of groundwater quality to an imposed contaminant load, which is determined by the intrinsic characteristics of the aquifer.

Thus defined, vulnerability is distinct from pollution risk. Pollution risk depends not only on vulnerability but also on the existence of significant pollutant loading entering the subsurface environment. It is possible to have high aquifer vulnerability but no risk of pollution, if there is no significant pollutant loading; and to have high pollution risk in spite of low vulnerability, if the pollutant loading is exceptional. It is important to make clear the distinction between vulnerability and risk. This because risk of pollution is determined not only by the intrinsic characteristics of the aquifer, which are relatively static and hardly changeable, but also on the existence of potentially polluting activities, which are dynamic factors which can in principle be changed and controlled.

Considerations on whether a groundwater pollution episode will result in serious threat to groundwater quality and thus to its (already developed, or designated) water supply are not included in the proposed definition of vulnerability. The seriousness of the impact on water use will depend not only on aquifer vulnerability to pollution but also on the magnitude of the pollution episode, and the value of the groundwater resource.

1.4. SUGGESTED SYSTEM OF VULNERABILITY EVALUATION AND RANKING FOR EEC MEMBER-STATES

Given the definition of vulnerability proposed in Section 1.3, it is important to recognise that the vulnerability of an aquifer will be different for different pollutants. For example, groundwater quality may be highly vulnerable to the loading of nitrates at the surface, originated in agricultural practices, and yet be little vulnerable to the loading of pathogens.

In view of this reality, it is scientifically most sound to evaluate vulnerability to pollution in relation to a particular class of pollutant, such as nutrients, organics, heavy metals, pathogens, etc., i.e. to create specific vulnerability maps. This point of view has been expressed by other authors (e.g. Foster [9]), and some work has been done in specific vulnerability mapping. An example is the work of Canter *et al.* [10] for nitrate pollution of agricultural origin. Alternatively, vulnerability mapping could be performed in relation to groups of polluting activities [9], such as unsewered sanitation, agriculture, and particular groups of industries. This has been attempted for some activities. An example is the work of le Grand [11] for waste disposal.

Although we recognise that specific vulnerability mapping is scientifically more sound, we must realise that there will generally be insufficient available data to perform specific vulnerability mapping. Therefore, it is necessary to adopt a mapping system that is simple enough to apply the data generally available, and yet is capable of making best use of those data in a technically valid and useful way. Various such systems of vulnerability evaluation and ranking have been developed and applied in the past. Examples are Albinet and Margat [12], Haertle [13], Aller *et al.* [14], and Foster [9].

Some of the systems for vulnerability evaluation and ranking include a vulnerability index which is computed from hydrogeological, morphological and other aquifer characteristics in some well-defined way. The adoption of an index has the advantage of, in principle, eliminating or minimising subjectivity in the ranking process. Given the multitude of authors and potential users of vulnerability maps in EEC countries, Lobo-Ferreira and Cabral [15] suggested that a vulnerability index be used in the vulnerability ranking performed for Community maps.

Such a standardised index has been adopted in the U.S., Canada and South Africa, and is currently used in those countries: the index DRASTIC, developed by Aller *et al.* [14] for the U.S. EPA. This index has the characteristics of simplicity and usefulness that we think to be necessary. Another index that has the desired characteristics was developed by Foster [9]: the index GOD. These two indices are briefly reviewed below.

1.4.1. *The index of vulnerability DRASTIC*

The index of vulnerability DRASTIC [14] corresponds to the weighted average of 7 values corresponding to 7 hydrogeologic parameters:

1 - **D**epth to the water table (D)
2 - Net **R**echarge (R)
3 - **A**quifer material (A)
4 - **S**oil type (S)
5 - **T**opography (T)
6 - **I**mpact of the unsaturated zone (I)
7 - Hydraulic **C**onductivity (C)

We attribute a value between 1 and 10 to each parameter, depending on local conditions. High values correspond to high vulnerability. The attributed values are obtained from tables, which give the correspondence between local hydrogeologic characteristics and the parameter value. Next, the local index of vulnerability is computed through multiplication of the value attributed to each parameter by its relative weight, and adding up all seven products. Thus, each parameter has a predetermined, fixed, relative weight that reflects its relative importance to vulnerability. The most significant factors have weights of 5; the least significant a weight of 1. A second weight has been assigned to reflect the agricultural usage of pesticides. In the following table the factors are presented together with the weights respectively for normal DRASTIC applications and for DRASTIC pesticide applications:

DRASTIC parameters	normal DRASTIC	pesticide DRASTIC
(a)- Depth to the water table	5	5
(b)- Net Recharge	4	4
(c)- Aquifer material	3	3
(d)- Soil type	2	5
(e)- Topography	1	3
(f)- Impact of the vadose zone	5	4
(g)- Hydraulic Conductivity	3	2

The minimum value of the DRASTIC index is therefore 23 and the maximum value is 226. Such extreme values are very rare, the most common values being within the range 50 to 200.

Examples of the application of DRASTIC may be seen in Figures 2 and 3. Figure 2 presents a DRASTIC groundwater vulnerability map of Indiana, USA, developed by Copper and extracted from http://pasture.ecn.purdue.edu/~frankenb/watershedmap.html. Navalur and Engel [16] compare a DRASTIC vulnerability assessment with groundwater quality data sampled across Indiana State. Figure 3 presents a DRASTIC groundwater vulnerability map of Southern Africa, developed by Lynch *et al.* [17], extracted from http://www.ccwr.ac.za/~lynch2/drastic.html.

1.4.2. *The index of vulnerability GOD*

According to Foster [9] the GOD index of vulnerability considers the following three factors:

1 - Groundwater occurrence, i.e., whether the aquifer is unconfined, semi-confined, confined, etc. (if there is an aquifer);

2 - Overall aquifer class in terms of degree of consolidation and lithological character;

3 - Depth to groundwater table or strike.

(The name GOD is formed by the first letter of each factor).

The aquifer is ranked on a scale up to unity in relation to each of the three factors. The index is computed through the product of three scores. The maximum index value is 1.0, representing high vulnerability, and the minimum value is 0.016 if there is an aquifer (zero if there is not). The ranking of each score is easy to obtain, as explained in Foster (1987). A qualifying suffix indicating the degree of fissuring and the attenuation capacity is added to the numerical index. This suffix provides a qualitative measure of the relative tendency for lateral pollutant transport in the saturated zone.

Figure 2. **DRASTIC groundwater vulnerability map of Indiana, USA, developed by Copper and extracted from http://pasture.ecn.purdue.edu/~frankenb/watershedmap.html**

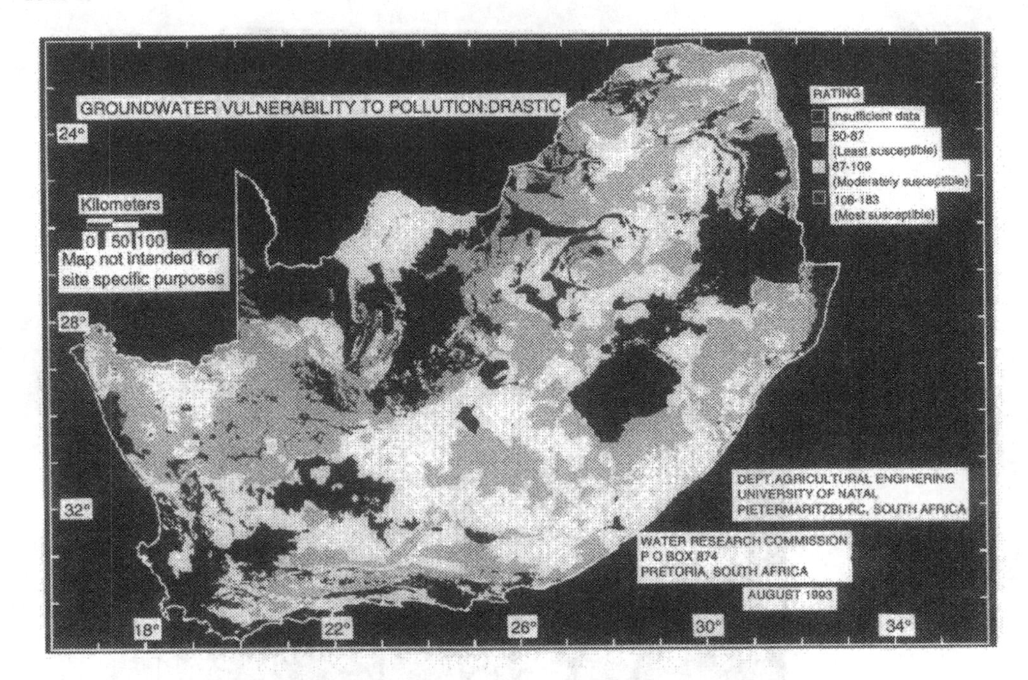

Figure 3. **DRASTIC groundwater vulnerability map of Southern Africa, developed by Lynch *et al.* [17] and extracted from http://www.ccwr.ac.za/~lynch2/drastic.html**

1.5. SUMMARY PROJECTS ON THE DEVELOPMENT OF AN INVENTORY OF THE GROUNDWATER RESOURCES OF PORTUGAL INCLUDING THE VULNERABILITY ASSESSMENT OF GROUNDWATER TO POLLUTION USING THE DRASTIC INDEX METHOD

In 1995, Laboratório Nacional de Engenharia Civil (LNEC) published a series of three volumes, [18], containing a collection of 11 Reports developed by LNEC. The study was developed for the European Commission's General-Directorate on the Environment, Nuclear Safety and Civil Protection (DGXI), for several Portuguese authorities, e.g. Direcção-Geral do Ambiente (DGA), Secretaria Regional de Habitação e Obras Públicas of Região Autónoma dos Açores and Laboratório Regional de Engenharia Civil of Região Autónoma da Madeira.

The main objective of the Project was the development of "An Inventory of the Groundwater Resources of Portugal", including:

- the characterisation of the hydrogeological systems;
- the assessment of groundwater resources, both the renewable ones, i.e. the groundwater recharge of aquifers, and the groundwater reserves, i.e. the exploitable part of the groundwater reserves;
- the assessment of groundwater uses;
- the mapping on a 1:500 000 scale, using a Geographical Information System (GIS), of the hydrogeological systems, the groundwater recharge, and the vulnerability of

groundwater to pollution using the DRASTIC index method, including the seven different parameters needed for its evaluations; and

- the analysis of the international legislation for groundwater zoning and protection. Further, in the framework of the Project, the first operational data-base on Portuguese groundwater quantity and quality parameters, called INVENTAR, was developed in LNEC, programmed and filled-in with data from more than 6000 groundwater well-point locations.

In relation to the DRASTIC mapping of aquifer vulnerability to pollution the information sources and steps developed for the application of the method in Portugal were the following:

- The variable D (depth to the water table) was extracted from the Data-Base INVENTAR, developed in LNEC by Lobo-Ferreira *et al.* [18], (for more information *cf.* http://www-dh.lnec.pt/gias/estudos/inventE.html)
- The variable R (net recharge) was calculated by the method presented in Vermeulen *et al.* [19], based on the BALSEQ sequential daily water balance model [20].
- The evaluation of variables A (aquifer material), I (impact of the vadose zone) and C (hydraulic conductivity) were based on data reports from the Geological Mapping of Portugal, on scale 1:500 000 [21].
- The variable S (soil type) was obtained in accordance with the method that is presented in [18].
- The variable T (topography) was obtained from elevation points, using the triangulation method in ARC/INFO system. The main sources of information were the Geological Map of Portugal by SGP [21] on scale 1:500 000 and the map of hydrographic basins of the "Atlas do Ambiente de Portugal", on scale 1:1 000 000, which has the elevation of all points where water lines start and end.
- Using the Geographic Information System ARC/INFO, LNEC and DGA treated the parameters above described, using the formulas and weights presented in the DRASTIC method.

The maps of the aquifer systems, hydrogeological parameters, aquifer's recharge and the final map of DRASTIC aquifer's vulnerability of Portugal, all in scale 1:500 000 were developed in ARC/INFO (the maps are presented in a 1:1 500 000 scale in Lobo-Ferreira *et al.* [18]. In Figure 4 we present, developed by Lobo-Ferreira and Oliveira [22] (also in [18]) a one page black-and-white map of the DRASTIC index vulnerability assessment of Portuguese groundwater.

Several other studies, that included DRASTIC groundwater vulnerability assessment, were developed in Portugal following the methodology presented [22]. Among those study we highlight the "Study for evaluation of the vulnerability of the reception capacity of coastal zone water resources in Portugal. The receiving water bodies: groundwater systems", [23]. In this study, the aquifers of the coastal areas of Portugal and the situation concerning groundwater exploitation were characterised. The main pollution problems affecting these areas, with a special attention to the salt water intrusion phenomenon, were described. Finally, an application of the DRASTIC method for evaluation of the vulnerability to pollution of the aquifer formations of the coastal areas was presented and the vulnerability mapping at the scale 1:100 000 was made. In Figure 5 an example of a vulnerability map at the scale 1:100 000 is presented. This map was developed by Lobo-Ferreira *et al.* [23] for the Peniche area, in Portugal's central coastal zone, and was extracted from http://www-dh.lnec.pt/gias/gias.html.

1.6. CONCLUSIONS OF SECTION 1

We have presented a brief review of three vulnerability ranking systems: (1) the system used in the previous EEC groundwater inventory (based on the pollutant travel time in the unsaturated zone), (2) the system GOD developed by Foster [9], and (3) the system DRASTIC developed by Aller *et al.* [14] and adopted by the U.S. EPA for vulnerability mapping in the U.S. and also applied in Canada.

It is our opinion that the vulnerability evaluation procedure should correspond to a well-defined computation of an index, in order to minimize subjectivity involved in the ranking. The system previously applied to the EEC countries (reviewed above) represented a pioneer effort, and regardless of its merit however allowed for subjectiveness in the evaluation process. This limitation was emphasized in the conference on "Vulnerability of Soil and Groundwater to Pollutants" held in The Netherlands in 1987. After Anderson and Gosk [24], the main factors reducing the applicability of the general type vulnerability maps are: incorporation of percolation time, subjective nature, and composite nature of the vulnerability concept.

To overcome these limitations and in order to guarantee the compatibility and coherence of the various national databanks, we suggest that vulnerability ranking be made through a well-defined computation leading to a final index. Such an index should meet the requirements of being relatively simple, given the limitations of generally available data, while being technically sound and valid for vulnerability classification. Two existing indices that meet these requirements, reviewed above, are the index GOD (by Foster [9]), and the index DRASTIC (by Aller *et al.* [14]), the latter adopted by the U.S. EPA and used in Canada as well.

We recommend that the index DRASTIC be selected for the elaboration of the EEC groundwater vulnerability maps. The index DRASTIC presents, in our opinion, advantages over the index GOD, which however also presents the basic desired characteristics of simplicity and soundness. Namely, DRASTIC considers a larger number of hydrogeologic factors than does GOD. An important advantage offered by DRASTIC is the amount of existing experience on its application, in the U.S. and Canada.

Most data required for computation of the index GOD are also required for DRASTIC. DRASTIC requires additional data, and data collection is therefore more lengthy and costly relatively to GOD. Lobo-Ferreira and Cabral [15] suggested that the data required by both systems be collected first, so that a preliminary assessment of vulnerability can be made, through the (faster) computation of the GOD index.

Research based on the aquifer vulnerability concept and the corresponding data acquisition process, such as the one required by the DRASTIC index method, exemplified in this section for Portuguese conditions, allows a sounder application of mathematical groundwater flow and mass transport models.

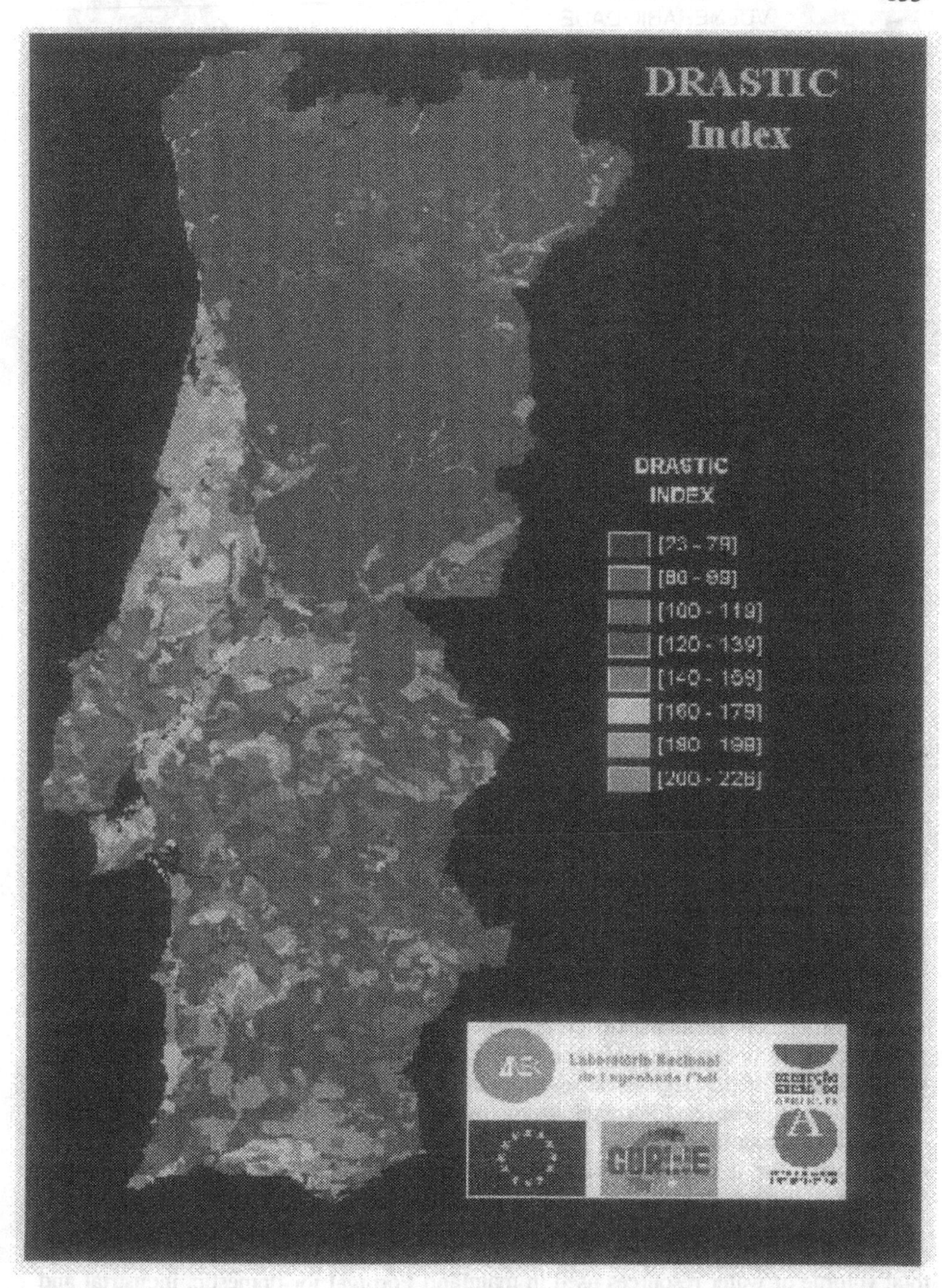

Figure 4. **DRASTIC groundwater vulnerability mapping of mainland Portugal developed by Lobo-Ferreira and Oliveira [22] (also in [18], see also http://www-dh.lnec.pt/gias/novidades/drastic_e.html)**

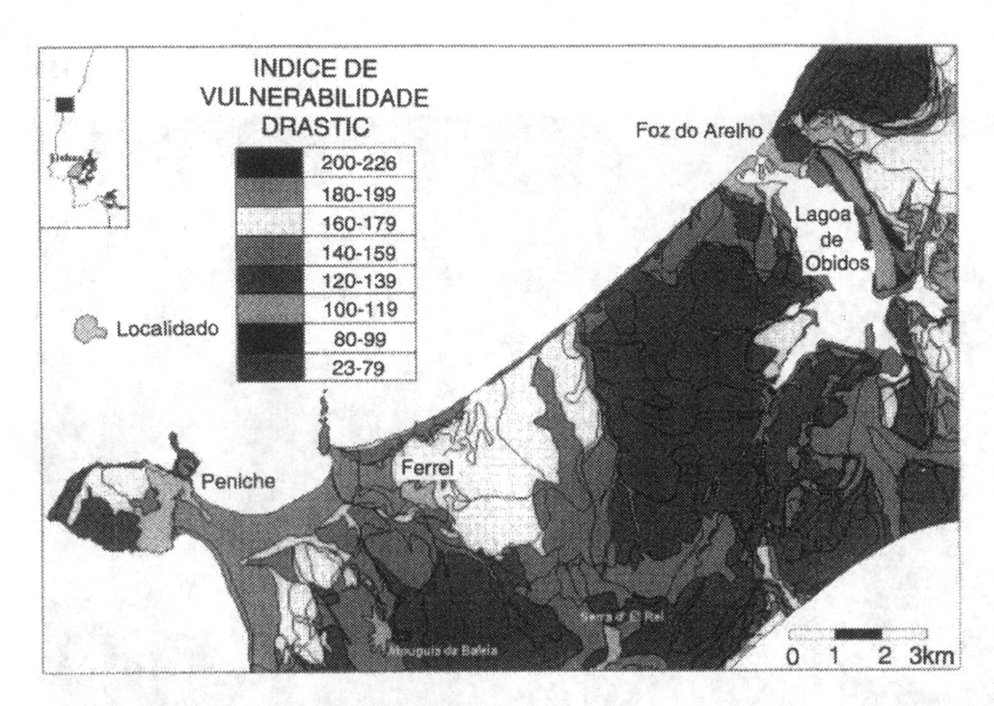

Figure 5. **Vulnerability map at the scale 1:100 000 of Peniche coastal zone, in central Portugal, developed by Lobo-Ferreira *et al.* [23], extracted from http://www-dh.lnec.pt/gias/evaluation.htm**

2. Assessment Of Legislation Concerning Geographical Zoning For Groundwater Protection

2.1. GENERAL COMMENTS

This section summarises a more comprehensive study developed under the DGXI/B/1 project on the "Inventory of the Groundwater Resources of Portugal", co-ordinated by Laboratório Nacional de Engenharia Civil, Lisbon, Portugal [18]. The study was also developed under the European Commission's DGXII Human Capital and Mobility Programme, WEEL - Water Environment European Laboratories Project [25]. The following topics were addressed:

- Legislation on methods of groundwater geographical protection zoning from selected European Union Member-States and the United States of America;
- Application of this legislation to the pumping well field of the unconfined porous aquifer of Estarreja, Portugal, using mathematical groundwater flow and mass transport modelling;
- Proposals for a revision of Portuguese legislation on groundwater geographical protection zoning.

According to Lobo-Ferreira and Oliveira [22], groundwater represents 73,3% of the total volume of water resources used in mainland Portugal for domestic, industrial and agricultural purposes. It represents therefore a strategic resource whose management and protection, both from quantitative and from qualitative points of view, should be appropriately considered at national and local authorities levels.

The protective and cleaning effects of aquifers assure, in most cases, good protection of groundwater from direct contamination caused by superficial sources of pollution. Nevertheless, groundwater vulnerability to pollution is an important issue in all areas in which topographical, geological and hydrogeological conditions are not favourable [18].

Contaminants can reach the aquifers from different sources, such as the handling, storage or treatment of hazardous substances, municipal solid wastes, industrial processes, fuels, mineral oils, radioactive materials, etc. Groundwater pollution can also come from other common human activities and facilities, such as: fertilisers and pesticide use in agricultural activities, sewer systems, septic tanks, waste water releases, whether to surface water, to the ground, or worst of all, into the ground. According to the Environmental Protection Agency, in the United States, more than 200 different chemical compounds have been identified in groundwater [26]. Some of these are considered to be extremely hazardous for human health.

Previous experience in the field, showed that the key to solving the problem of qualitative management of groundwater is pollution prevention. This is due to the very high costs of setting up and developing aquifer rehabilitation programs and, eventually, of setting up research programs for the assessment of new water supply sources.

The development of a groundwater protection program, in the context of the latest land use control policies, has the objective of avoiding problems concerning groundwater pollution. It makes possible the achievement of two fundamental goals: (1) to ensure the availability of a high quality water supply source and (2) to save large amounts of financial resources (which can then be used in other areas of public utility), through avoiding the need for groundwater rehabilitation.

In the framework of the European Union, most of the Member-States already have specific legislation which establishes protection zones around wells, defining the polluting activities that should be banned in each zone, and having as the main objective the preservation of groundwater quality.

Nevertheless, the application of that legislation is still difficult. Recently (December 1993), General Direction XI of the European Commission organised a meeting of experts to discuss an action program concerning groundwater (cf. new DGXI Action Programme for Integrated Groundwater Protection and Management, approved by Member-States in Summer 96). Problems concerning geographical zoning for the protection of groundwater were also discussed.

In Portugal, although the most recent legislation (D.L. 84/90, 86/90 and 90/90 of March 16, 1990, and D.L. 45/94 e 46/94 of Feb. 22, 1994) has, once more, pointed out the importance of defining protection zones as an instrument of groundwater protection, the size of those zones is still defined by *Norma Portuguesa Definitiva NP836*, established in 1971. *NP836* defines two zones, called "near protection zone" and "far protection zone", with extensions depending on the aquifer type and on its filtration capacity. In this norm the size of protection zones is still defined in a very simplified way. Moreover, its application is still not complete all over the country. According to Bicudo *et al.* [27], 64% of Portuguese wells surveyed in their study, used for public water supply, did not yet have defined protection zones (in 1989).

2.2. GROUNDWATER GEOGRAPHICAL PROTECTION ZONING CRITERIA USED IN THE EUROPEAN UNION AND IN THE UNITED STATES OF AMERICA

According to Roux [28] more than 50% of the European Member-States water supply comes from groundwater. In some European countries, such as France, Belgium, Germany, Italy, Portugal and The Netherlands, groundwater supplies more than 60% of the total drinking water. These values show the fundamental role played by groundwater in satisfying European water needs.

In most of the European Union Member-States, legislation concerning limitation of hazardous activities endangering groundwater was established in the beginning of the '70s. Legislation concerning groundwater pumping systems protection was also set up.

The former European Economic Community, and today the European Union, has always taken carefully into account the problem of natural resources protection. This special sensibility has lead to a remarkable legislative production. All the Member-States followed with their own legislation, in some cases improving some aspects specifically concerning groundwater protection.

According to Margat [29], groundwater protection in the European Member-States is provided by two kinds of measures:

- general regulatory measures, often linked to environmental impact studies;
- definition of two or three protection zones around groundwater pumping systems.

This last measure has been or is being established in most Community countries and has proved a success so far.

Ciabatti and Lobo-Ferreira [25] summarised most aspects concerning groundwater geographical protection methods used in some European countries and in the United States of America. The following aspects were addressed:

- general legislation for groundwater protection;
- authorities responsible for monitoring activities in protected zones and for ensuring that controls are observed;
- geographical criteria for regional groundwater protection zoning;
- geographical criteria for groundwater pumping systems protection zoning;
- economic aspects of protection (average and limiting costs of protection, who bears the cost of protection, etc.);
- number or percentage of groundwater pumping systems protected by regulation.

Generally speaking, it is possible to specify two main criteria for the definition of the sizes of protection zones: the horizontal distance criterion and the horizontal travel time criterion.

The *horizontal distance criterion* consists of defining a circular area around the well, with a radius value which is determined without taking into account the hydrogeological and hydrodynamic characteristics of the aquifer to be protected.

The *horizontal travel time criterion* is based on the definition of travel time of polluted particles in the aquifer flow field. It must be considered more advantageous because it permits the definition of the residence time of contaminants in the aquifer and their arrival time at the pumping well, depending on the hydrogeological and hydrodynamic characteristics of the aquifer and the location of the pollutant source.

The problem of confined aquifer protection is not specifically addressed in any of the legislation analysed. Confined aquifer protection is achieved simply by applying the same methodologies used for unconfined aquifers, reducing the size of protection areas when a superficial impervious or semi-impervious layer is present. This approach is often combined with measures for avoiding the removal of the covering impervious layers protecting the underlying aquifer. No hydrogeological criteria are used to justify this very simplified approach.

The various items of legislation show great differences concerning the degree of specification of banned and/or controlled activities inside protection zones. The innermost area (i.e. the operational courtyard) must be purchased by the well owner and fenced in. All the activities not strictly related to water abstraction must be forbidden inside this area. In the more external area, related to the micro-biological protection of the well, the following activities, processes and installations are generally forbidden or put under control: building sites, waste sites, pesticide and fertilisers use, animal feedlots, paddocks and breeding, new roads or railways, transport or storage of any pollutants. In the outer protection zone, in most cases corresponding to the recharge area of the well, some activities are allowed but a certain amount of control is maintained.

2.3. APPLICATION OF GROUNDWATER PROTECTION LEGISLATION TO THE ESTARREJA UNCONFINED AQUIFER

The application of legislative measures for the protection of groundwater pumping systems to the unconfined aquifer of Estarreja, in central Portugal, is described in this chapter. The case study of the municipal well field of Estarreja is a very interesting one for the application of the different protection zoning methods described above. The reasons for this choice were as follows:

- the unconfined aquifer presents a shallow water table;
- in the vicinity of the well field, several activities and facilities are located, having a potential negative impact on the quality of the pumped groundwater;
- a large amount of geological and hydrogeological data is available for setting up the conceptual and numerical models of flow and mass transport.

These conditions (which imply high vulnerability of the aquifer to pollution) make the Estarreja case study ideal for the application of protection legislation used in the European Union and in the United States of America. An assessment of potential sources of contamination located in the vicinity and internally in the groundwater recharge area of the pumping well field, was also carried-out.

The municipal well field of Estarreja is located 1.5 km north of the small city with the same name, and 500 m south-east of a very large chemical industrial complex. Drinking water for the city of Estarreja is supplied by eight wells, located in an area of about 10 hectares, that pump water from the Quaternary unconfined aquifer. The average discharge of each well is 2 l/s.

According to the methodology usually used in Europe and in the United States for this kind of study, land use in the vicinity of the well field was assessed and, with the co-operation of the Municipality of Estarreja, an inventory of the potential sources of contamination was developed, with the objective of plotting the sources on a map.

In the area of interest eight wells are more or less aligned with and within a distance of 50 to 150 m of Highway 109 (i.e. E.N. 109, in Figure 6).

The area located to the east of the pumping wells (i.e. east of Pinhal da Cardosa), is a woodland in which no harmful activities or potential sources of contamination for the groundwater are located.

The area located to the north of the pumping wells is farm land (i.e. north of Rio Sardinha). The local crops are essentially vines and corn. According to information provided by the Municipality, pesticides and fertilisers are not used in large quantities. In this area some large superficial wells used for irrigation are located, drawing water from the same aquifer used for public drinking water supply.

Residential areas are mainly small rural communities. Some isolated houses are located along Highway 109. They are nowadays connected to the municipal sewers system, which came into service only in the first months of 1994. In the other cases, organic wastes are probably discharged using septic systems or cesspools.

Highway 109 is the major road of the area, connecting the town of Estarreja with the chemical industrial complex and the village of Avanca, in the North. The traffic on this road is very high, with a high percentage of trucks and vehicles transporting hazardous chemical substances. The probability of discharge of these products and substances to the ground, in the case of road accidents, is therefore, very high.

The area west of Highway 109 is occupied by a very large chemical industrial complex, whose activities have a very harmful and negative impact on the soil and groundwater used for drinking water supply and agricultural uses in the whole area of the Estarreja region [30].

In the area between the municipal pumping wells and Highway 109 several activities and facilities are located (Figure 6), having a potential negative impact on the quality of pumped water.

To each potential source of groundwater contamination a reference number was given, which also identifies the kind of activity (Table 1).

Table 1. Inventory of activities, processes, facilities and installations representing potential contamination sources of groundwater in the Estarreja area

N.	Activity developed	N.	Activity developed
1	Prêt-à-porter clothing manufacturing and storage	7	Gas portable containers storage facilities
2	Drinks manufacturing and storage	8	Fuel station with car washing facilities
3	Drinks manufacturing and storage	9	Metalworking repair shop
4	Clothes manufacturing and storage	10	Auto repair shop
5	Abandoned sand extraction site	11	construction commercial areas
6	Isolated houses	12	construction commercial areas

The risk of pollution that each of these sources poses, depends essentially on two factors: the contaminant substances produced and their ability to infiltrate and reach the saturated zone. In the light of this, the activities and facilities that pose major threats to groundwater are the following:

- the gasoline service area (n. 8) with car washing facilities, located 100 m west of well p8 (i.e. possibility of infiltration into the aquifer of fuels stored in underground tanks, or other substances such as solvents, soaps, detergents, waxes and miscellaneous chemicals, [26];
- the metalworking factory (n. 9) located 150 m west of well p8, at which the principal activity is metal structures manufacturing for construction uses and gas portable containers washing. Waste waters produced in this process are discharged directly into the ground. According to EPA [26], this kind of industrial process may cause very hazardous substances to infiltrate to the aquifer, such as: heavy metals, sodium and hydrogen cyanide, metallic salts, several acids and hydroxides, etc.;
- the auto repair garage (n. 10) located 150 m west of well p8, in which wrecked automobiles are also stored. In this case, the main hazardous infiltrating substances are waste oils, solvents, acids, paints and fuels [26];
- some isolated houses (n. 6) are now connected to the municipal sewer system, which came into service only in the first months of 1994. Before 1994, a septic tank system was used for the discharge of waste water, which posed serious risk of contamination to the aquifer due to their content in coliform and noncoliform bacteria, viruses, nitrates, detergents, solvents, elevated concentration levels of chloride, sulphate, calcium, magnesium, potassium and phosphate [26].

Others activities and storage facilities do not seem to represent a threat to the aquifer, even if EPA [26] points out the possibility of infiltration of solvents, paints, glues and other adhesives, epoxy waste, etc., from construction commercial areas and materials (n. 11 and n. 12).

A particular problem is posed by the abandoned sand extraction site (n. 5) located 800 m East of the pumping wells. The extraction activity does not imply direct discharge of hazardous materials and contaminants onto or into the ground, but the removal of protective superficial unsaturated layers increases aquifer vulnerability. In fact, the protective and cleaning effects of these superficial layers afford good protection to the aquifer from direct contamination by superficial sources of pollution.

The ASM - Aquifer Simulation Model code, [31], was used for the setting up and solving of a 2-D finite difference numerical model of the Estarreja aquifer and the groundwater pumping systems. A 2-D model was considered adequate on account of the shallowness of the aquifer. The total modelled area is about 2 km^2 large (Figure 7). The model is set up of 768 square cells of 50 x 50 m^2 each. The model was specifically aimed at the study of the flow field in the vicinity of the pumping wells, for computing the horizontal travel time and the isochrones. It was applied for a selection of groundwater protection zoning methods used in the European Union and in the United States of America. The ASM model uses the "random-walk" method for its dispersive transport assessment and for the evaluation of the isochrones. Porosity was considered constant in all cells of the simulated aquifer, and equal to 0.1. Adsorption and diffusion/dispersion phenomena were not taken into account.

Figure 7 shows groundwater piezometric distribution and flow directions in the municipal groundwater pumping system of Estarreja, as calculated by the calibrated numerical model. The recharge area of the wells extends as far as Beduido and Outeiro.

Based on the model calibration (for steady state conditions) isochrones were calculated and geographical groundwater protection zones were defined for a selection of European and American legislation. The evaluation of the protection scenarios is shown in Figures 8 to 14.

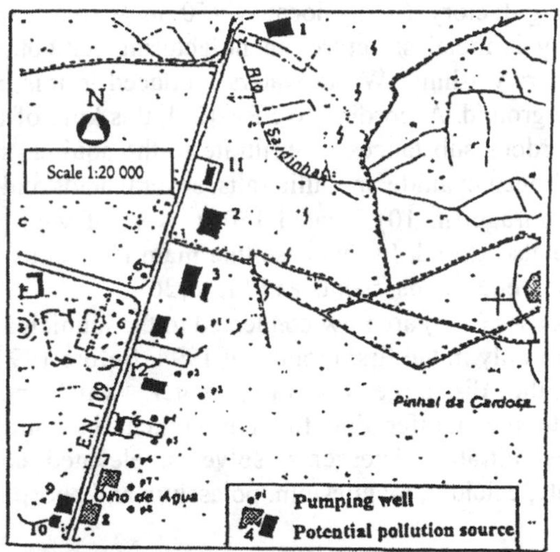

Figure 6. **Main potential pollution sources near the Estarreja municipal groundwater pumping system**

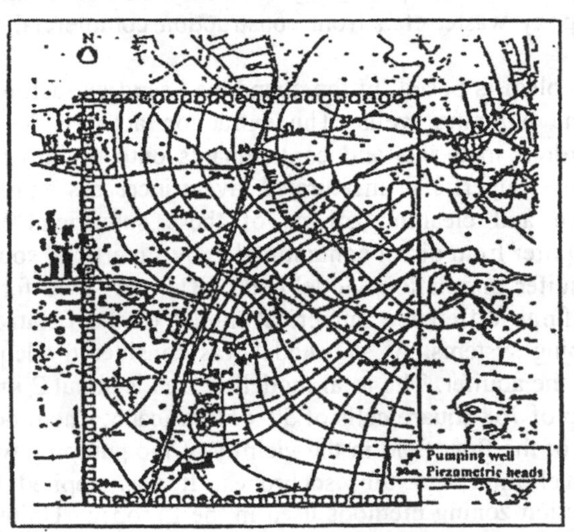

Figure 7. **Piezometric and flow lines computed with the mathematical model**

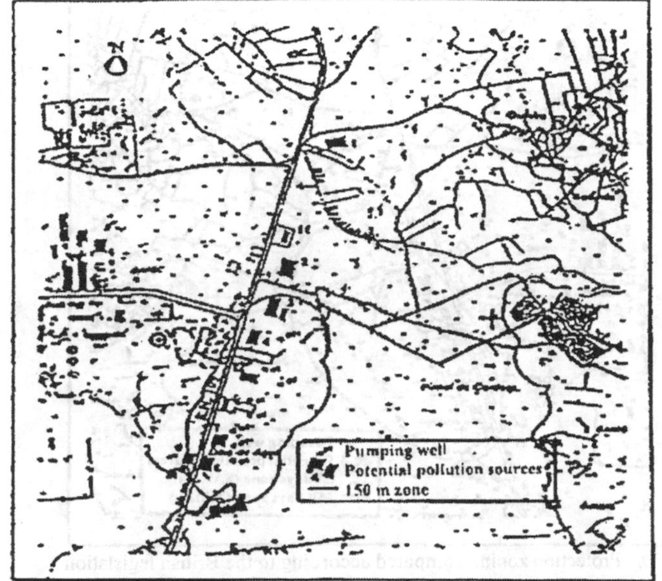

Figure 8. Protection zoning computed according to the Portuguese legislation

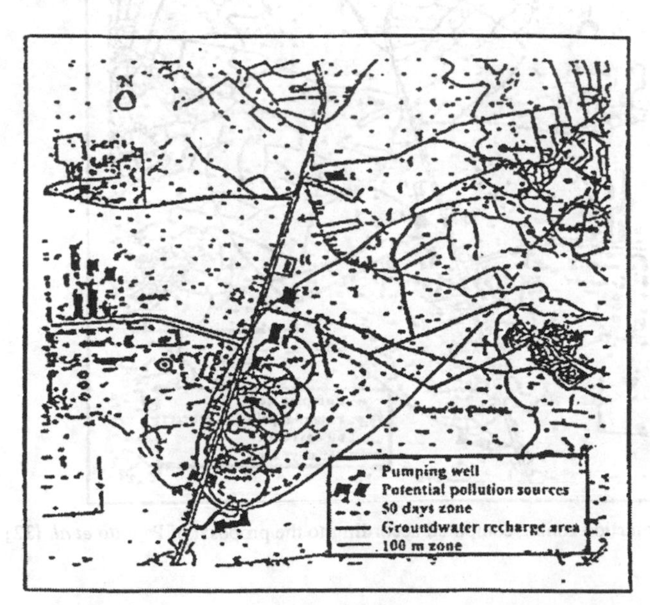

Figure 9. Protection zoning computed according to the German legislation

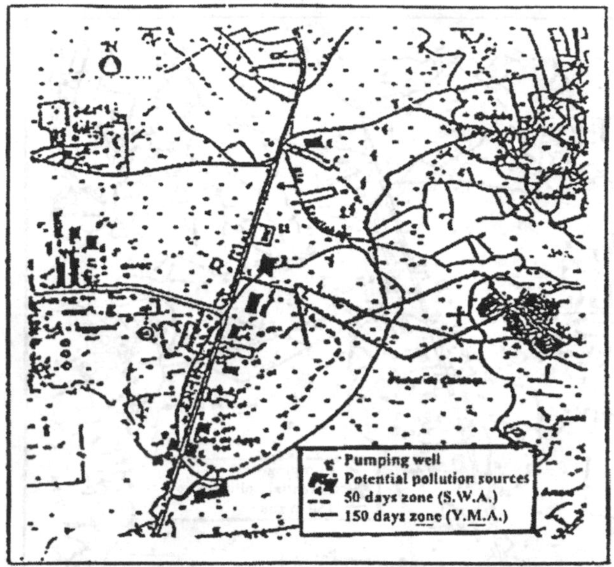

Figure 10. Protection zoning computed according to the British legislation

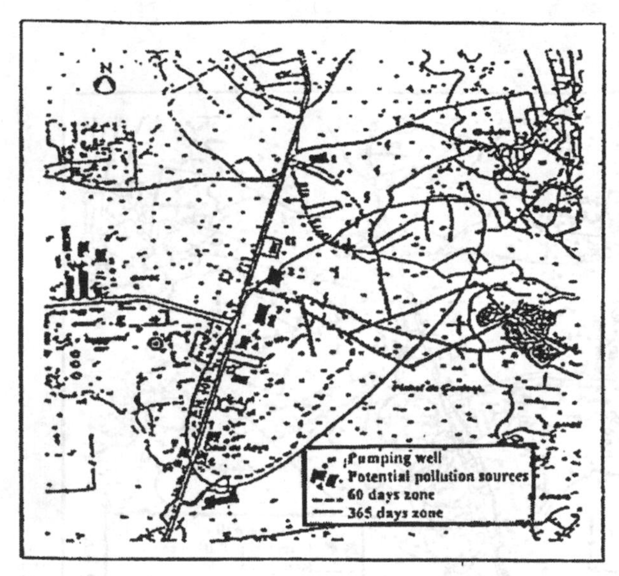

Figure 11. Protection zoning computed according to the proposal of Pagotto *et al.* [32]

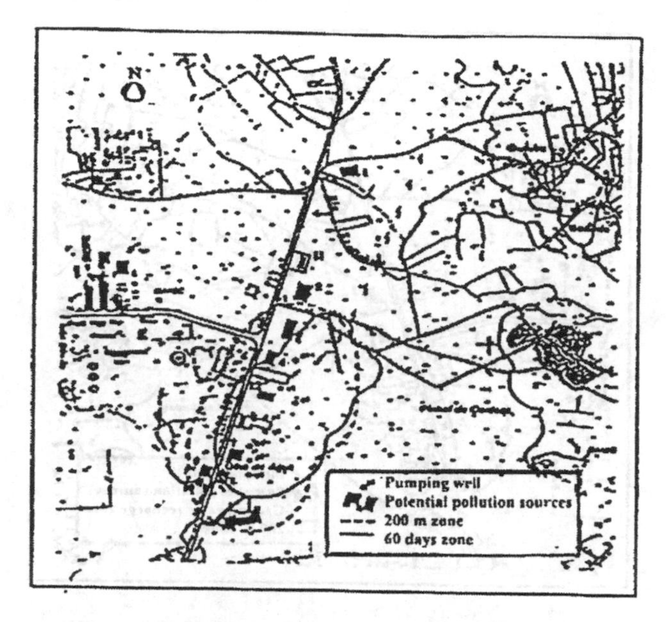

Figure 12. Comparison of the "rispetto" (near field) zones, computed according to the Italian legislation and according to the proposal of Pagotto *et al.* [32]

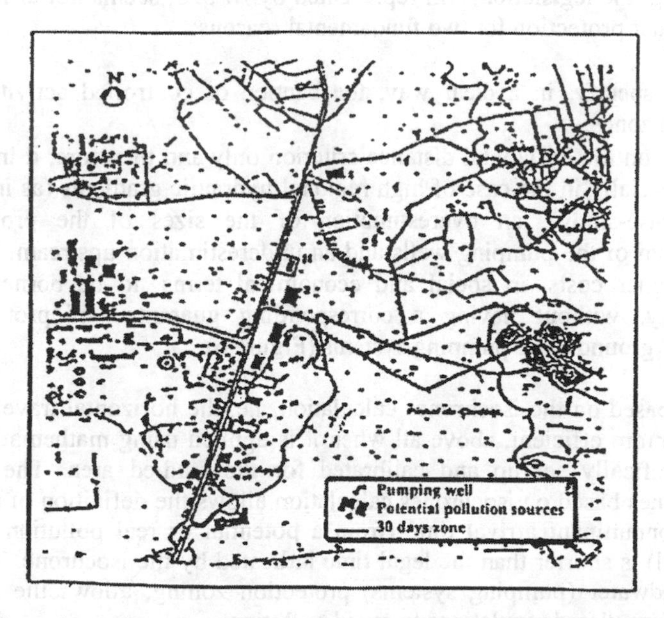

Figure 13. Protection zoning computed according to the Dead County (Florida, USA) legislation

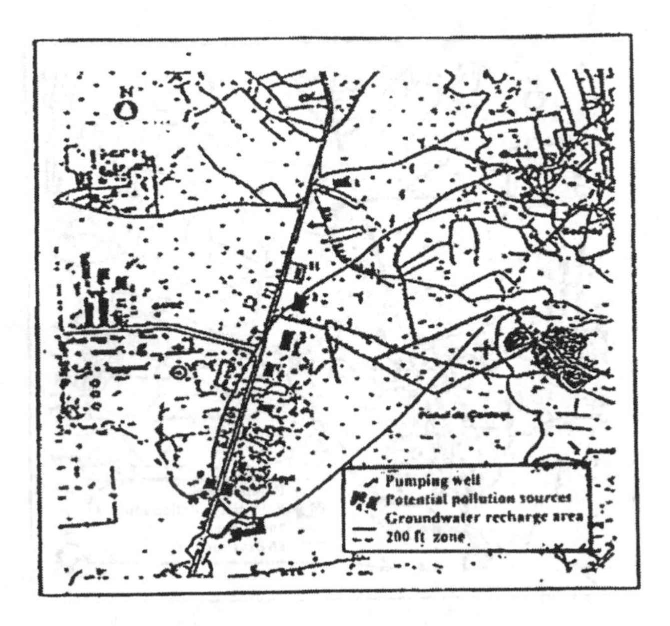

Figure 14. Protection zoning computed according to the State of Vermont (USA) legislation

The Portuguese legislation, still represented by *NP836*, seems not to fully guarantee the groundwater protection for two fundamental reasons:

- it doesn't specify, in a clear way, the banned or controlled activities inside the protection zones;
- it is based on the horizontal distance criterion only and therefore, it implies in most cases, especially in the case of high regional hydraulic gradients (as in the Estarreja aquifer case-study), an overestimation of the sizes of the protection zones downstream of the pumping wells and an underestimation upstream. This probably means higher costs, in social and economical terms, to be borne by the local community, without having a corresponding guarantee of protection of the municipal groundwater pumping systems (Figure 6).

The criteria based on the isochrones calculation, i.e. the horizontal travel time criteria, seems to be more efficient, above all when it is applied using mathematical numerical models, specifically set up and calibrated for the studied area. The definition of protection zones based on isochrones calculation allows the definition of the area inside which the contaminant arrival time (from a potential or real pollution source to the pumping well) is shorter than the legal time indicated by the isochrone. This approach, to the groundwater (pumping systems) protection zoning, allows the application of specifically formulated regulations to avoid pollution.
The protection scenario which leads to more strict land use restrictions is the one based on the application of the Federal Republic of Germany legislation [33], see Table 2).

2.4. SUMMARY AND CONCLUSIONS OF SECTION B

The implications of the foregoing analysis for land use control in general (and for the case-study site in particular) are listed below. These will be useful for a revision of Portuguese legislation on groundwater geographical protection zoning:

- The zoning approach which seems to guarantee the best protection for unconfined porous aquifers is based on the definition of the following three zones around pumping wells:
  - an operational courtyard (zone I), immediately around the well, with the goal of protecting the well and its immediate environment from any kind of contamination;
  - an inner protection zone (zone II), which is related to the micro-biological protection of abstracted waters and protection against contamination which would be hazardous a short distance from the well;
  - an outer protection zone (zone III), which, is related to the protection of groundwater from contamination affecting water over long distances or due to chemicals which are not decomposed or not decomposed easily. Eventually, this zone could be divided into two sub-zones (IIIa and IIIb) for very large aquifers.
- For the delimitation of the zone II and of the eventual zone IIIa the time of travel criteria must be considered the most suitable, because it is able to define the residence time of contaminants in the aquifer and their arrival time at the pumping well taking into account the hydrogeological and hydrodynamic characteristics of the aquifer and the location of the pollutant source. The horizontal distance criteria, in contrast, may cause inaccuracies in the definition of the size and shape of the protection areas. This is because, in the case of high regional hydraulic gradient it may lead to an overestimation of sizes of protection zones downstream of the wells and an underestimation upstream. This would lead to a lower degree of protection and a higher implementation cost/benefit ratio.
- The horizontal distance criteria is more suitable for the definition of zones I and III (or the eventual zone IIIb) for the following reasons:
  - concerning zone I, the dynamic hydraulic head distribution in the vicinity of a pumping well is only slightly influenced by the regional hydraulic gradient;
  - zone III (or zone IIIb) represents (following the approach to groundwater protection of the Federal Republic of Germany) a general protection area which ensures a sufficient degree of protection for all groundwater resources.
- The definition of the 50 days isochron may be considered reasonable for the delimitation of zone II. Nevertheless, it is important to consider that (as in the Estarreja case-study) the area delimited by the 60 days isochron does not imply a great increase in land use limitation and assures a higher degree of protection against pathogenic contamination.
- For the delimitation of zone III it seems to be better to define two kinds of approach to protection zoning:
  - for very large aquifers, a zone IIIa (the extent of which corresponds to the 1 year isochron) and a zone IIIb (the extent of which coincides with the recharge area of abstractions) may be defined;

Table 2. Activities, processes and installations not acceptable in wellhead protection areas, according to the German legislation

zone I	veichle and pedestrian traffic agriculture manure and pesticides (and all mentioned bans or restrictions of zones II, IIIA and IIIB)
zone II	constructions, plants and workshop farms, stables and sheds building sites and stock of building material roads and railway transfer points and parking lots sport facilities and camping sites tenting and bathing establishments at surface waters car washing and oil change cemeteries removal of surface layers mining and quarries intensive grazing allotments fuel storage and transport of water endangering substances waste water pipes fishponds (and all mentioned bans or restrictions of zones IIIA and IIIB)
zone IIIA	commercial use of water endangering substances mass livestock open storage and water endangering pesticides waste water treatment hospitals, sanatoriums and urbanisation storage of water endangering substances airports and associated facilities military facilities and manoeuvres waste sites sewage treatment plants injection of cooling water essential removal of surface layers new cemeteries shunting stations road construction with water endangering substances drilling (and all mentioned bans or restrictions of zone IIIB)
zone IIIB	oil refineries and smelting works chemical plants and nuclear reactors waste water injection deposition and underground storage of water endangering substances pipelines for water endangering substances

- for aquifers of limited extension, only a zone III might be defined, corresponding to the recharge area of abstractions.
- The banned activities inside the protection zones may be identified with those listed in Table 2. This list is based on the Federal Republic of Germany legislation.

The implementation of a wellhead protection program is very difficult if not supported by measures for:

- assuring a good level of compensation for whoever incurs extra costs, resulting from the imposition of the groundwater protection measures;
- establishing an adequate fees or charges policy specifically aimed at forcing consumers and anyone else endangering the groundwater quality to bear the cost of protection;
- avoiding diluted responsibilities or conflicts between ministries, state or local departments, state or local agencies, water and/or basin authorities, etc., in the implementation of such programs;

assuring social support from local communities for groundwater and wellhead protection programs.

3. References

1. Parascandola, M.-F. (1979) - Carte de classification des terrains en fonction de leur vulnérabilié - vallée du Rhône entre Valence et Orange, du Viverai, du Vercors, du Dévoluy et des Baronnies. Travaux de thése Université Science et Technique de Montpellier. Edition B.R.G.M.
2. Aust, H., H. Vierhuff and Wagner, W. (1980) *Grundwasservorkommen in der Bundesrepublik Deutschland.*
3. Lobo-Ferreira, J.P. and Calado, F. (1989) *Avaliação da Vulnerabilidade à Poluição e Qualidade das Águas Subterrâneas de Portugal*, Laboratório Nacional de Engenharia Civil, Lisboa.
4. Villumsen, A., Jacobsen, O.S. and Sonderskov, C. (1982) *Mapping the vulnerability of groundwater reservoirs with regard to surface pollution*, Geologic Survey of Denmark, Yearbook 1982, Copenhagen.
5. Fried, J.J. (1987) *Groundwater resources in the European Community, 2nd phase: Vulnerability - Quality* (not published).
6. Rodrigues, J.D., Lobo-Ferreira, J.P., Santos, J.B, Miguéns, N. (1989) Caracterização sumária do recursos hídricos subterrâneos de Portugal, *Memória No. 735*, Laboratório Nacional de Engenharia Civil, Lisboa and in *Groundwater in Western Europe*, Natural Resources Water Series, No. 27, 1992, United Nations, New York.
7. Hashimoto, T., Stedinger, J.R. and Loucks, D.P. (1982) Reliability, Resiliency, and Vulnerability Criteria for Water Resource System Performance Evaluation, *Water Resources Research*, 18(1), p14-20.
8. Duijvenbooden, W. van and Waegeningh, H.G. van (1987) *Vulnerability of Soil and Groundwater to Pollutants*, Proceedings and Information No. 38 of the International Conference held in the Netherlands, in 1987, TNO Committee on Hydrological Research, Delft, The Netherlands.
9. Foster, S.S.D. (1987) Fundamental concepts in aquifer vulnerability, pollution risk and protection strategy, in W. van Duijvanbooden and H.G. van Waegeningh (eds.), *Vulnerability of Soil and Groundwater to Pollution*, Proceedings and Information No. 38 of the International Conference held in the Netherlands, in 1987, TNO Committee on Hydrological Research, Delft, The Netherlands.
10. Canter, L., Knox, R. and Fairchield, D. (1987) *Ground Water Quality Protection*, Lewis Publishers, Inc., Chelsea, Mi.
11. le Grand, H.E. (1983) *A standardized system for evaluating waste disposal sites*, NWWA, Worthington, Ohio.
12. Albinet, M. and Margat, J. (1970) Cartographie de la vulnérabilité a la pollution des nappes d'eau souterraine, *Bull. BRGM* 2me Series 3 (4).
13. Haertle, A. (1983) *Method of working and employment of EDP during the preparation of groundwater vulnerability maps*, IAHS Publ. 142(2).
14. Aller, L., Bennet, T., Lehr, J.H. and Petty, R.J. (1987) *DRASTIC: a standardized system for evaluating groundwater pollution potential using hydrogeologic settings*, U.S. EPA Report 600/2-85/018.

15. Lobo-Ferreira, J.P. and Cabral, M. (1991) Proposal for an Operational Definition of Vulnerability for the European Community's Atlas of Groundwater Resources, in *Meeting of the European Institute for Water, Groundwater Work Group Brussels, Feb. 1991.*
16. Navalur, K.C.S. and Engel, B.A. (1997) Predicting spatial distribution of vulnerability of Indiana State Aquifer system to nitate leaching using a GIS, in http://ncgia.ucsb.edu/conf/SANTA_FE_CD-ROM/sf_papers/navulur_kumar/my_paper.html
17. Lynch, S.D., Reynders, A.G. and Schulz, R.R. (1993) Preparing input data for a national-scale groundwater vulnerability map of Southern Africa, in *6th National Hydrological Symposium, SANCIAHS*, Sept. 1993 and Water S.A. (1994) 20, 239-246, Pietermaritzburg, South Africa.
18. Lobo-Ferreira, J.P., Oliveira, M. Mendes and Ciabatti, P.C. (1995*) Desenvolvimento de um Inventário da Águas Subterrâneas de Portugal. Volume 1.* Laboratório Nacional de Engenharia Civil, Lisboa.
19. Vermeulen, H., Lobo-Ferreira, J.P. and Oliveira, M. M. (1994) A Method for Estimating Aquifer Recharge, in DRASTIC Vulnerability Mapping, in*Second European Conference on Advances in Water Resources Technology and Management*, held in Lisboa, June 1994, A.A. Balkema and E.W.R.A., Rotterdam.
20. Lobo-Ferreira, J.P. and Delgado-Rodrigues, J. (1988) BALSEQ - A model for the estimation of water balances, including aquifer recharge, requiring scarce hydrogeological data, in E. Simmers (Ed.) *Estimation of Natural Groundwater Recharge.* Dordrecht, D. Reidel, NATO ASI Series Vol. 222.
21. SGP (1972) *Carta Geológica de Portugal à Escala 1/500 000,* Serviços Geológicos de Portugal, Lisboa.
22. Lobo-Ferreira, J.P. and Oliveira, M.M. (1993) *Desenvolvimento de um Inventário das Águas Subterrâneas de Portugal. Caracterização dosrecursos Hídricos Subterrâneos e Mapeamento DRASTIC da Vulnerabilidade dos Aquíferos de Portugal,* Laboratório Nacional de Engenharia Civil, Relatório 179/93 - GIAS, Lisboa.
23. Lobo-Ferreira, J.P., Oliveira, M.M., Moinante, M.J., Theves, T. andDiamantino, C. (1995) *Mapeamento das Águas Subterrâneas da Faixa Costeira Litoral e da Vulnerabilidade dos seus Aquíferos à Poluição. Relatório Específico R3.3,* LNEC, Relatório 237/95 - GIAS, 585 pp., Laboratório Nacional de Engenharia Civil, Lisboa.
24. Andresen, L.J. and Gosk, E. (1987) Applicability of vulnerability maps, in W. vanDuijvanbooden and H.G. van Waegeningh (eds.), *Vulnerability of Soil and Groundwater to Pollution,* Proceedings and Information No. 38 of the International Conference held in the Netherlands, in 1987, TNO Committee on Hydrological Research, Delft, The Netherlands.
25. Ciabatti, P., and Lobo-Ferreira, J.P. (1994) *Desenvolvimento de um Inventário das Águas Subterrâneas de Portugal. Análise de legislação sobre zonamento de protecção de captações de águas subterrâneas. Aplicação a dois casos de estudo portugueses,* Laboratório Nacional de Engenharia Civil, Relatório 247/94 - GIAS, Lisboa.
26. Environmental Protection Agency (1993) *Wellhead Protection: A Guide for Small Communities,* Seminar Publication, EPA/625/R-93/002, Washington, DC.
27. Bicudo, J. R., Alegre, H., Albuquerque, A., David, L. and Bartolomeu, F. (1993) *Inquérito effectuado pelas Administrações Regionais de Saúde para Caracterização dos Sistemas de Abastecimento da Água - Síntese de resultados,* Laboratório Nacional de Engenharia Civil e Direcção - Geral da Saúde, Lisboa.
28. Roux, J.C. (1992) Legislations et Pratiques de Protection des Eaux Souterraines, notamment des Zones de Captages d'Eau Destinée a la Consommation Humaine dans les Pays de la Communauté Européenne, in *Final Report of the Atelier "Le Concept de Zonage Geographique en Matière de Protection des Eaux Souterraines"* coordinated by J. Margat of European Institute of Water, Bruxelles.
29. Margat, J. (1992) Régionalisation de la Protection des Eaux Souterraines - Vue d'Ensamble, in *Final Report of the Atelier "Le Concept de Zonage Geographique en Matière de Protection des Eaux Souterraines"* coordinated by J. Margat of European Institute of Water, Bruxelles.
30. Leitão, T. Eira, Lobo-Ferreira, J.P. and Inácio, M. (1994) *Metodologias para a Recuperação de Águas Subterrâneas e Solos Contaminados. Partes C, D, E e F. Relatório Final.* Direcção-Geral da Qualidade do Ambiente/ Laboratório Nacional de Engenharia Civil, Relatório 145/94 - GIAS, Lisboa.
31. Kinzelbach, W. e Rausch, R. (1991) *ASM - Aquifer Simulation Model - Version 3.1 - Operating Manual,* Kassel/Stuttgart.
32. Pagotto, A., Marino, L. and Barelli, G. (1990) Esperienze di Perimetrazione di Aree di Salvaguardia attorno alle Captazioni Idropotabili nell'Alta Pianura Modenese, in *Studi sulla Vulnerabilitá degli Acquiferi, Vol. 2, a cura di N. Paltrinieri, M. Pellegrini e A. Zavatti.*
33. Schleyer, R., Milde, G. and Milde K. (1991) Development of Aquifer Protection Policy in Germany, in *Groundwater Pollution and aquifer Protection in Europe*, Proceedings of the Annual Symposium 1991 of The Institution of Water and Environment Management, London, UK.

MODELLING OF OIL AND CHLORINATED HYDROCARBONS IN SATURATED AND UNSATURATED ZONES AT THE MILITARY BASE IN MILOVICE

VÁCLAV BENEŠ
Aquatest a.s.
Geologická 4
14100 PRAHA 5
Czech Republic
VÁCLAV ELIÁŠ
Institute of Hydrodynamics, ASCR
Pod Pa ankou 5
166 12 Praha 6

Abstract

Two transport models of oil and chlorinated hydrocarbons were constructed using the use of the computing code SUTRA. The models were applied to solve of environmental problems at the military base Milovice in the Czech Republic. The first model describes pollutant transport in an unsaturated zone and the second one the transport in an aquifer. The models were used for optimisation of remedial processes on this site and for the optimisation of pollutants capture in this massively contaminated area. The objective of remediation of this site is to protect the groundwater at Sojovice-Káraně which is one of the three main water resources for Prague.

1. Introduction

One of many military bases of the Soviet Army in (former) Czechoslovakia was garrisoned at Milovice some 35 km Northeast of Prague. The contaminated locality is situated in the Jizera river basin which has great significance for the drinking water supply of Prague. A principal water source for Prague, Káraný, with capacity of about 2000 l/s lies ashore of the Jizera river. The water is drawn partly from a series of wells adjacent to the stream and partly from wells situated around recharge basins which are filled with mechanically treated water of the Jizera river.

The model solution presented in this paper is focused on the decontamination problems of one of the most polluted parts of the military base, i.e. the fuel stores. This area is contaminated predominantly by oil hydrocarbons. Prior to the clean-up, these products formed a continuous layer several decimetres thick on the groundwater level.

F. Fonnum et al. (eds.),
Environmental Contamination and Remediation Practices at Former and Present Military Bases, 171–179.

The contamination originated from leaky storage tanks and fuel pipelines that had been irresponsibly manipulated.

2. Basic Site Description

The geological structure of the modelled region is formed by chalk rocks (Upper Cretaceous - Middle Turonian) and partly by quaternary sediments. The ground waters discharge from the contaminated region in the Southwest direction and are drained by the Jizera and Labe (Elbe) rivers. A part of the groundwater penetrates into the Mlynaøice creek which flows along the military base limits. At higher water level the surface water of the Mlynaøice creek can infiltrate into the groundwater about 5 km downstream of the base and thus jeopardise the water quality of the near Káraný. The extent of oil derivatives contamination in the groundwater has been determined by means of chemical analysis of water samples collected in 21 wells. The maximum pollutants concentration was 3.3O mgl^{-1}. The measured concentrations were used as initial values for the model calculations of pollutant transport in the aquifer.

3. Model SUTRA Application

The model SUTRA was developed by the US Geological Survey in 1984 [4]. The model is continuously updated, the 1992 version was used for the simulation.The model solves dissolved substance transport in a heterogeneous and unisotropic porous medium. The materials may be sorbed on a rock medium and their concentration may be influenced by chemical , microbial or other reactions. The model can also be used in a vertical cross-section also for the system unsaturated-saturated zone.
Model calculations objectives of the pollutants transport in the unsaturated zone were:

- to provide information about the pollutants residence time in this zone,
- to permit the assessment of the chosen remedial methods influence on time development of the substance concentrations in this zone,
- to determine the amount pollutants that will discharge into the aquifer under the individual chosen variants.

The model calculations were carried out in three variants:

- Variant No.1: continuous pollutants inflow through the soil surface
- Variant No.2: ceasing of pollutants inflow through the soil surface and substance transport accumulated in the unsaturated zone - without considering biodegradation
- Variant No.3: ceasing of the pollutants inflow including consideration of biodegradation

Model of the pollutants transport in the aquifer was the basis for the design of the optimal remedial system of wells, their quantity and situation and, for the assessment of the pumped or injected volume. The objective of the remedial pumping, aside from ground water cleaning, was to prevent leakage of the pollutants from massively contaminated area.For the definition of the flow regime, namely with respect to the

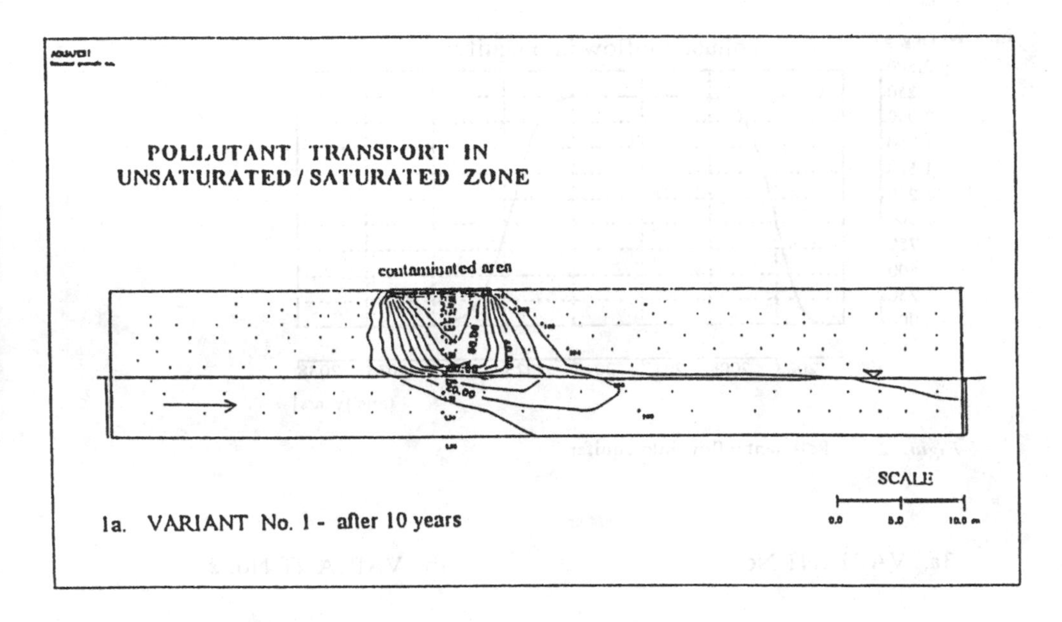

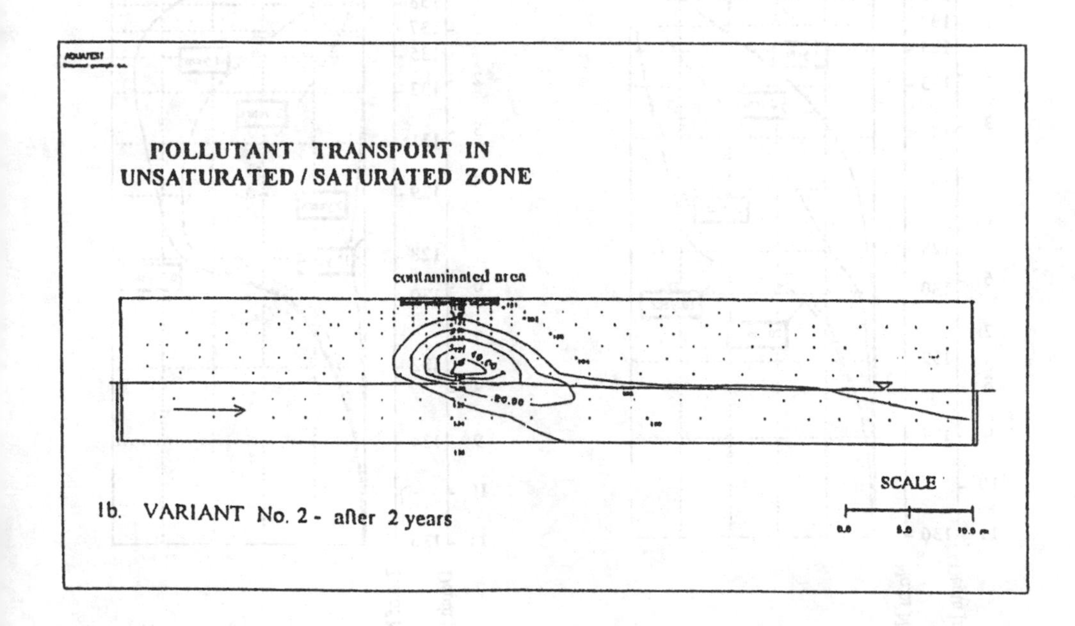

Figure 1. Isolines of the pollutant concentration in the unsaturated zone

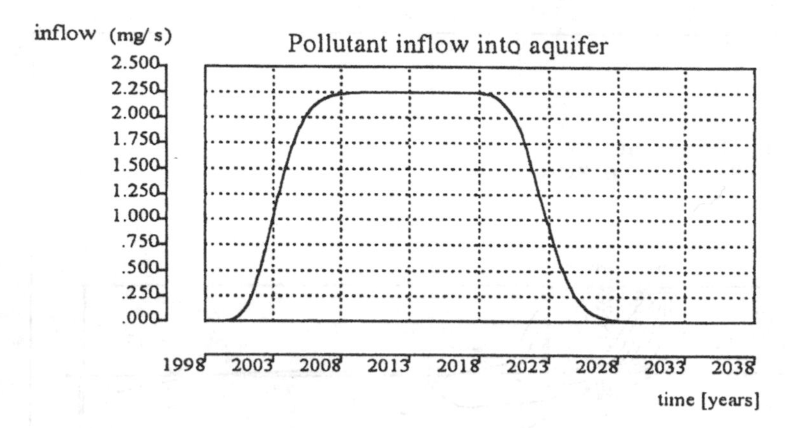

Figure 2 Pollutant inflow into aquifer

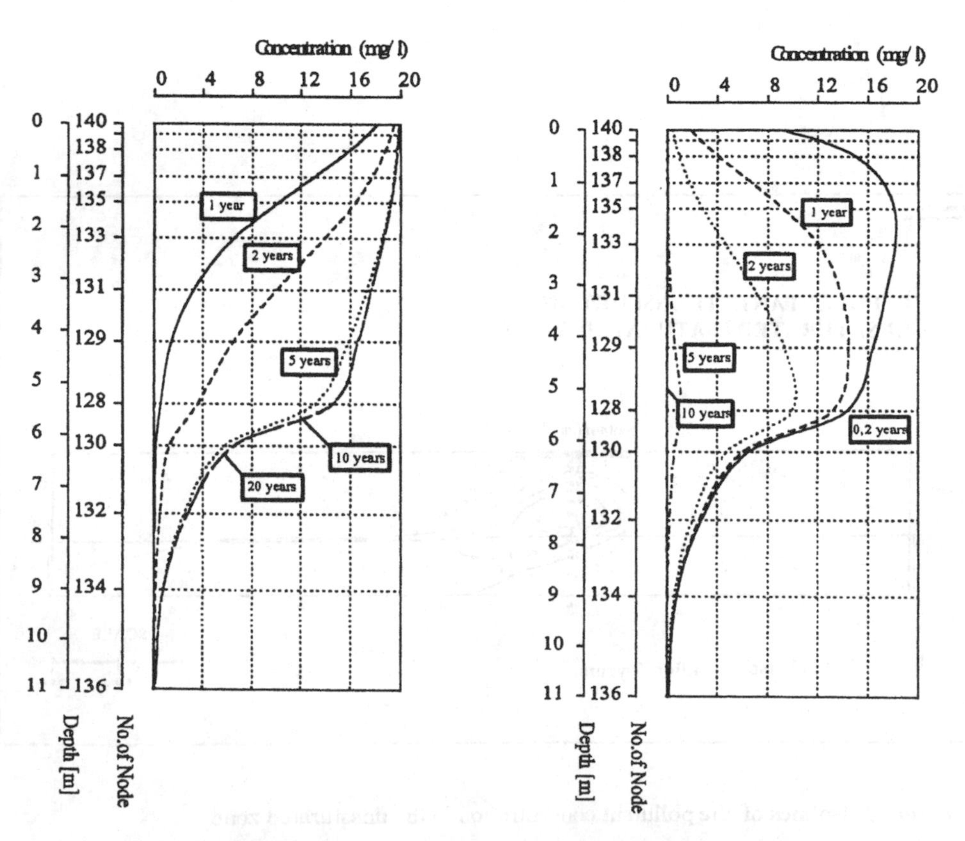

Figure 3 Pollutant concentration profile in the unsaturated/saturated zone

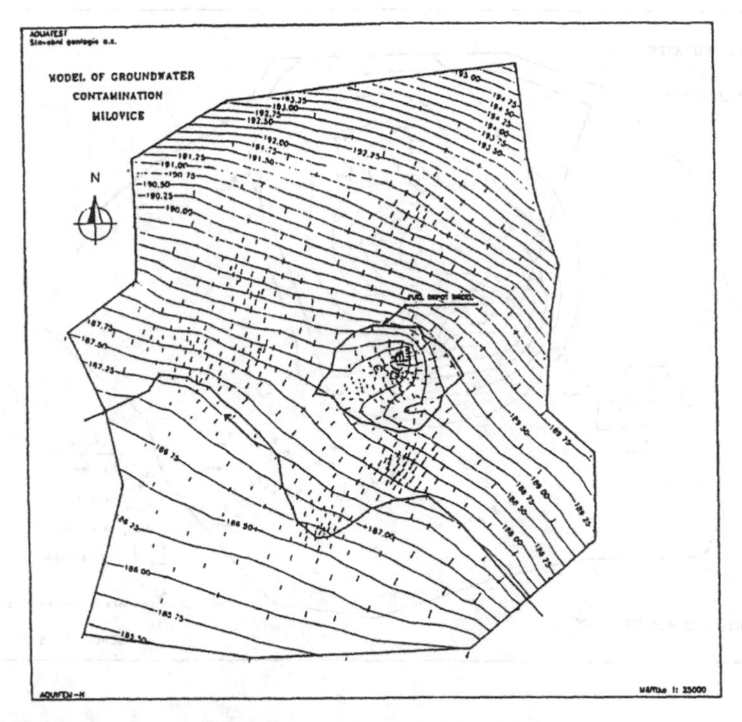

Figure 4. Groundwater head contours - regional model

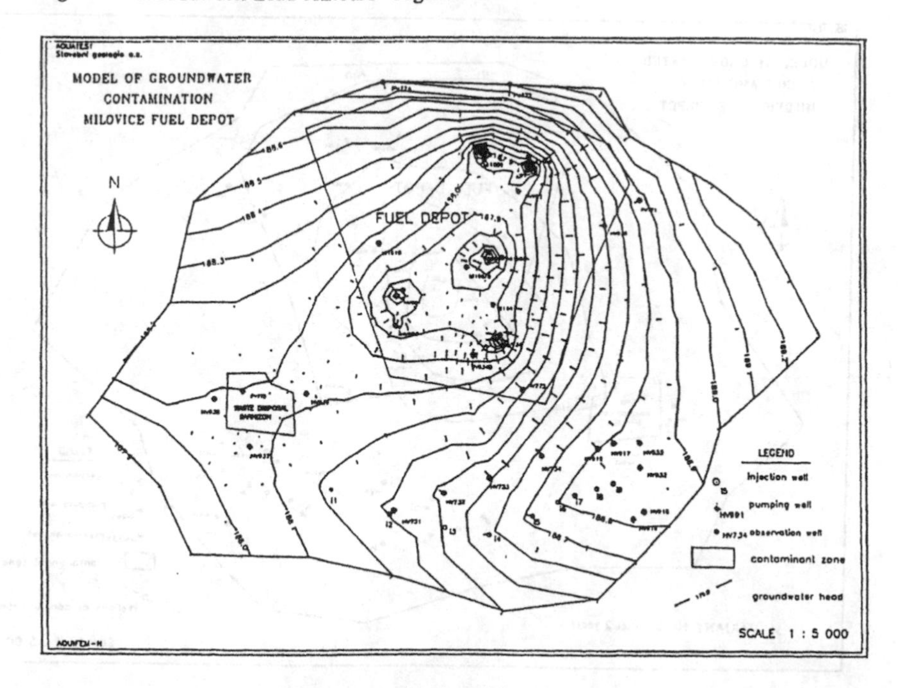

Figure 5. Groundwater head contours - detail model

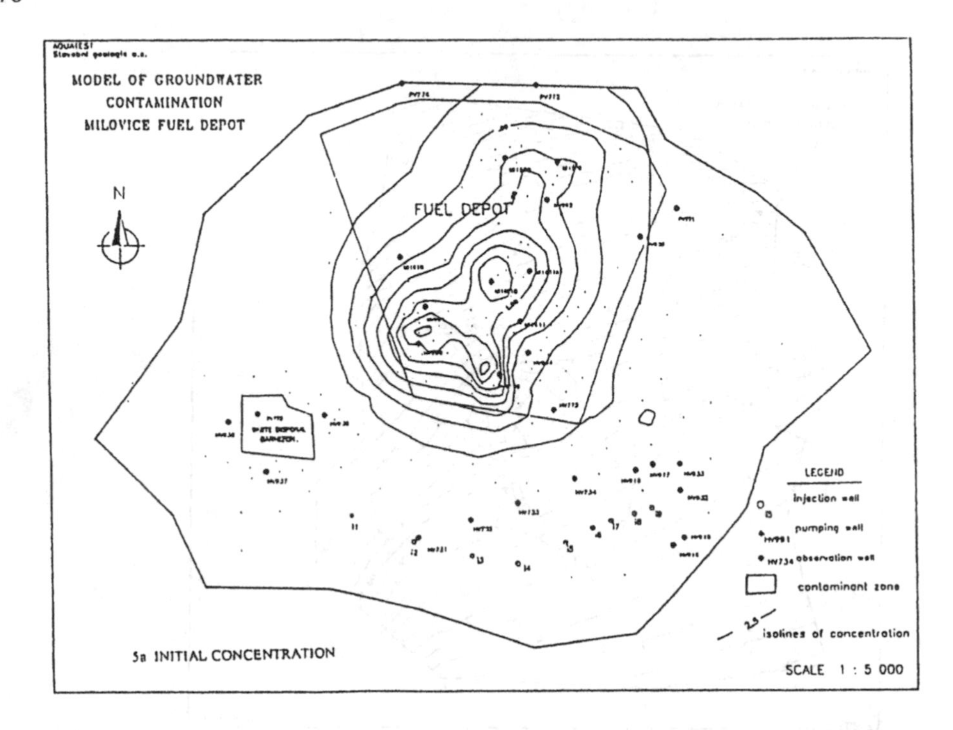

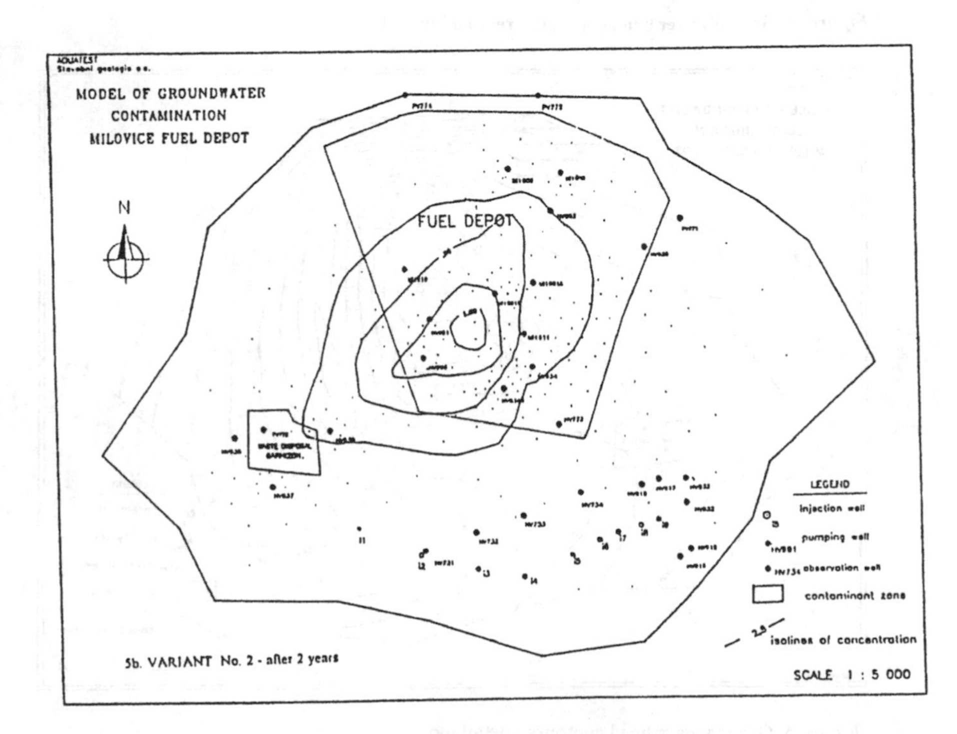

Figure 6. Pollutant concentration contours

specific boundary conditions of the transport model, a regional flow model was set up in the first phase. The proper pollutants transport was solved by a detailed model (see Fig.4).

4. Input Data

Hydrogeological investigation at Milovice has been in progress since 1992 and during this time a great deal of input data have been collected which are useful for the model calculations.

Unsaturated zone: Thickness of the unsaturated zone was 6.00 m, and for hydraulic purposes was divided into three layers. The characteristics are sumarized in Table.1.

TABLE 1. Hydraulic characteristics of the unsaturated zone

Layer	Thickness (m)	K_s ($m\,s^{-1}$)	ε (-)	α (cm^{-1})	n (-)
1	0.60	3.95 E-05	22.64	1.00 E-04	3.00
2	1.20	1.14 E-04	15.04	1.00 E-04	3.00
3	4.20	1.60 E-04	16.23	1.00 E-04	3.00

Other parameters were identical for the whole cross-section of the unsaturated zone.
Coefficient of the hydrodynamic dispersivity: longitudinal $\alpha_L = 0.40$ m
transversal $\alpha_T = 0.20$ m
Linear distribution coefficient of the sorption isotherm: $K_d = 0.045$ g cm^{-3}. The biodegradation was considered as first order kinetics with the half period: $t_{1/2}$ = 2.65 years.
Initial distribution of the soil moisture was considered constant across the whole profile. The moisture saturation coefficient was taken: $S_w = 0.45$.
The pollutants inflow through the soil surface into the unsaturated zone was calculated as a product of the infiltrated precipitated water flow and of the solubility of the pollutant in the water. The surface infiltration was estimated as 2.5O $l.s^{-1}km^{-2}$ and the oil hydrocarbons solubility was considered to be 20,0 mgl^{-1}.
Aquifer: The thickness of the aquifer was in the range of 5 -35 m. The filtration coefficients were determined by means of pumping tests and were in the range of 1.20 E-04 to 2.80 E-04 ms^{-1}. The surface distribution of the filtration coefficient was adjusted at the calibration of the regional model. The effective porosity was considered as a constant value on the whole model area: $n_0 = 0.12$.
The boundary conditions of the detailed transport model were considered to be of the first type, i.e. h=const., and their values were assumed from the regional groundwater flow model which had covered larger territory and in which the interaction of the ground- and surface water in streams (Mlynaøice creek) was considered as well.
The pollutants inflow rate was assumed from the results of the model solution of the transport in the unsaturated zone.

Coefficient of the hydrodynamic dispersivity: longitudinal $\alpha_L = 42.00$ m
transversal $\alpha_T = 14.00$ m

Linear distribution coefficient of the sorption isotherm (determined as a product of the partial distribution coefficient and of the organic carbon content): $K_d = 0.045$ cm^3 g^{-1}.
The half period - biodegradation : $t_{1/2} = 2.64$ years.

Initial distribution of the pollutants in aquifer was determined on the basis of the chemical analysis of water samples collected from individual monitoring and pumping wells.

5. Results of the Model Calculations.

With respect to the paper limitations the results are presented only in the form of selected examples of some graphical outputs.

Unsaturated zone: The spatial distribution of the pollutant's concentration in the unsaturated zone and its inflow into the collector of the ground water represent a typical output of the model calculations.The concentration development in a cross-section profile is presented in Fig.1. Fig.1a shows a vertical profile of the oil hydrocarbons concentration after 10 years at a continuous inflow, Fig 1b shows the concentration distribution 2 years after the ceasing of the inflow into the unsaturated zone. In Fig.2 the time development of the pollutant inflow into the aquifer is depicted.A permanent inflow over 20 years and consequent washing-out of the pollutant by the clean infiltrating water (after the removal of the contaminated source) is considered here. Fig. 3 shows the development of the oil hydrocarbons vertical distribution, in Fig.3a at a continuous inflow (curves representing 10 and 20 year periods coincide - resulting in formation of a stationary transport regime), and in Fig 3b after the ceasing of the inflow.

Aquifer: The groundwater flow in the regional model is illustrated in Fig.4 with the map of groundwater head contours. The groundwater table in the territory of the detailed model for the optimal variant of the remedial pumping and injection of the clean water is presented in Fig.5. Migration of the oil hydrocarbons is depicted in Fig.6: Fig.6a shows the initial surface distribution of the oil hydrocarbons, Fig.6b shows the same situation 5 years after the the inflow into the aquifer ceased.

6. Selection of Remedial Actions and Conclusion

Site remediation is essential, without it a considerable decrease in water quality in the source area Kárané could occur. Implementation of remediation (venting, bioventing etc.) in the unsaturated zone will significantly shorten the time needed for the cleanup of the former military base Milovice. From model calculations, remedial pumping from 10 wells was proposed with a total discharge of 32 l s^{-1}. After cleanup the water will be infiltrated back into 9 wells with a total discharge of 16 l s^{-1}. Implementation of the proposed variant of remedial pumping and injection of clean water into the wells will reduce the concentration of pollutants in the centre of contamination after 5 years to the value of max. 0.150 mg l^{-1}, i.e.below 5% of the initial value After 10 years the concentration will decrease to 0.05 mg l^{-1}. The proposed optimal variant of the remedial pumping will prevent the leakage of pollutants.

7. References

1. Bear,J. (1979) *Hydraulics of groundwater,* Mc Graw Hill, New York
2. Beneš,V. (1995) *Hydrodynamics of transport and transformation process of pollutants in groundwater,* 175 p, Academia Prague, (in Czech)
3. Kinzelbach, W. (1986) *Groundwater modelling,* Elsevier, Amsterdam
4. Voss,C.J. (1984) SUTRA - *a finite-element simulation model for saturated-unsaturated fluid density dependent groundwater flow with energy transport or Chemically reactive single species solute transport,* US Geological Survey, Report No84-4369 Reston, Virginia USA

HOW TO MAP HYDROCARBON CONTAMINATION OF GROUNDWATER WITHOUT ANALYSING FOR ORGANICS

A Study of Oil Contamination of Soil and Groundwater at Viestura Prospekts Former Military Fuel Storage Depot, Riga, Latvia.

D. BANKS[1], L. BAULINS[2], A. LACIS[3], G. SICHOVS[2] & A. MISUND[1]
[1]*Norges geologiske undersøkelse*
Postboks 3006 Lade, N7002 Trondheim, Norway
[2]*Geo-Konsultants*
3 K. Ulmana Gatve, Riga, LV-1004, Latvia
[3]*Geological Survey of Latvia*
5 Eksporta Street, Riga, LV-1010, Latvia

Abstract

The Geological Surveys of Latvia and Norway and the Norwegian Defence Research Establishment have undertaken a risk-based investigation and assessment of a former Soviet military oil depot at Viestura Prospekts, Riga, using the firms Dames & Moore (U.K.) and Geo-Konsultants as subcontractors.

The site lies on homogeneous fine-medium grained Baltic Ice Lake sands containing a shallow unconfined water table. Georadar, calibrated against monitoring boreholes, has been used to map the north-westward gradient of the water table.

16 monitoring boreholes have been employed to assess groundwater contamination. Free phase oil has been proven in boreholes near fuel storage bunkers and at the oil-transfer railhead.

Vertical electric sounding profiling (VES), a geophysical technique measuring the apparent resistivity of the sediments, has been employed at the site. VES showed two plumes of low resistivity leading away from the bunker area and the railhead in a NW direction. These may represent plumes of groundwater with high ionic content.

Oil degradation releases CO_2, which enhances weathering of carbonate and silicate mineral phases in the aquifer, releasing alkalinity and several cations and elevating the water's electrical conductivity (EC), explaining the VES map. Additionally, degradation consumes O_2 and other electron acceptors, promoting reducing conditions and releasing dissolved iron. Thus, a combination of high Fe, alkalinity, ion content, EC and low oxygen, sulphate and nitrate, may be used as low-cost inorganic indicators of hydrocarbon contamination.

F. Fonnum et al. (eds.),
***Environmental Contamination and Remediation Practices at Former and Present Military Bases,* 181–196.**

1. Introduction

The Geological Survey of Norway (NGU), the Geological Survey of Latvia (GSL) and the Norwegian Defence Research Establishment (FFI) are currently carrying out a risk-based site investigation and assessment of a former Soviet military fuel storage depot at Viestura Prospekts, Riga, Latvia. The main partners in the project have also contracted out parts of the project to sub-consultants, namely; Dames and Moore (U.K.) who have carried out the risk assessment of the site, and Geo-Konsultants (Latvia) who have carried out the geophysical investigations. The project has been financed by the Ministry of Foreign Affairs in Norway and the Ministry of Environment in Latvia.

1.1. SITE LOCATION AND TOPOGRAPHY

The site is located on the eastern side of Viestura Prospekts in the Meza Parks area of northern Riga. The site has a total area of ca. 20.5 Ha. A small part of the north of the site is currently in use by a private fuel management firm for daily operations including vehicle refuelling and repair. The remainder of the area is used by the firm for oil off-loading (from rail), pumping and storage, but is fenced off and declared a customs zone.

The site has an elevation of between 5 and 12 m above sea level (a.s.l.). The terrain is hummocky. While some of these hummocks may be related to excavated materials during installation of the fuel bunkers, the terrain in the surrounding Meza Parks is similar, suggesting that most of the hummocks are natural. They may represent wind blown dunes in the sandy deposits of the Baltic Ice Lake, which constitute the underlying sediments. The topography is dominated by the two fuel storage bunker complexes in the east of the site. These consist of groups of steel tanks encased in concrete bunkers. Low bunds of natural materials have been constructed around the bunkers. Within the bunds the ground is natural (i.e. sandy). While the bunds may have been effective at hindering spreading of surficial oil contamination, they would not have been effective at preventing ground and groundwater contamination.

1.2. SITE HISTORY

The Viestura Prospekts site was in use from 1941 until 1992 as a fuel storage depot for the Soviet (later, Russian) Army (Figure 1). Estimates of the total volume of fuel storage at the site vary from 4,400 m^3 (GSL) to 13,200 m^3[1]. There exist a railhead, a road tanker loading bay and a pipeline to Riga port (following the course of Viestura Prospekts). It is believed that fuel arrived by rail and was distributed throughout the Soviet Union and Warsaw Pact lands by road and sea. At a similar, though larger and more strategically central, locality in Lithuania (Valciunai [2]), fuel also entered by rail. In the event of a critical situation, fuel could be freighted by road to a number of sub-depots and recipients.

Following the withdrawal of the Russian/Soviet Army in 1992, the site has been in use by a commercial fuel handling company who operate along the same lines as the Soviet Army, but on a much smaller scale. The site is bordered on the north and east by Meza Parks, one of Riga's main park areas.

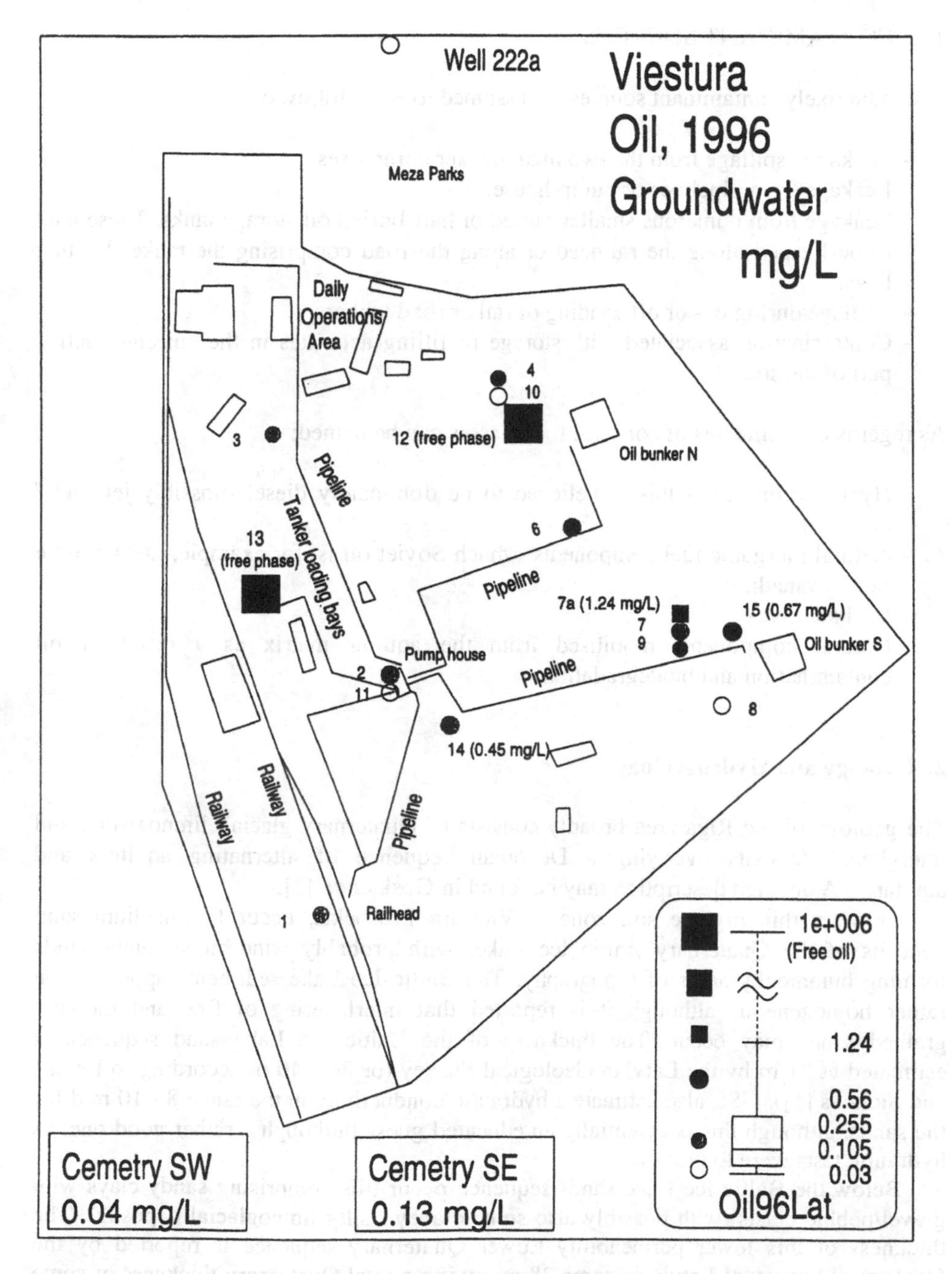

Figure 1. **Map of the Viestura Prospekts site, showing concentrations of oil in groundwater (mg/L) in 1996. Numbers on the map are borehole numbers.**

1.3. CONTAMINANT SOURCES

The main likely contaminant sources are assumed to be as follows:

- Leakage / spillage from the two main bunker complexes.
- Leakage from pipelines or pump-house.
- Leakage from numerous smaller buried or half-buried oil storage tanks. These tend to be located along the railhead or along the road comprising the tanker loading bays.
- Spillage during on- or off-loading of rail or road tankers.
- Contamination associated with storage or filling activities in the currently active part of the site.

As regards contaminants of concern, four classes can be named:

- Hydrocarbon fuel - this is believed to be dominantly diesel (possibly jet fuel / paraffin).
- Natural inorganic fuel components - much Soviet oil is, for example, known to be rich in vanadium
- Fuel additives.
- Natural components mobilised from the aquifer matrix as a result of oil contamination and biodegradation.

2. Geology and Hydrogeology

The geology of the Riga area broadly consists of Quaternary glacial, limnoglacial and interglacial deposits overlying a Devonian sequence of alternating aquifers and aquitards. A detailed description may be found in Gosk *et al.* [3].

Below a thin organic soil zone at Viestura Prospekts, occur fine-medium sand deposits of the Quaternary Baltic Ice Lake, with probably wind-blown dune sands forming hummocky areas of topography. The Baltic Ice Lake sediments appear to be rather homogeneous although it is reported that interlayering of fine and medium grained sands may occur. The thickness of the Baltic Ice Lake sand sequence is estimated at 21 m by the Latvian Geological Survey (or 30 - 40 m according to Levins and Sichovs [1]). GSL also estimate a hydraulic conductivity in the range 8 - 10 m/d for the sands, although this is essentially an educated guess (although a rather good one, as hydraulic tests were to prove).

Below the Baltic Ice Lake sands sequence occur tills comprising sandy clays with gravel/pebble clasts (with possibly also some clayey / silty limnoglacial deposits). The thickness of this lower permeability Lower Quaternary sequence is reported by the Geological Survey of Latvia as some 28 m, giving a total Quaternary thickness of some 49 m.

The Quaternary deposits overlie sedimentary rocks of the Upper Devonian Gauja Formation comprising sandstone with silty interlayers. This is an important aquifer for

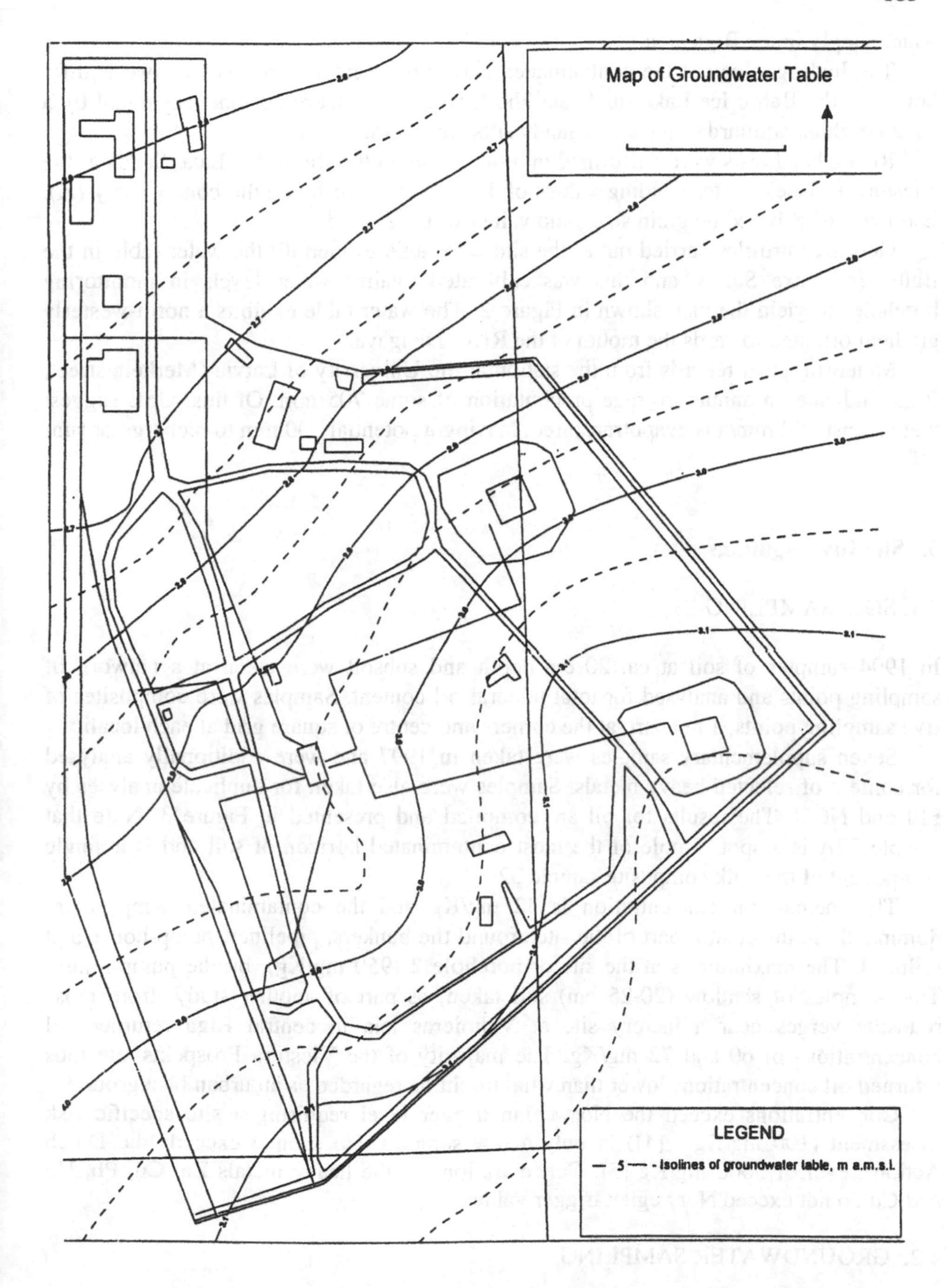

Figure 2. **Map of Viestura Prospekts site showing contours on water table deduced by georadar (m above sea level). Scale bar is 100 m.**

water supply in the Riga area.

The hydrogeology at the contaminated site thus essentially consists of two aquifer horizons, the Baltic Ice Lake sands and the Devonian Gauja Sandstone, separated by a ca. 28 m thick aquitard sequence of sandy silts and clays.

Rising head tests were performed in 5 boreholes in the Baltic Ice Lake Sands at the Viestura Prospekts site, yielding values of 1 - c. 8 m/d for hydraulic conductivity (K). Estimates of K based on grain size yield values of 6 - 20 m/d.

Georadar profiles carried out at the site were able to identify the water table in the Baltic Ice Lake Sands, and this was calibrated against water levels in monitoring boreholes to yield the map shown in Figure 2. The water table exhibits a north-westerly gradient oriented towards the mouth of the River Daugava.

Meteorological records from the station at the University of Latvia (Merkela street, Riga) indicate an annual average precipitation of some 705 mm. Of this, GSL suggest that at least 400 mm/a is evapotranspired, leaving a potential 300 mm to recharge or run-off.

3. Site Investigations

3.1. SOIL SAMPLING

In 1994 samples of soil at ca. 20 cm depth and subsoil were taken at a network of sampling points and analysed for total mineral oil content. Samples were composites of five sampling points, 1 m apart, at the corners and centre of square grid at each locality.

Seven supplementary samples were taken in 1997 and were additionally analysed for content of selected heavy metals. Samples were also taken for duplicate analyses by FFI and NGU. The results for oil are compiled and presented in Figure 3. Note that sample 32A is a spot sample of the most contaminated horizon of soil and is a single component of the bulk composite sample 32.

The median oil concentration is 32 mg/Kg and the contaminated samples are dominantly in the central part of the site around the bunkers, pipelines, pump-house and railhead. The maximum is at the site of borehole 2 (950 mg/Kg), by the pump-house. Two samples of shallow (20-25 cm) soil taken, as part of another study, from grass roadside verges near a factory site at Valmieras iela in central Riga returned oil concentrations of 60 and 72 mg/Kg. The majority of the Viestura Prospekts site thus returned oil concentrations lower than what might be regarded as an urban background.

Concentrations exceed the Norwegian trigger level requiring a site specific risk assessment (100 mg/Kg - [4]) in only a few samples. No sample exceeds the Dutch Action Level of 5000 mg/Kg [5]. Concentrations of the heavy metals Zn, Cu, Pb, Hg and Cd do not exceed Norwegian trigger values.

3.2. GROUNDWATER SAMPLING

Site investigations of contamination at the site were undertaken in 1994, 1996 and 1997. In 1994, boreholes 1 to 8 were drilled with (typically) a 1 m long well-screen fully

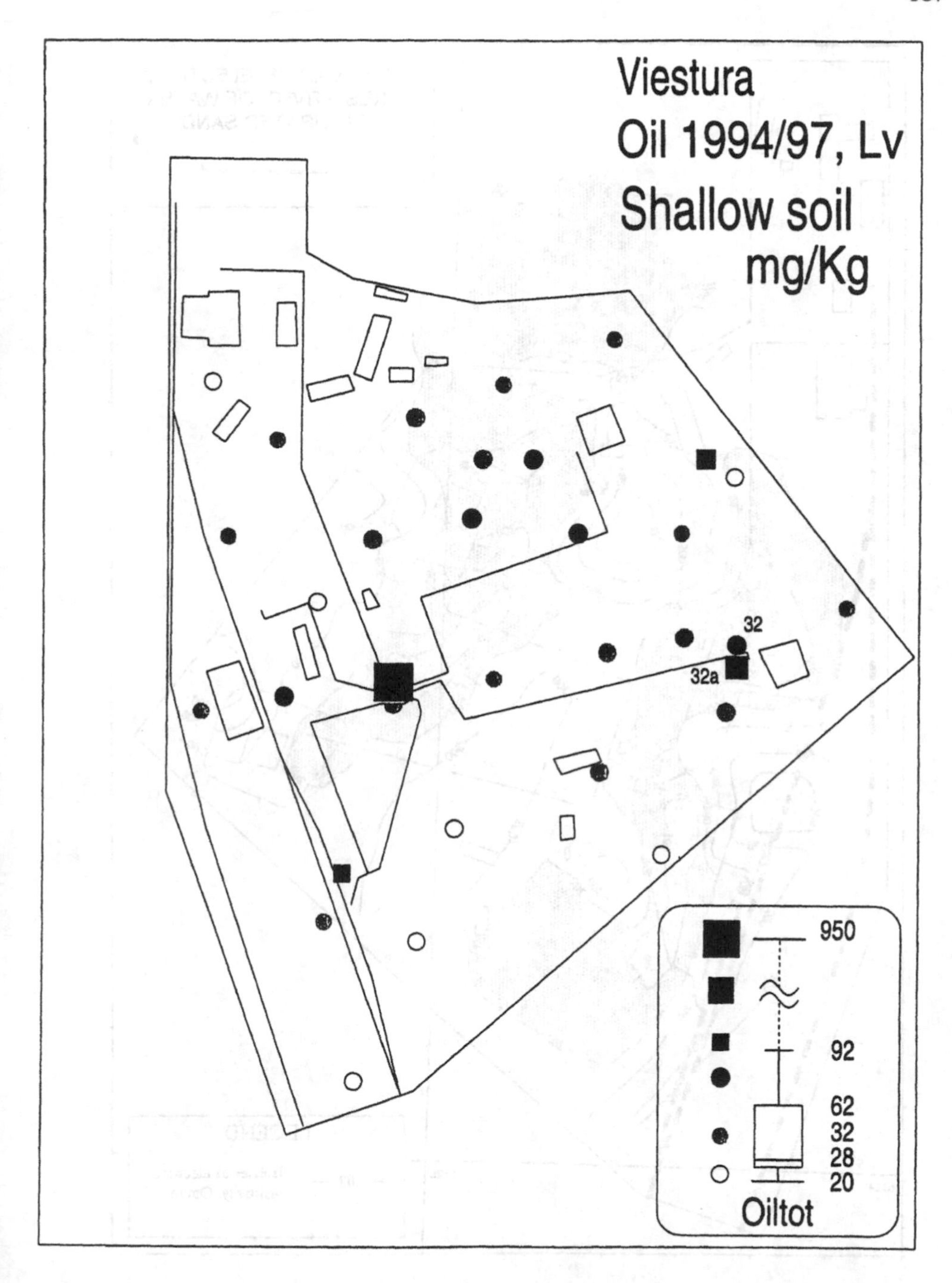

Figure 3. **Concentrations of mineral oil in topsoil (ca. 20-25 cm depth) at the Viestura Prospekts site, according to samples collected in 1994 and 1997.**

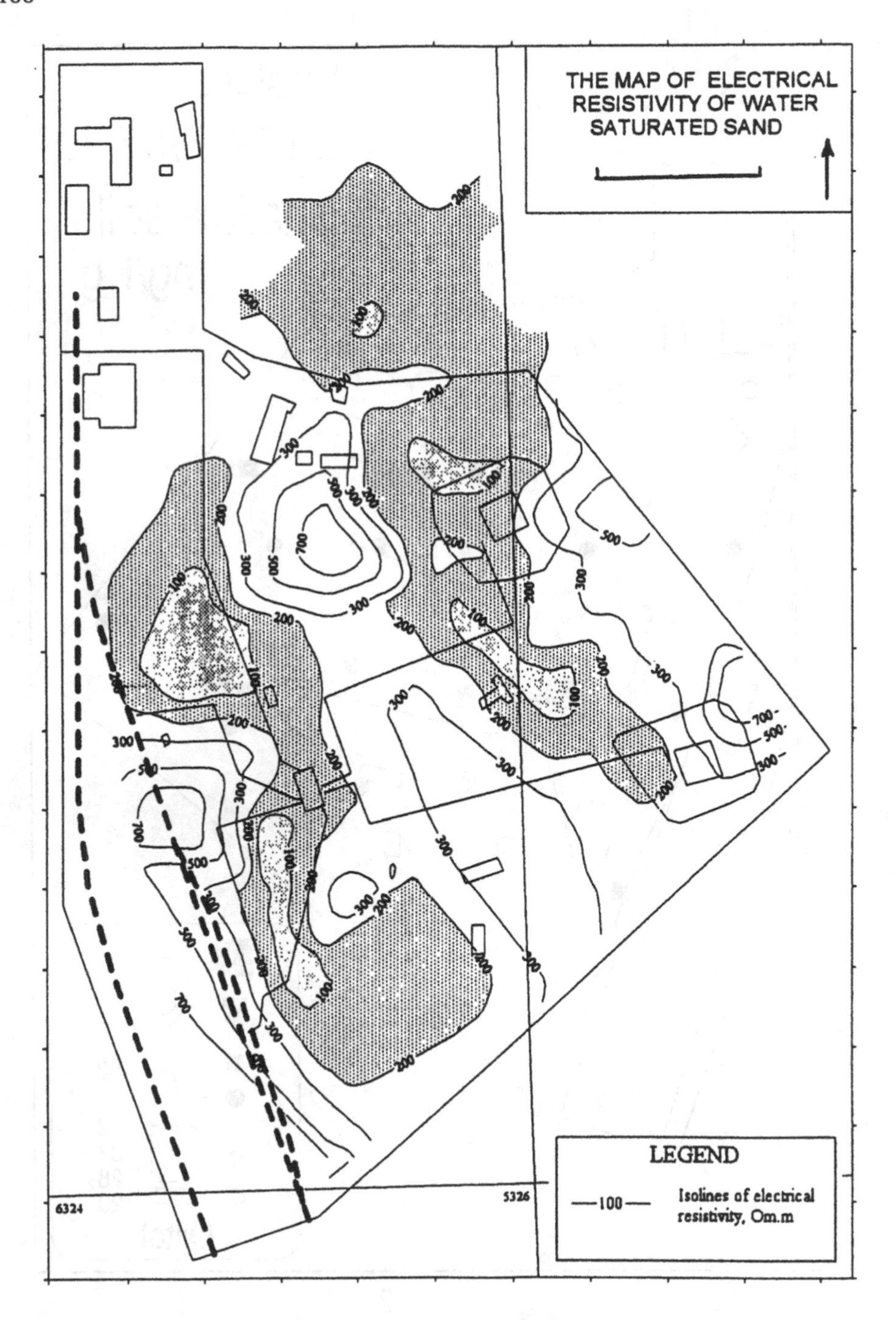

Figure 4. **Electrical resistivity of water-saturated sand (ohm.m) deduced by VES profiling. Shaded areas show low resistivity. Scale bar = 100 m.**

below the water table. At three sites (2, 4 and 7), a second borehole was drilled with a filter at a deeper level in the sand aquifer (at depths of around 8-10 m). These deep boreholes were named 9, 10 and 11. In all boreholes, the well screen was fully submerged: the wells were thus designed to sample dissolved contamination, rather than light non-aqueous phase liquid (LNAPL - i.e. free oil) contamination.

To supplement this network, four new wells were installed in 1996 (numbers 12-15), with well screens straddling the water table, rendering them suitable for LNAPL monitoring. The casing in well 7 was pulled up so that it too straddled the water table (referred to as well 7a). A pre-existing monitoring borehole (number 222a - with well screen at 8.6-10.6 m) was also incorporated into the network. Additionally, two wells at cemeteries (used for tending plants at graves) some 500-800 m SSW and SSE of the oil depot were sampled in an attempt to establish "background" groundwater chemistry.

Sampling rounds were carried out in 1994, 1996 and 1997. Samples were analysed for organic parameters in Latvia, while duplicate control samples were taken for inorganic analysis at NGU and organic analysis at FFI. Latvian laboratories yielded the most reproducible values for organic parameters, while NGU's laboratories returned the most reproducible values for the majority of inorganic parameters.

Free phase oil was found in only two wells; well 12, near the northernmost fuel bunker, and well 13, near the railhead.

4. Geophysical Investigations

Ground penetrating radar (Georadar), calibrated against water level measurements in existing boreholes, was used to produce a map of the water table, showing a north-westerly groundwater gradient.

Vertical electrical resistivity sounding (VES) was also used to produce a map of electrical resistivity of the aquifer below the site. In such flat, relatively homogeneous geology, variations in resistivity will be mainly attributable to variations in groundwater conductivity. The resistivity map is presented in Figure 4. It appears to show two elongated areas of low resistivity. These could be interpreted as plumes of high conductivity groundwater moving north-westward away from the fuel bunker area and the railhead area. Initially, these results were perplexing: the high conductivity water might represent contamination, but hydrocarbons are not themselves electrically conductive, nor do they dissolve to yield high concentrations of charged species.

5. Results: Groundwater Chemistry

Free phase oil was found in wells 12 and 13, both of which lie within the high-conductivity plumes detected in the VES survey.

Figure 1 shows the distribution of dissolved (or possibly emulsified) oil concentrations found in groundwaters during the 1996 sampling round. For reference, the Norwegian drinking water limit for oil is 0.01 mg/L [6], while the Dutch background and action levels are 0.05 and 0.6 mg/L, respectively [5].

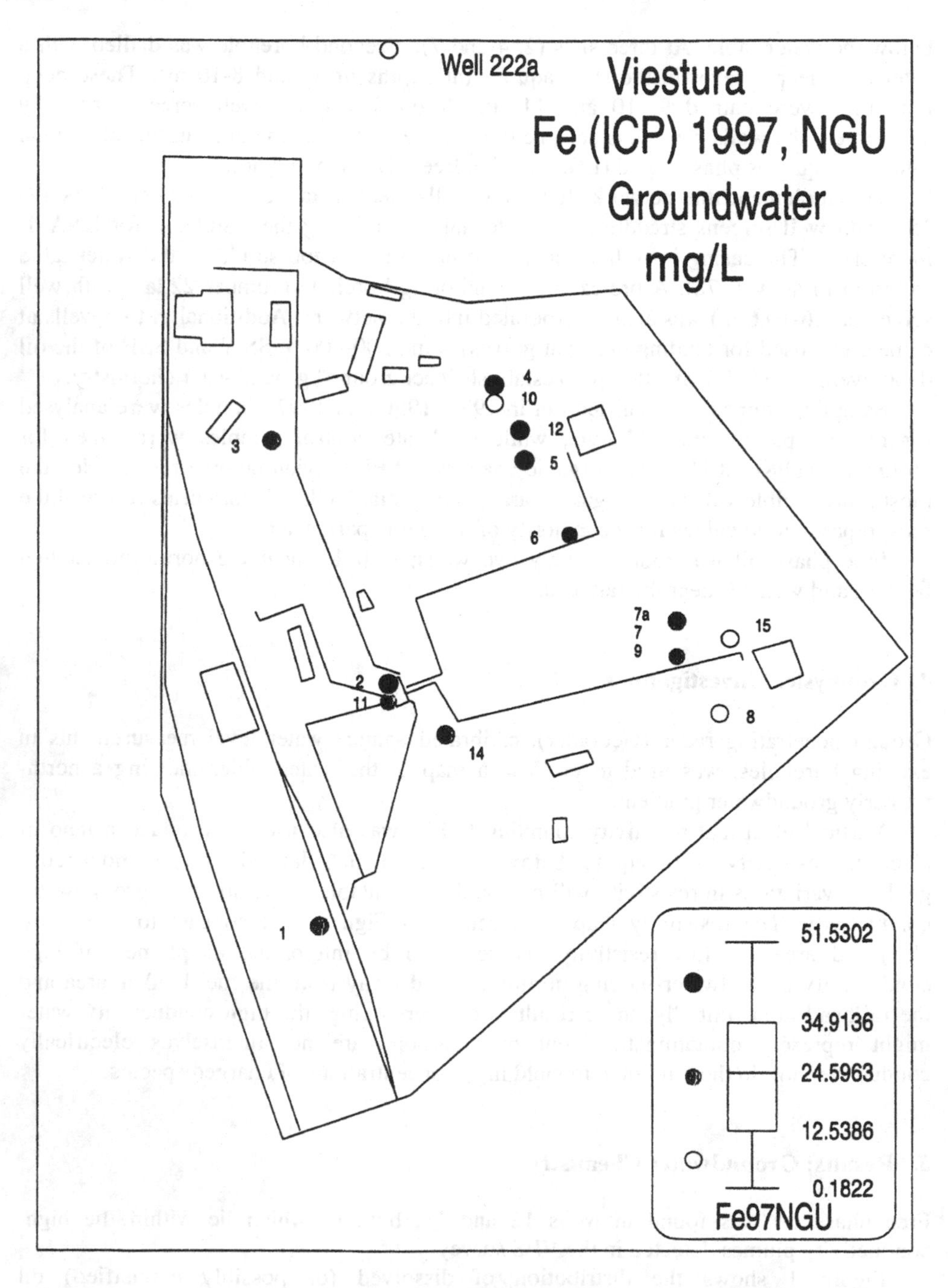

Figure 5. **Concentrations of dissolved iron in groundwater (mg/L) at the Viestura Prospekts site (1997). Numbers on map refer to monitoring wells.**

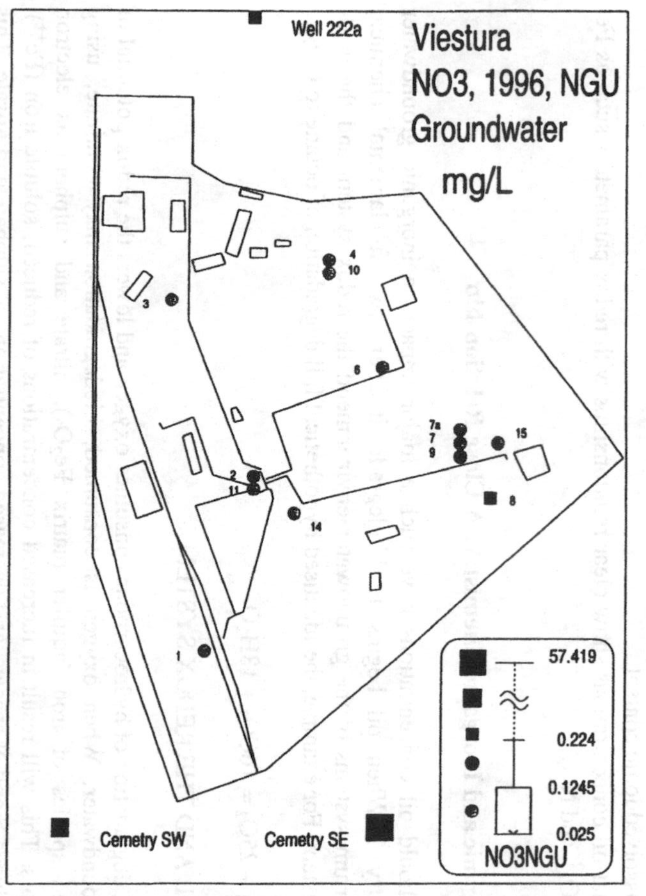

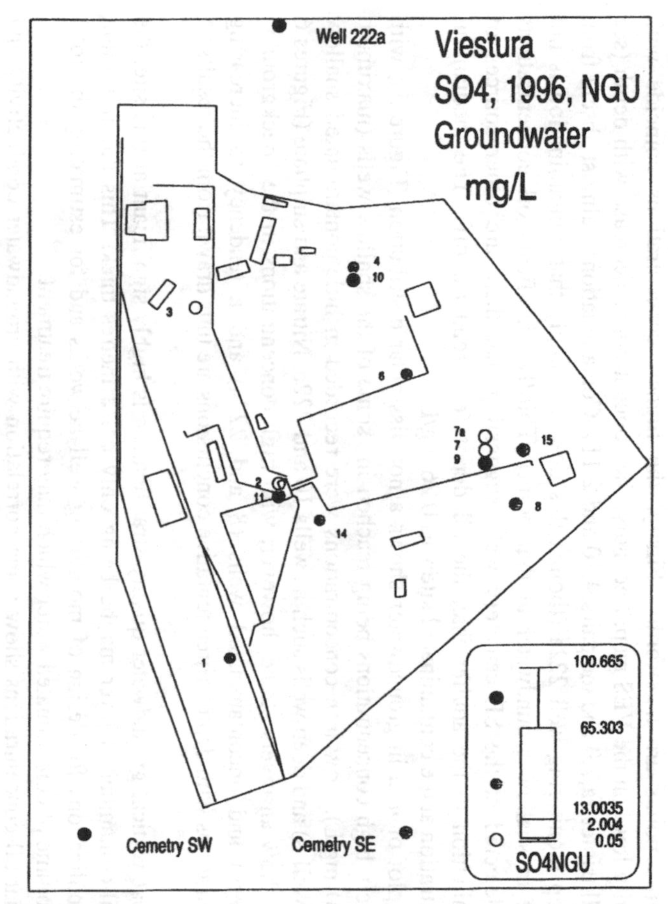

Figure 6 (left) and Figure 7 (right). **Concentrations of nitrate and sulphate in groundwater at the Viestura Prospekts site (mg/L), as measured in 1996.**

Three points are noteworthy. Firstly, the highest concentrations coincide with resistivity lows on the VES map. Secondly, the concentrations decrease with depth (see the well triplet 7a,7,9 and the pairs 4,10 and 2,11). Contamination is almost absent from the deep monitoring well 222a (though this does not imply that contamination is not present in shallow groundwater at that point). Thirdly, significant oil concentrations were detected at the SE cemetery well, suggesting that there are other sources of contamination to the aquifer than the oil depot (e.g. road run-off). The median oil concentration at the contaminated site was 0.26 mg/L.

A plot of iron in groundwater shows a not-dissimilar distribution (Figure 5), with extremely high concentrations being reached in some of the shallow wells (maximum over 50 mg/L). Low iron concentrations were recorded in little-contaminated shallow wells (well 8) and deep wells such as wells 10 and 222a. Nitrate and sulphate (Figures 6 and 7) show an inverse trend, however, with high concentrations in the "background" (cemetery) and uncontaminated wells (8 and 222a) and a tendency to increasing concentrations with depth. Three tentative conclusions are thus drawn from the results:

- that vertical groundwater quality stratification is highly significant at this site, the oil-contaminated layer maybe being only ca. 2 metres thick. This has important implications for design of monitoring well networks and for estimating the total volume of contaminated water which may require treatment.
- that oil concentrations show some correlation with groundwater conductivity and dissolved ionic content.
- that oil concentrations show clear relationships with redox parameters such as Fe, SO_4^{2-} and NO_3^-.

6. Organic and Inorganic Chemistry: A Close Relationship

Why should oil concentrations have such a major impact on inorganic groundwater chemistry ? When oil begins to biodegrade it alters two fundamental chemical equilibrium systems of the groundwater environment: the redox system and the acid-base system. For example, the idealised hypothetical full degradation of octane to CO_2:

$$2C_8H_{18} + 25O_2 = 16CO_2 + 18H_2O$$

6.1. OIL AND THE REDOX SYSTEM

Firstly, degradation of hydrocarbons consumes oxygen and lowers the redox potential of the groundwater. When oxygen is consumed, biodegradation may continue, using oxidised species of iron (aquifer matrix Fe_2O_3), nitrate and sulphate as electron acceptors. This will result in increased concentrations of reduced, soluble iron (Fe^{++}), and reduced S and N-species, and decreased concentrations of sulphate and nitrate. This is indeed observed: Figures 8 and 9 illustrate the positive correlation between iron and oil concentrations and the negative one between iron and sulphate/nitrate. Previous investigations also suggest a correlation between oil and ammonium concentrations [1].

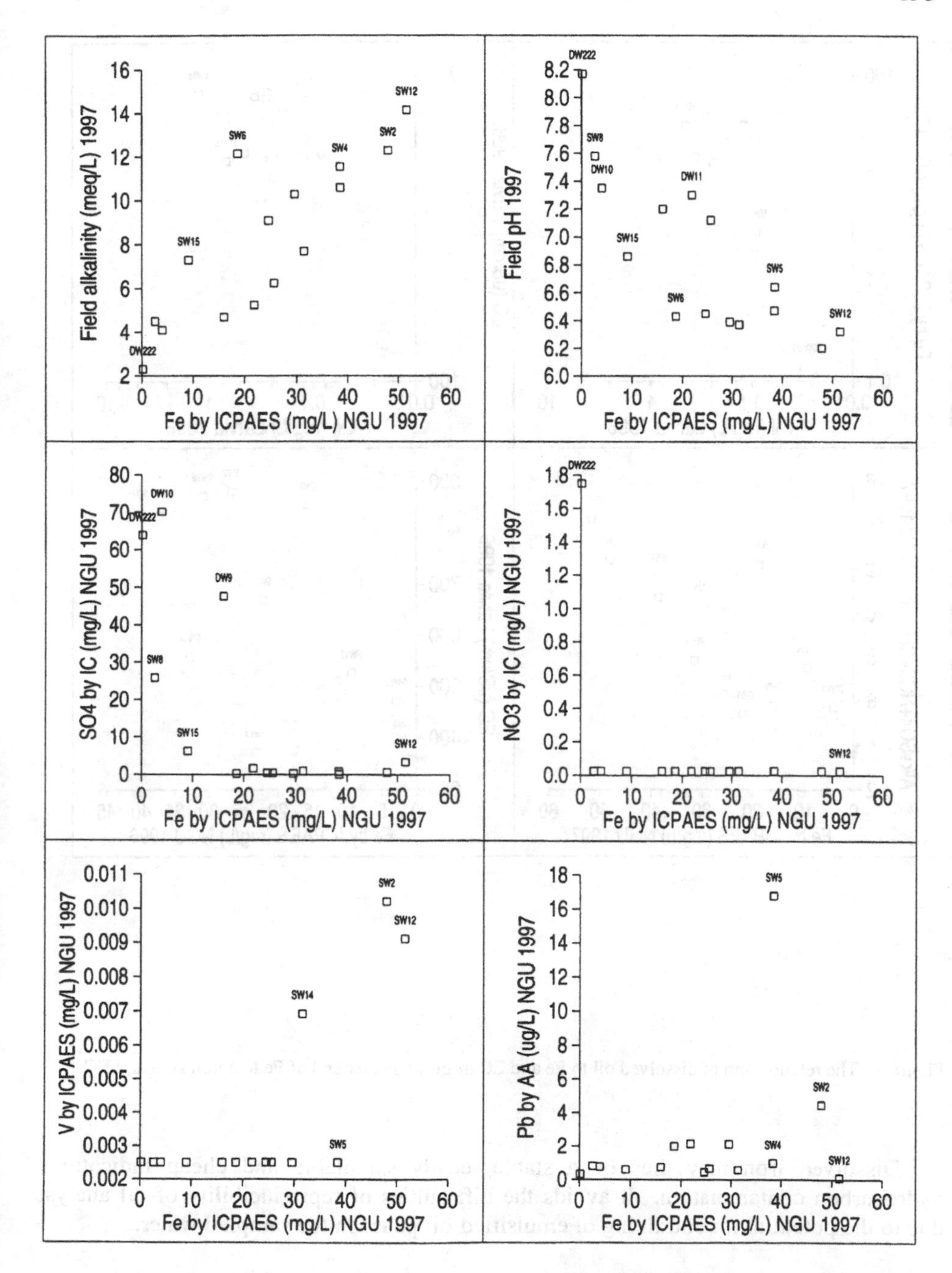

Figure 8. **Relationship of alkalinity, pH, sulphate, nitrate, vanadium and lead to dissolved iron in groundwater at Viestura Prospekts. 1997 samples. Concentrations less than analytical detection limit are plotted at half the detection limit. SW = shallower well screen, DW = deeper well screen (wells 9,10,11, 222).**

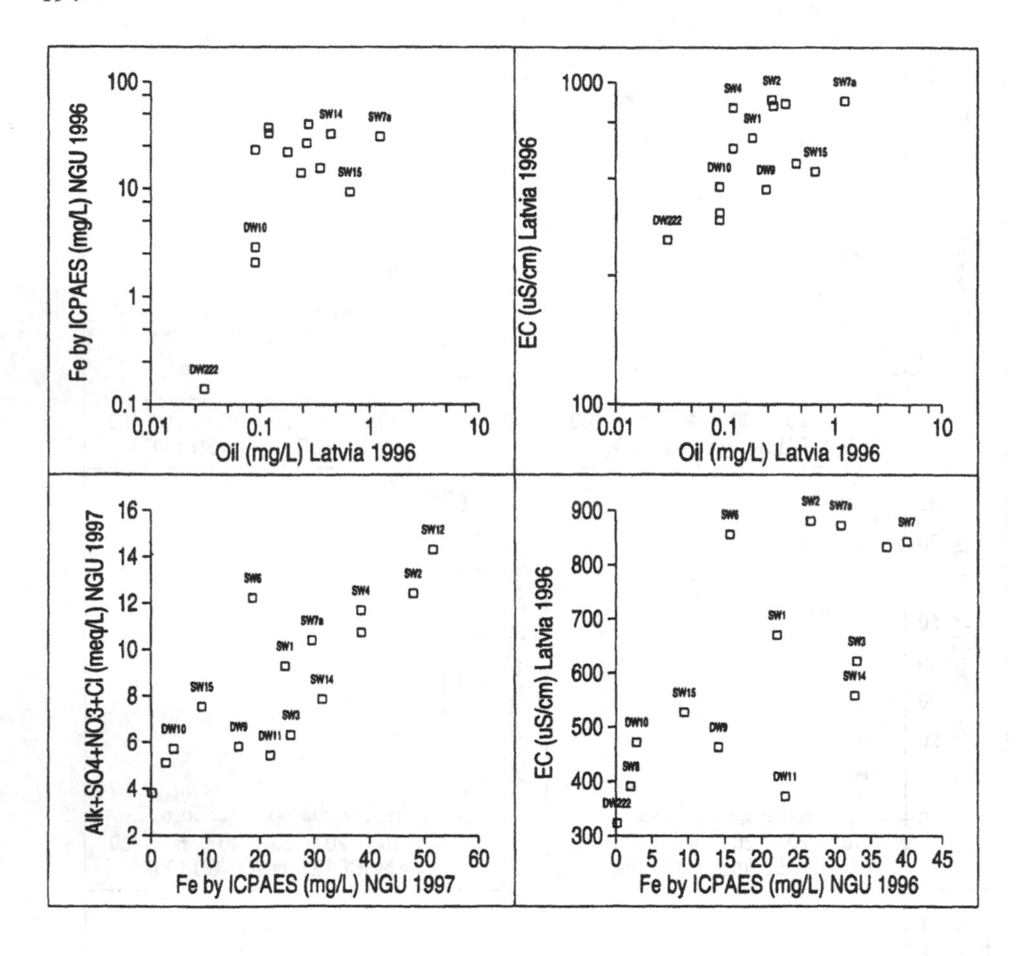

Figure 9. The relationship of dissolved oil to Fe and EC in groundwater and of Fe to anion sum and EC.

Dissolved iron may thus be a stable, easily samplable and cheap indicator of hydrocarbon contamination. It avoids the difficulties of reproducibility of oil analyses due to the potential for entraining of emulsified oil phase in the sampled water.

6.2. OIL AND THE ACID-BASE SYSTEM

Secondly, the reaction generates carbon dioxide which, when dissolved in water, acts as an acid and can attack and weather bothe carbonate and feldspar minerals, such as calcite:

$$H_2O + CO_2 + CaCO_3 = Ca^{21} + 2HCO_3$$

or sodium feldspar:

$$2NaAlSi_3O_8 + 2CO_2 + 3H_2O = 2Na^+ + 2HCO_3^- + 4SiO_2 + Al_2Si_2O_5(OH)_4$$

These reactions release alkalinity, dissolved silica and cations. Figures 8 and 9 confirm that positive correlations do indeed exist between oil and electrical conductivity (a measure of ionic content) and between Fe and ionic sum and alkalinity.

6.3. TRACE METALS - A CURIOSITY

Figure 9 indicates that, while the majority of sampled groundwaters display concentrations of < 5 μg/L vanadium (here, as on all figures, concentrations less than the analytical detection limit are plotted at a value of half the detection limit), several of the most iron rich (and oil-contaminated) groundwaters display concentrations of up to 10 μg/L. It is conceivable that the vanadium has been mobilised from the aquifer matrix by pH/redox changes during oil degradation, but it is also known that many Soviet oil resources are typically rich in vanadium [7].

The same observation applies, although less clearly, to lead. It is conceivable that lead may have been used as an additive to some fuels formerly stored on the site.

7. Conclusions

Contamination of groundwater due to hydrocarbon spillages, far from being merely the province of the organic chemist, offers much food for thought for the inorganic hydrochemist. This is not a new observation, but has been tackled in detail by the work of Mary-Jo Baedecker and her colleagues [8,9,10,11]. This study provides clear indications that oil contamination in groundwater may be positively correlated with total ion content, alkalinity, electrical conductivity and iron, and negatively correlated with redox indicators such as sulphate and nitrate. These species thus offer low cost alternatives to hydrocarbon analyses for monitoring of oil contamination. Organic analyses are expensive, crave extremely high standards of purity in sampling techniques and are often less reproducible than inorganic analyses due to problems with entrainment of emulsified oil phase. Such inorganic parameters may also allow some assessment to be made of the contamination's degradation status.

Secondly, the correlation of oil concentrations with electrical conductivity appears to allow contamination plume migration to be identified with geophysical techniques such as VES.

Thirdly, it has been shown that the distribution of oil contamination in the aquifer at Viestura Prospekts is highly vertically stratified and confined to within the upper few metres below the water table. This has important implications for design of monitoring networks.

The message is clear: contamination specialists should avoid a philosophy which involves peppering a site with monitoring boreholes and analysing sampled waters merely for the contaminants under question.

Rather, the use of geophysics can assist in reducing the number of necessary observation boreholes, and in locating them most efficiently. A monitoring network should not just aim to achieve maximum areal coverage, but to obtain information on the three-dimensional distribution of contaminants. Analytical programs should place more emphasis on characterising major ionic chemistry. Such a philosophy may save money on costly organic chemical analyses.

8. References

1. Levins, I. and Sichovs, G. (1994) Ecological evaluation of the oil storage facility belonging to "Lat-East-West" Co. (2 Viestura Prospekts, Riga). Report by the consultancy firms Urbsanas Centrs Ltd. and Geo-Konsultants Ltd., Riga.
2. Paukstys, B., Misund, A., Tucker, C., Banks, D., Segar, D., Kadunas, K. and Tørnes, J.Aa. (1996) Assessment methodologies for soil/groundwater contamination at former military bases in Lithuania. Part 1: chemical investigation. *Norges geologiske undersøkelse report* **96.146**, Trondheim, Norway.
3. Gosk, E., Levin, I. and Jacobsen, A.R. (1996) Groundwater as a source of drinking water for Riga city. *Geological Survey of Denmark and Greenland (GEUS) report* **1996/52**, Copenhagen.
4. Statens forurensningstilsyn (1995) Håndtering av grunnforurensningssaker. Foreløpig saksbehandlingsveileder. *Statens forurensningstilsyn report* **95:09**, Oslo.
5. Ministerie van Volkshuisvesting, Ruimtelijke Ordening en Milieubeheer (1994) *Leidraad bodembescherming. Aflevering 9, oktober 1994.* Sdu Uitgeverij Koninginnegracht, Den Haag, Netherlands.
6. Sosial- og helsedepartementet (1995) Forskrift om vannforsyning og drikkevann m.m. *Sosial- og helsedepartmentet* **nr. 68, I-9/95**, Oslo.
7. Rankama, K. and Sahama, T.G. (1950) *Geochemistry.* The University of Chicago Press.
8. Cozzarelli, I.M., Eganhouse, R.P. and Baedecker, M.J. (1990) Transformation of monoaromatic hydrocarbons to organic acids in anoxic groundwater environment. *Environmental Geology and Water Science* **16**, 135-141.
9. Bennett, P.C., Siegel, D.E., Baedecker, M.J. and Hult M.F. (1993) Crude oil in a shallow sand and gravel aquifer - I. Hydrogeology and inorganic geochemistry. *Applied Geochemistry* **8**, 529-549.
10. Eganhouse, R.P., Baedecker, M.J., Cozzarelli, I.M., Aiken, G.R., Thorn, K.A. and Dorsey, T.F. (1993) Crude oil in a shallow sand and gravel aquifer - II. Organic geochemistry. *Applied Geochemistry* **8**, 551-567.
11. Baedecker, M.J., Cozzarelli, I.M., Eganhouse, R.P., Siegel, D.I. and Bennett, P.C. (1993) Crude oil in a shallow sand and gravel aquifer - III. Biogeochemical reactions and mass balance modeling in anoxic groundwater. *Applied Geochemistry* **8**, 569-586.

MONITORING A DIESEL SPILL AT SKODDEBERGVATN MILITARY SITE IN NORTHERN NORWAY

A. MISUND
Norges geologiske undersøkelse
Postboks 3006 Lade, N7002 Trondheim, Norway

1. Introduction

On behalf of the Norwegian Defence Construction Establishment, the Geological Survey of Norway (NGU) has conducted a two year (1995-96) monitoring survey of a diesel spillage at the Lake Skoddebergvatn military site in Northern Norway, described in [1] and [2]. The construction of the site started in 1991, and the first sign of diesel pollution was reported in February 1992.

1.1. AREAL SETTING

The investigated area is located on the south-western side of Lake Skoddebergvatn in Troms County of Northern Norway. The area that surrounds the site is characterised by steep topography with an annual precipitation of approximately 1000 mm. The area has a thin Quaternary cover of till/moraine overlying crystalline bedrock. The thin and poorly permeable till cover causes most of the precipitation to run off as surface water via a number of small streams towards Lake Skoddebergvatn. The flat, lower parts of the stream' catchment areas next to Lake Skoddebergvatn are dominated by peaty bog deposits, due to the poorly draining underlying till. The groundwater levels in the bog and till fluctuate throughout the year at around 0.1 - 0.5 m below ground level, generally reflecting variations in stream levels.

Lake Skoddebergvatn is not used for public drinking water purposes, but a small number of cottages may use it for water supply.

1.2. PROBLEM DESCRIPTION

On the 26th October 1992 a thin film of oil and the smell of hydrocarbons was observed in the streams entering Lake Skoddebergvatn. It is believed that this oil was released during construction work being carried out for the Defence Department within the stream catchments. At low lake water levels, hydrocarbons could also be observed in the exposed bottom sediments in Skoddebergvatn.

F. Fonnum et al. (eds.),
Environmental Contamination and Remediation Practices at Former and Present Military Bases, 197–207.

Today, construction work is completed and there are no activities that can cause pollution in the area between Lake Skoddebergvatn and the former construction area. It thus seems most likely that the detected pollution has its origin in one or more spills from the former construction area. There is some indication that there may have been a large spill in the area for refilling of diesel in generators / compressors used for the construction work. This is likely to have happened at a time of extensive snow cover during the winter 1992, such that the oil was transported towards Skoddebergvatn during the spring snow-melt. Most of the spill was transported in the streams, and can be observed in the peat and sediments in the stream banks today. As diesel is lighter than water, most of the contamination is found in the uppermost part of the bank side sediments. As the streams overtop their banks during times of heavy rainfall, contamination may be expected in the uppermost peat/till sediments of the flat, lower part of the stream catchments, at some distance from the streams.

1.3. AIM OF THE STUDY

The aim of the study was to determine the concentrations of hydrocarbons remaining in the peat and till streamside deposits and to evaluate the water quality of the main streams entering Lake Skoddebergvatn from the former construction area.

As part of an ongoing monitoring programme, the Geological Survey of Norway (NGU) re-sampled some locations after one year (i.e. in 1996) to (1) examine if there is a degradation of diesel in the sediments by studying the nnC-17/pristane relationship, (2) determine the degree of pollution in the bottom sediments in Skoddebergvatn and (3) determine if there is any significant input of 'diesel-like' hydrocarbons via the streams to Lake Skoddebergvatn.

2. Methodology

In the autumn of 1995 a total of 28 soil samples and 14 water samples were collected. There had been heavy rainfall in the period prior to the sampling in 1995, which caused a high water level in Lake Skoddebergvatn and flooding in the streams. Most of the soil samples were collected 20 - 30 cm into the stream bank. In addition, 3 samples were collected in the bog area, approximately 10 m from the main stream channel. The samples from the bog area had a higher content of organic material (peat). The soil samples were collected in sealed 0.5 L glass Kilner jars.

As part of a monitoring programme 13 soil samples and 10 water samples were collected in the summer and autumn of 1996. Five of the 1996 soil samples were collected from the bottom sediments of Lake Skoddebergvatn.

The stream water samples were all collected prior to soil sampling in order to avoid problems with mobilisation of diesel due to disturbance of the contaminated soil. The water samples were collected in 1 L brown glass bottles.

3. Results

3.1. SOIL CONTAMINATION

Of a total of 41 soil samples collected in 1995 and 1996, it was possible to detect hydrocarbon in 36. As shown in Table 1, the total hydrocarbon content (THC) in the samples from 1995 varied between 17 and 16400 mg/kg (related to dry mineral soil), and the distribution of contaminants along stream 1 and 2 are shown in Figure 1.

TABLE 1. Results of oil analyses of soil samples in the study area at Skoddebergvatn, 1995.

Locality No.	Date	Dry content % of "wet" original sample	THC (mg/kg dry material)	nC-17/pristane ratio
Standard diesel				1.92
2	07.09.95	37	1000	0.1
1	12.10.95	46.7	11.4	-
2	12.10.95	48.5	750	0.08
4	12.10.95	37.9	1940	0.1
5	12.10.95	20.2	5470	0.11
6	12.10.95	54.1	22.1	-
7	12.10.95	70.9	38.7	0.14
8	12.10.95	50	60.5	0.13
9	12.10.95	48.7	7660	0.29
10	12.10.95	49.6	1670	0.2
11	12.10.95	49.1	11000	0.63
12	12.10.95	36.3	5120	0.06
13	12.10.95	24.4	69.1	-
14	12.10.95	22	4800	0.19
15	12.10.95	32.9	67.5	-
16	12.10.95	53.8	18.8	0.19
17	12.10.95	47.8	5020	0.19
18	12.10.95	29.1	16400	0.23
19	12.10.95	58.8	17.3	0.26
20	12.10.95	60	146	0.07
21	12.10.95	52.9	1070	0.4
22	12.10.95	71.1	5.14	-
23	12.10.95	56.7	9.35	-
24	22.10.95	72.6	447	0.24
25-1	22.10.95	69.8	9600	0.94
25-2	22.10.95	69	9090	0.85
25-3	22.10.95	73.6	5620	0.78
25-4	22.10.95	74.8	4350	0.77

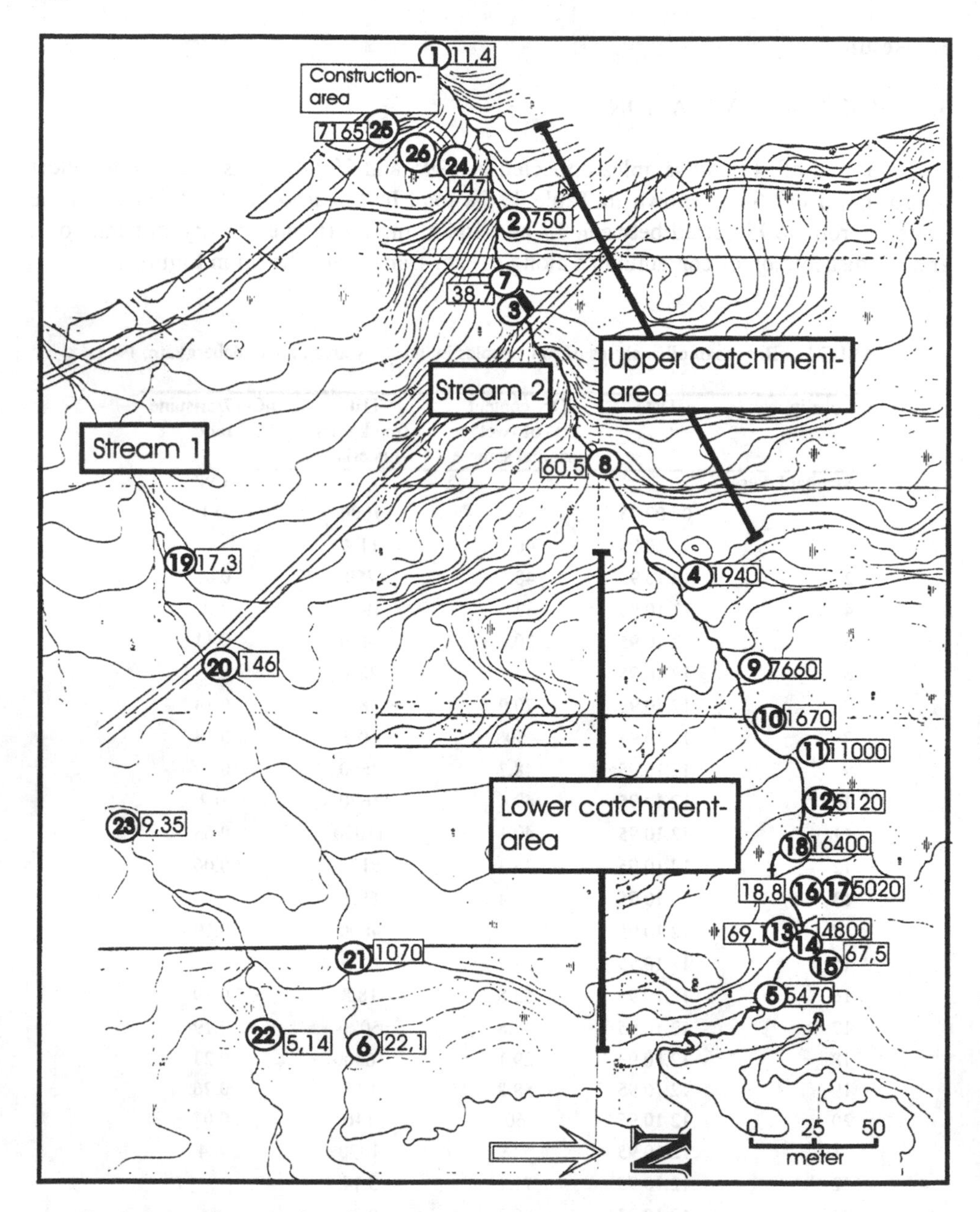

Figure 1: Sampling point for soil and groundwater 1995. Concentrations of THC in mg/kg dry soil.

There were no signs of contamination by hydrocarbons in the bottom sediments in Skoddebergvatn. The appearance of the GC-traces as described in [3] can be used to determine type of hydrocarbons in the sample. The GC traces in this study indicated that, in only 19 cases, the hydrocarbons appeared to be degraded from of diesel/heating-oil (but these represent the most contaminated samples). Examples of gas chromatograms are shown in Figure 2. In the other samples, the GC trace was characteristic of other types of mineral oil and non-mineral hydrocarbons. In Table 2, the results from 8 selected sites are compared with the results from 1996.

Table 2 Results of oil analyses of soil samples from the study area at Lake Skoddebergvatn, Streams 1 and 2. *Sample-1* is taken on 12.10.1995, *Sample-2* one year later on 22.10.1996.

Locality	Dry content-1 %	Dry content-2 %	THC-1 (mg/kg dry content)	THC-2 (mg/kg dry content)	Change in THC %	nC17/pristane-1	nC17/pristane-2
Standard diesel						1.92	1.92
2	37	38.3	750	1360	+81	0.08	0.11
5	20.2	38.1	5470	3670	-33	0.11	0.27
9	48.7	51.3	7660	1390	-82	0.29	0.12
11	49.1	54.2	11000	7010	-36	0.63	0.34
16	53.8	57.1	18.8	10.7	-43	0.19	0.65
17	47.8	44.6	5020	3910	-22	0.19	0.2
18	29.1	19.9	16400	18900	+15	0.23	0.25
21	52.9	47.4	1070	173	-84	0.4	0.31
Mean %					-25		
Dutch C-value [4]			5000	5000			

In the former construction area, an excavator was used to examine if there had been any major spill of hydrocarbons. The work revealed a substantial pollution of hydrocarbons at sampling site 25 (Figure 1). This was not a great surprise as this area had been used for refilling of diesel and other oil products into vehicles and/or other equipment.

The investigation also indicated that the contaminant source was concentrated in the area adjacent to stream 2. The amount of hydrocarbons in the sediments probably reflects the contact time between the sediments and the diesel as the pollution was carried along the stream towards Lake Skoddebergvatn. The stream catchment can be divided into two regimes: (upper) a steep part from the former construction area to sampling point 4, and (lower) a flatter area dominated by bog. In order to estimate the amount of diesel remaining in the sediments, we have used a 'rough' calculation assuming the main volume of diesel pollution to be 0.5 m wide and 0.5 m deep on each side of the stream. From investigations at Trandum military landfill [5] and many other publications, it is known that pollution in many cases will spread out as a plume from a

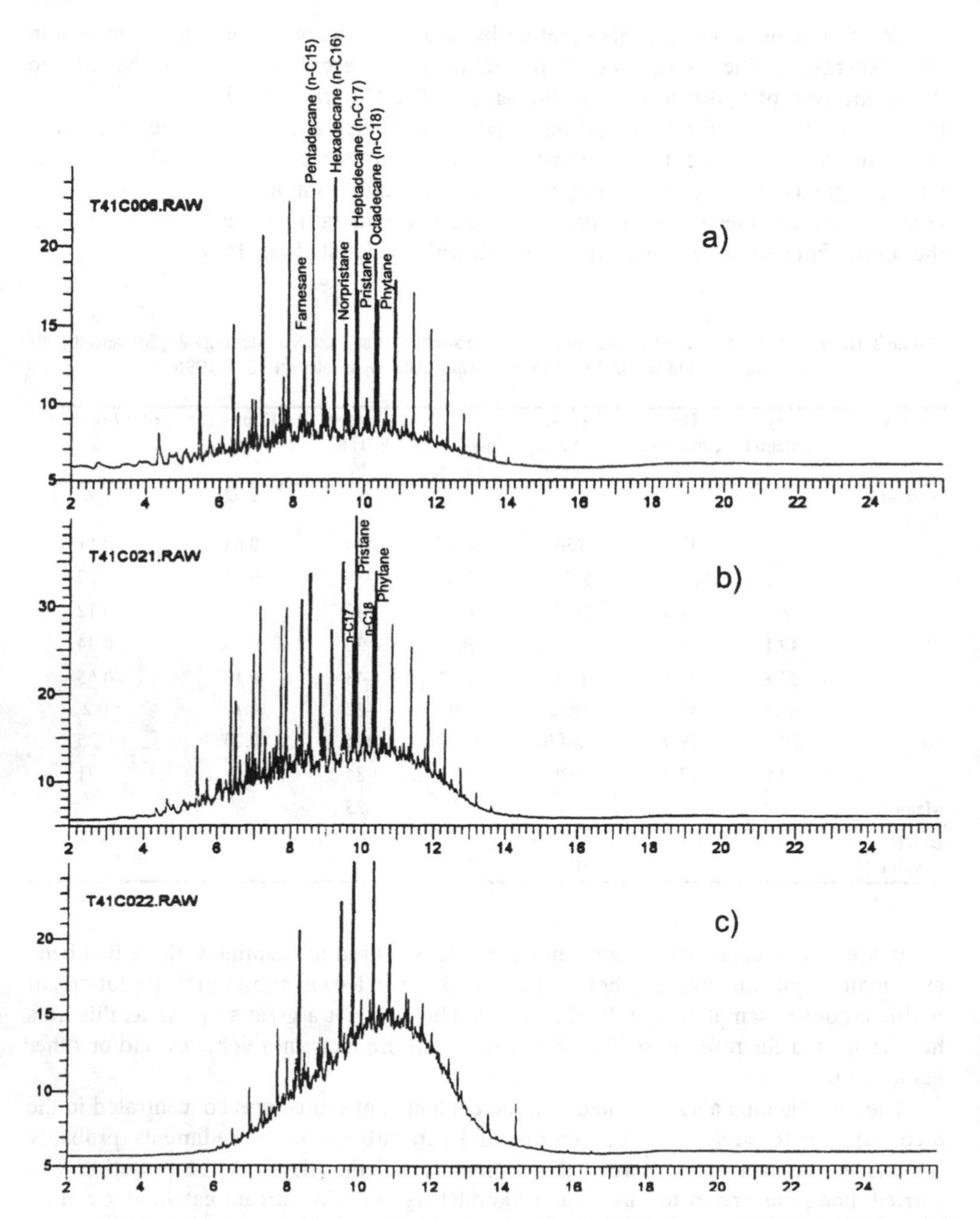

Figure 2: GC-chromatograms showing a) standard marine diesel nC-17/pristane = 1.92, b) site 11 degraded diesel nC-17/pristane = 0.63, c) site 12 strongly degraded diesel nC-17/pristane = 0.06.

point source. The pollution will also gradually be degraded as a function of the distance from the source, due to greater contact time with water, air and nutrients. In the source area, the degradation process is slower, due to the pollution creating toxic conditions for bacteria and poorer access for air. Our 'rough' estimate of hydrocarbon content in the catchment of stream 2 uses the values shown in Figure 1 for each segment of the stream, and results in a total of c. 660 kg diesel in the sediments in 1995. As shown in Table 2 there is on average a 25 % reduction of THC in the sediments from 1995 to 1996, implying an estimated 1996 total of c. 500 kg diesel in the sediments.

A rather high concentration of hydrocarbon was measured in soils at sampling point 21 in the sediments of stream 1(1070 mg THC/kg dry soil), but other samples from stream 1 do not confirm this high value. We thus regard the major pollution in the area to be associated with stream 2's catchment.

TABLE 3. Results of oil analyses of stream-water samples from the investigated area near Skoddebergvatn.

Location	Date	THC (mg/L)	nC-17/pristane
Standard diesel			1.92
23	07.09.95	0.350	
1	12.10.95	< 0.1	
2	12.10.95	< 0.1	
3	12.10.95	< 0.1	
4	12.10.95	< 0.1	
5	12.10.95	< 0.1	
6	12.10.95	< 0.1	
3	21.11.95	< 0.1	
3	21.11.95	< 0.1	
25	22.11.95	5.14	0.50
3	22.11.95	< 0.2	
3	12.12.95	0.21	
3	12.12.95	0.23	
1	12.06.96	0.025	
3	12.06.96	< 0.005	
5	12.06.96	< 0.005	
6	12.06.96	< 0.005	
22	12.06.96	< 0.005	
1	22.10.96	-	
3	22.10.96	-	
5	22.10.96	-	
6	22.10.96	-	
22	22.10.96	-	

< : trace of small amounts of hydrocarbons, but below detection (i.e. quantification) limit
- : not detected

3.2. CONTAMINATION OF WATER

The stream water samples can be divided into two groups: (1) those contaminated by free oil phase due to disturbance of soil in the stream banks and (2) those sampled prior to sampling (and disturbance) of soil. For the first group, THC-values as high as 15 mg/litre have been measured (Table 3), in contrast to the second group of samples which were all below the analytical detection limit of 0.1 mg/L, except for two samples taken in December 1995. Some analytical problems exist, since the Norwegian drinking water standard is as low as 0.010 mg/L for THC in water [6].

During the sampling in October 1995, stream-flow was very high and one might have expected that this would lead to a leaching of hydrocarbons from the sediments, which was, in fact, not the case.

3.3. AGE AND DEGRADATION OF DIESEL

To evaluate the age and degree of degradation of diesel, the nC-17/pristane ration was determined from the GC-chromatogram. In fresh marine diesel this ratio is typically 1.92 but, due to pristane's higher degree of resistance to microbiological degradation, the relative concentration of pristane will increase with time with respect to nC-17 (Figure 2) . Pristane is a branched-chained alkane (isoprenoid).

Studies in Denmark and the Netherlands [7] of a number of well-documented diesel spills have demonstrated a clear relationship between the nC-17/pristane ratios and the number of years since the spill took place. The oldest spills were more than 20 years old, with a nC-17/pristane ratio close to 0. Important preconditions for this published study were the selection of sites where the polluted soil had an impermeable cover (e.g. asphalt), and the soil sample was collected above the water table.

In the soil samples collected at Skoddebergvatn, the nC-17/pristane ratio varies between 0.06 and 0.94. As shown in Tables 1 and 2, most of the sampling sites have a nC-17/pristane ratio below 0.3. According to Christensen & Larsen [7], this indicates an age of spill of 16 years, although we know that the spill must have happened during the construction period commencing in Winter 1991. This discrepancy can be explained by considering the many factors that contribute to the degradation of diesel: including exposure to air and sunlight, oxygen content, permeability of soil, bacterial content in soil, bacterial availability of nutrients and water, and the content of organic matter in the soil. Christensen & Larsen [7] used only soil samples with poor exposure to sunlight, water and nutrients (i.e. below cover and above the water table). This may have resulted in a slower rate of degradation than at Skoddebergvatn, resulting in generally higher nC-17/pristane ratios. At Skoddebergvatn, the situation is quite different. The pollution mainly occurs very close to the surface, and is exposed to abundant quantities of water, air and nutrients from surrounding vegetation, yielding conditions ideal for rapid degradation of the diesel. The importance of water and nutrients for degradation of hydrocarbons is documented in many research projects, e.g. during the remediation of a site contaminated with heating oil [8].

Figure 3 shows the correlation between the nC-17/pristane ratio and THC content for the samples collected in 1995. The overall correlation coefficient (r) for the data set is 0.5 i.e. poor correlation. Figure 3 can also be interpreted as two separate data groups, represented by the upper steep and lower flat parts of stream 2 (Figure 1), with correlation coefficients of 0.93 and 0.75 respectively. The difference might be explained in terms of the slower stream flow in the lower catchment of stream 2 giving possibilities for infiltration of more diesel into stream banks, resulting in low nC-17/pristane ratios despite high concentrations of THC in the sediments.

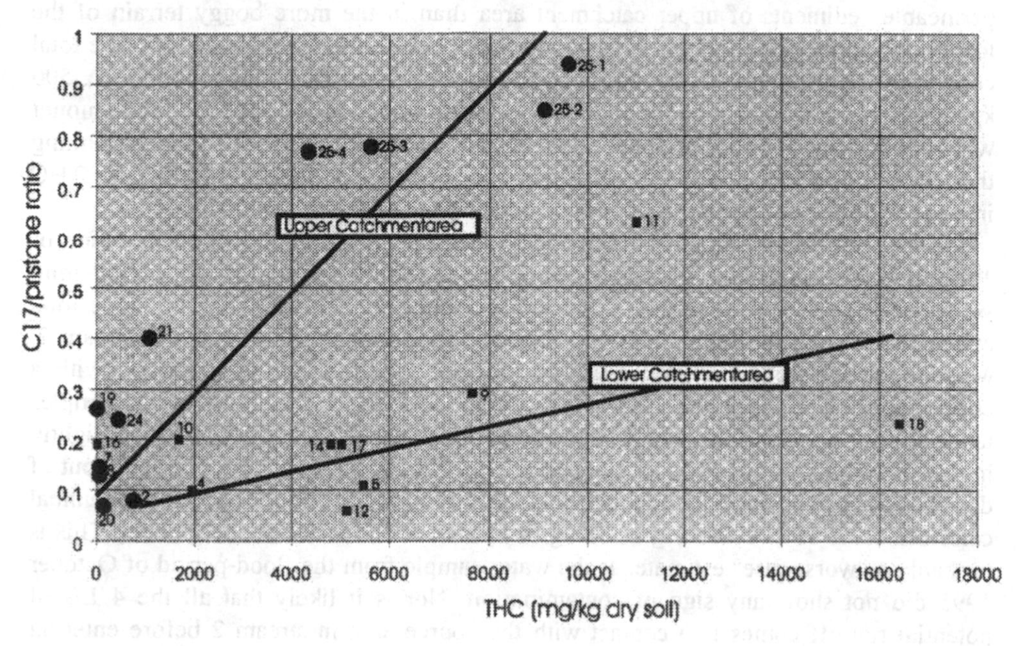

Figure 3: Relationship between the nC-17/pristane ratio and content of THC in soil (mg/kg dry soil) in samples from Skoddebergvatnet, Troms county. The steep line represents samples from upper catchment area (●), the flat line represents samples from lower catchmentarea (■). Correlation coefficient for all data is 0.5, but are for the upper- and lower catchment area, 0.93 and 0.75 respectively.

4. Risk and Consequences of the Diesel Spill

The investigation has shown that the soils along stream 2 are heavily contaminated with diesel. For the sampling points 5, 9, 11, 12, 17, 18 and 25, the measured values for THC are all above the Dutch C-level of 5000 mg THC/kg dry soil [4], which should indicate the necessity of remedial action. The investigation has also shown that many of the lighter diesel components have already been removed by degradation/volatilisation/leaching, based on the evidence of nC-17/pristane ratios, leaving the less soluble components in elevated concentrations in the soil. The greatest risk of leaching of hydrocarbons from soil to the stream and further to the Lake probably occurs at times of heavy rainfall (Autumn) and during snow-melt (Spring). Nevertheless, the water samples collected during these periods reveal hardly any sign of diesel contamination, supports the conclusion that most of the soluble diesel components have already been leached out.

For most of the sites, except sites 25 and 11, the diesel pollution appears to be highly degraded, and it might be expected that the remainder will gradually degrade due to natural processes. The bacterial activity at stream 2 will probably be higher in the permeable sediments of upper catchment area than in the more boggy terrain of the lower catchment area (Figure 1). There appeared to be a reduction of 25 % in the total content of hydrocarbons in the sediments from 1995 (c. 660 kg THC) to 1996 (c. 500 kg THC). If we assume a continuing 25 % annual reduction in THC, the total amount will be less than 10 kg (on average c.10 mg THC/kg dry soil) after 15 years, assuming that all the components of the diesel is degradeble. The Norwegian guideline for THC in areas with the most sensitive area use is set to 100 mg /kg dry soil [9].

The catchment area of stream 2 is approx. 0.2 km^2, with an annual precipitation of 1000 mm, and an effective rainfall (rainfall less evapotranspiration) of c. 700 mm, which may contribute to surface water and groundwater run-off. This implies a total annual runoff of 150 000 m3, corresponding to an average flow of 4 L/s in stream 2. Most of the precipitation takes place in the period September to February, with a correspondingly higher stream flow. The content of diesel in the stream water samples taken from the 'source-area' on 12th December 1995 was 0.23 mg THC/L, only slightly higher than the Dutch B-level (0.2 mg THC/L). If one assume that the average input of diesel to the stream catchment equals this concentration, this implies a potential annual contribution of 35 000 000 mg or 35 kg of hydrocarbons to Skoddebergvatnet. This is probably a "worst-case" estimate, as the water sample from the flood-period of October 1995 did not show any sign of contamination. Nor is it likely that all the 4 L/s of potential run-off comes into contact with the source area in stream 2 before entering Lake Skoddebergvatn.

Lake Skoddebergvatnet covers an area of 8 km^2 and, with an average depth of 5 m, has a volume of some 40 000 000 m^3. In the unlikely contingency that all the 500 kg of THC estimated to exist in the catchment of stream 2 was to be diluted into this body of water, the resulting concentration of THC would just reach the Norwegian drinking water standard of 0.01 mg/L THC [6].

5. Conclusions

The results of this study show that the detected pollution will gradually degrade due to natural processes. The ratio nC-17/pristane indicates that the most volatile and soluble components of the diesel have already decayed, vaporized or been leached out of the soil. The re-sampling of 8 soil samples in 1996 showed the same contaminant distribution as 1995, but there was on average a 25 % reduction in hydrocarbon content. It was also observed a further reduction in the nC-17/pristane ratio from 1995 to 1996. It should be borne in mind that there may be methodological problems related to sampling of the exactly same sample sites (and this can be demonstrated by the difference in dry matter content, which is in itself related to natural organic matter / peat content, in Table 2). No significant concentrations of hydrocarbons were detected in stream water and it appears that any attempt at remedial action involving physical disturbance of soils will probably do more harm than good, mobilizing the diesel. A re-sampling of the sites in 1998 is recommended as part of the monitoring programme.

6. References

1. Misund, A. og Lauritsen, T. 1995: Miljøtekniske undersøkelser ved Skoddebergvatnet, Skånland kommune, Troms. *NGU rapport. 95.146.*
2. Misund, A. 1996: Overvåkning av forurensning ved Skoddebergvatnet, Skånland kommune, Troms. *NGU rapport. 97.004.*
3. Zemo, D. A., Bruya, J. E. og Graf, T. E. 1995: The Application of Petroleum Hydrocarbon Fingerprint Characterization in Site Investigation and Remidiation. *Ground Water Monitoring Review Spring 1995, pp 147-156.*
4. Moen, J.E.T., Cornet, J.P. og Evers, C.W.A. 1986: Soil protection and remedial action: criteria and decision making and standarization of requirements. *In: Contaminated soil, ed. Assink J.W. og Vanderbrink W.J., Martinus Hijhoff Publishers, Dordrecht.*
5. Sæther, O. M., Misund, A., Ødegård, M., Andreassen, B. T. & Voss, A. 1992: Groundwater contamination at Trandum landfill, Southeastern Norway. *Norges geologiske undersøkelse Bull. 422*, s.83-95.
6. Sosial- helsedepartementet, 1995: Forskrift om vannforsyning og drikkevann m.v.. *Forskrift gitt 1. Januar 1995, Sosial- helsedepartementet, Oslo.*
7. Christensen, L. B. og Larsen, T. H. 1993: Method for Determing the Age of Diesel Oil Spills in the Soil. *Ground Water Monitoring Review Fall 1993, pp 142-149.*
8. Breedveld, G. D., Kolstad, P. Hauge, A., Briseid, T. og Brønstad, B. 1992: In situ bioremidiation of oil pollution in the unsaturated zone. 11. Nordiske Geoteknikermøte. Ålborg, Vol 1-3 mai 1992. Dansk geoteknisk forening.
9. Nordal, O., Andersen, S., Weholt, Ø. & Huse, A. 1995: Håndtering av grunnforurensningssaker. Foreløpig saksbehandlerveiledning. *SFT Rapport 95:09*

IMPACT OF MILITARY CONTAMINATION ON QUALITY OF POTABLE WATER RESOURCES

V. JUODKAZIS
Department of Hydrogeology and Engineering Geology
Vilnius University, Èiurlionio 21, LT-2009 Vilnius, Lithuania,
B. PAUKŠTYS
Hydrogeological Company "Grota"
Eiðiðkiø plentas 26, LT-2038 Vilnius, Lithuania

1. Introduction

Political changes following the collapse of the Soviet Union have caused large economical and environmental alterations in Europe. Among the first manifestations of these changes was the evacuation of millions of troops and military equipment from thousands of military sites. The environmental problems created by the military presence, however, remained in the host countries. Military activities are considered to be multi-faceted point source of contamination not only in ex-Warsaw pact countries but also in most NATO countries.

In the Baltic region countries, groundwater aquifers are important if not the sole sources of potable water. The density of military sites in these territories was rather high, and military installations were often built in the sanitary protection zones of the well fields. Water in its capacity, as an excellent solvent is able to transport dissolved contaminants from the soils of military sites into aquifers, which subsequently cannot be used for drinking purposes. Although water abstraction wells are usually closed before a population is affected, a loss of potable water resources may lead to large expenditures when alternative water extraction areas need to be explored, and new wells drilled and prepared for drinking water consumption.

The Soviet army operated with minimal concern for the environment. Military bases distributed throughout the occupied Baltic States, and the majority of East European countries were similar from the point of view of ecological culture. However, their number varied according to country, and natural conditions at the bases were also different. Therefore the effect of pollutants migrating from military bases to the environment and groundwater quality was uneven. The necessity to determine the scale of the environmental threat stimulated specialists to search for methodological solutions for evaluation of environmental damage, especially in relation to the protection of potable groundwater resources.

In this respect Lithuania, is a typical region, situated as it is along with Latvia, Estonia, the Kaliningrad enclave and a major part of Poland and Germany in a zone of

F. Fonnum et al. (eds.),
***Environmental Contamination and Remediation Practices at Former and Present Military Bases*, 209–226.**

continental glaciation with Quaternary glacial and aquaglacial deposits lying at the surface (Kiriuchin, 1978). From a structural-hydrogeological point, these states belong to the artesian systems known as the Baltic and Polish-German artesian basins (figure 1) with horizontally stratified porous (water-bearing) and compacted (low permeable or impermeable) soils (Grigelis and Juodkazis, 1980). Only groundwater is utilised as drinking water in Lithuania, and the situation is similar in other states within the above mentioned region (Grigelis, et al, 1985). Therefore in our opinion, experience gained in Lithuania, could be important in a wider theoretical methodological sense.

Evaluation of the impact of military bases on groundwater quality is important since the results obtained should facilitate an answer to the following questions: should contaminated soil and water in the military bases be cleaned or not? Methodologically such evaluation is rather conditional (at least in post-communist countries) due to small number investigations having been carried out on polluted areas. Nevertheless, methodological principles and results are very important since they can form a basis for planning further studies and remediation actions.

2. A strategy for analyses of military impact on groundwater resources

In order to evaluate the impact of military pollution on water resources, a specific methodology (or in military terms, a strategy) is necessary. However, it is not so easy to formulate such methodological principles since the areas of the military bases have not been sufficiently studied. Moreover, in former military bases of Lithuania even monitoring is not organised on contaminated sites. Thus, methodological principles are based on general assumptions of pollutant migration in natural environmental conditions.

Using the available information and accumulated experience it is nevertheless possible to discuss a general strategy for analysis of regional military impact on groundwater resources that is based on an evaluation of the interaction of human-induced and natural factors and specific methodology for the assessment of local environmental damage at a particular site.

The general approach is based on an evaluation of the following factors:

Anthropogenic load:

- Number of military bases on the territory of the country;
- Density of military bases in the area unit;
- Ratio of the military areas to the total area of the country;
- Classification of military bases from environmental and groundwater pollution viewpoints.

Natural hydrogeological conditions:

- Lithology of surficial soils;
- Geomorphology and landscape;
- Intensity of vertical groundwater circulation under natural conditions.

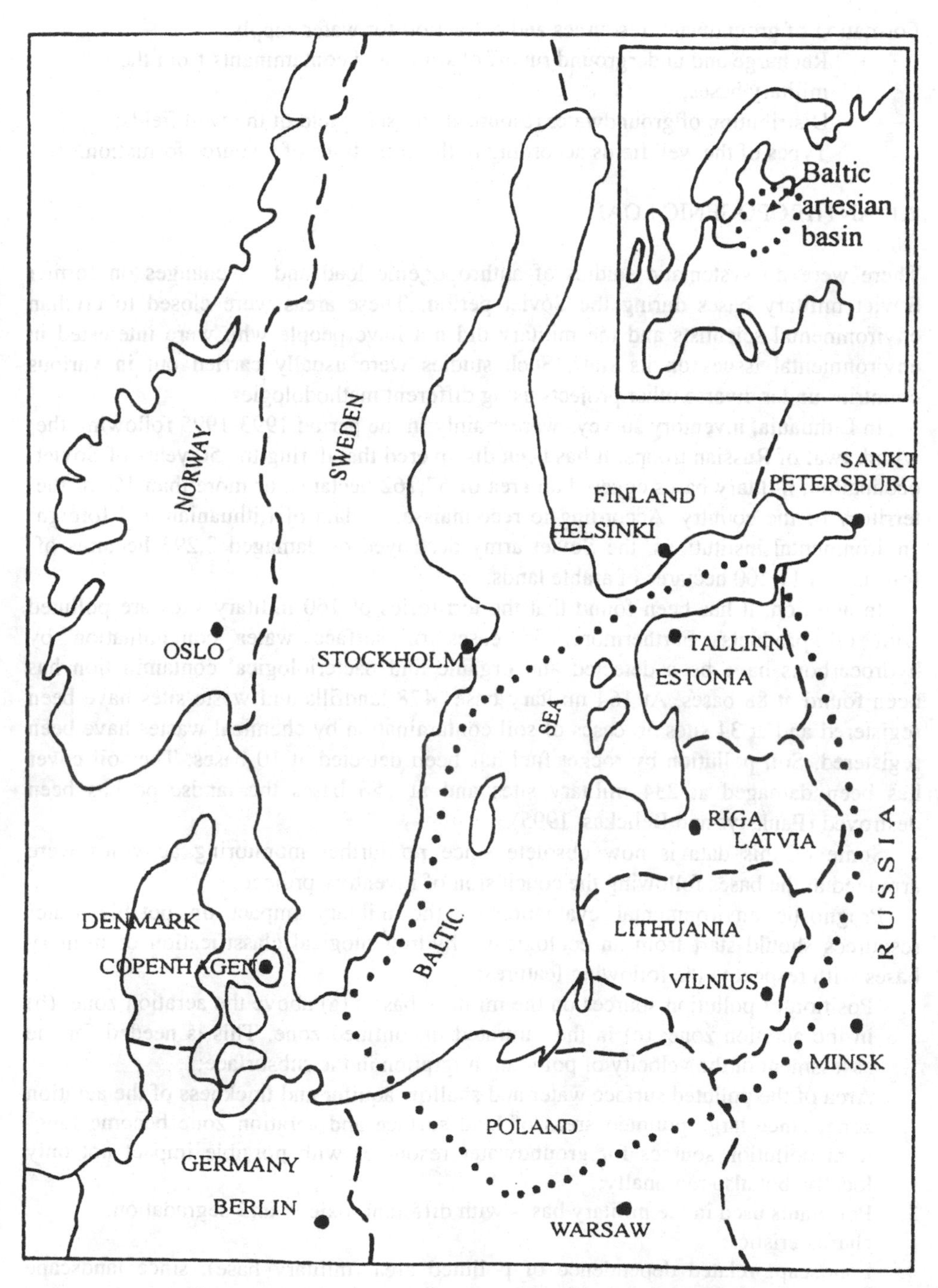

Figure 1. Situation of the Baltic States and the Baltic Artesian Basin

Formation of groundwater resources and extraction for water supply:

Recharge and underground runoff of water and contaminants from the military bases;

Distribution of groundwater resources and safe yield of the well fields;

Types of the well fields according to the conditions of resource formation.

2.1. ANTHROPOGENIC LOAD

There were no systematic studies of anthropogenic load and its changes on former Soviet military bases during the Soviet period. These areas were closed to civilian environmental scientists and the military did not have people who were interested in environmental issues on its staff. Such studies were usually carried out in various countries under diverse other projects using different methodologies.

In Lithuania, inventory surveys were mainly in the period 1993-1995 following the withdrawal of Russian troops. It has been discovered that during the 50 years of Soviet occupation, military bases covered an area of 67,762 hectares, or more than 1% of the territory of the country. According to reconnaissance data of Lithuanian and foreign environmental institutions, the Soviet army destroyed or damaged 3,293 hectares of forests and 11,400 hectares of arable lands.

In addition, it has been found that the territories of 160 military sites are polluted with oil products. Furthermore, 35 cases of surface water contamination by hydrocarbons have been detected and organic and bacteriological contamination has been found at 88 bases. At 164 military bases 478 landfills and waste sites have been registered and at 34 sites 56 cases of soil contamination by chemical wastes have been registered. Soil pollution by rocket fuel has been detected at 10 bases. The soil cover has been damaged at 234 military sites and at 156 bases the landscape has been destroyed (Paukstys and Belickas, 1995).

Some of this data is now obsolete since no further monitoring activities were arranged at the bases following the conclusion of inventory project.

Prognostic environmental evaluation of the military impact on potable water resources should start from an ecological - hydrogeological classification of military bases with respect to the following features:

1. Position of pollution sources on the military bases: (a) above the aeration zone; (b) in the aeration zone; (c) in the saturated unconfined zone. This is needed for the assessment of the velocity of pollutant migration in the subsurface;
2. Area of the polluted surface water and shallow aquifer and thickness of the aeration zone, since large polluted areas of land surface and aeration zone become long-term pollution sources for groundwater resources with possible impact not only locally, but also regionally;
3. Pollutants used in the military bases with different toxicity and degradation characteristics;
4. Landscape-related dependence of polluted area (military base), since landscape affects the dynamics of shallow groundwater, and its horizontal and vertical filtration;

5. Present economic and military objectives in the area determining the situation whether it hampers economic (military) development and whether economic units are able to fund the cleaning. In this regard, the Valèiûnai military base near Vilnius can be shown as an example where the Company "Lietuvos Kuras" (Lithuanian Fuel) funds works related to remediation of the polluted site.

2.2. NATURAL HYDROGEOLOGICAL CONDITIONS

The intensity of horizontal and vertical migration of pollutants under natural conditions in the subsurface is caused by two major factors: * the physical-chemical features of pollutants and their amount in the pollution plume and * the distribution of lithological and hydrodynamical barriers, limiting pollutant migration in the hydrogeological system containing the pollution source.

1. The first structural element of the artesian basins is the upper soil, where shallow groundwater is forms the unconfined aquifer. In Lithuania and other Baltic states, the surface layer consists of two-type soils (figure 2): * sand - mostly fluvioglacial, deltaic and alluvial and clay - mostly till loam and limnoglacial clay. Clayey deposits with hydraulic conductivity not exceeding n x 10^{-2} m per day is a good barrier that hinders both vertical and horizontal migration. The recharge (underground runoff index) in the areas covered with clayey deposits make up usually about 0.5-1.0 l/s from 1 km^2, whereas in sand-prevailing areas it is significantly higher - up to 3-5 l/s from 1 km^2 (Grigelis et al, 1985). It should be noted that the majority of former Societ military bases were mainly built in the areas underlain by sandy deposits, thus pasing a higher threat to environment and groundwater.
2. The second barrier is a hydrodynamic one; it determines the distribution of recharge and discharge areas. It is well known that there is a downward vertical infiltration (recharge) in hilly regions within the active groundwater exchange zone and an upward discharge in lowlands and river valleys (Juodkazis, 1979). Therefore, in the latter case (under natural conditions) the confined aquifers are protected from surface pollution since their piezometric level is higher than the free level of shallow groundwater. However, in recharge areas with vertical seepage via aquicludes, there is a threat that confined aquifers may be polluted. In this case, the intensity of vertical filtration plays the main role. In Lithuania almost half of the military bases are situated in the regional areas of groundwater recharge (hilly regions) (figure 3).

The third barrier is represented by low-permeable aquicludes which significantly hamper vertical exchange. Their hydraulic conductivity does not exceed n x 10^{-3}-n x 10^{-5} m per day. For this reason, the intensity of vertical recharge is rather low, and at depths of 50-100 m, it does not exceed 0.1-0.2 l/s to km^2 (Juodkazis, 1979). However, there are sites in the Quaternary cover where the aquicludes are either very thin or they are absent. Such sites, so-called "hydrogeological windows" (figure 4), show the infiltration index as being significantly higher. Nevertheless, analysis of ambient hydrogeological conditions shows that under natural conditions unaffected by exploitation, the probability of confined aquifer pollution is not high on a regional scale. About 90 percent of pollution is related to unconfined shallow groundwater that

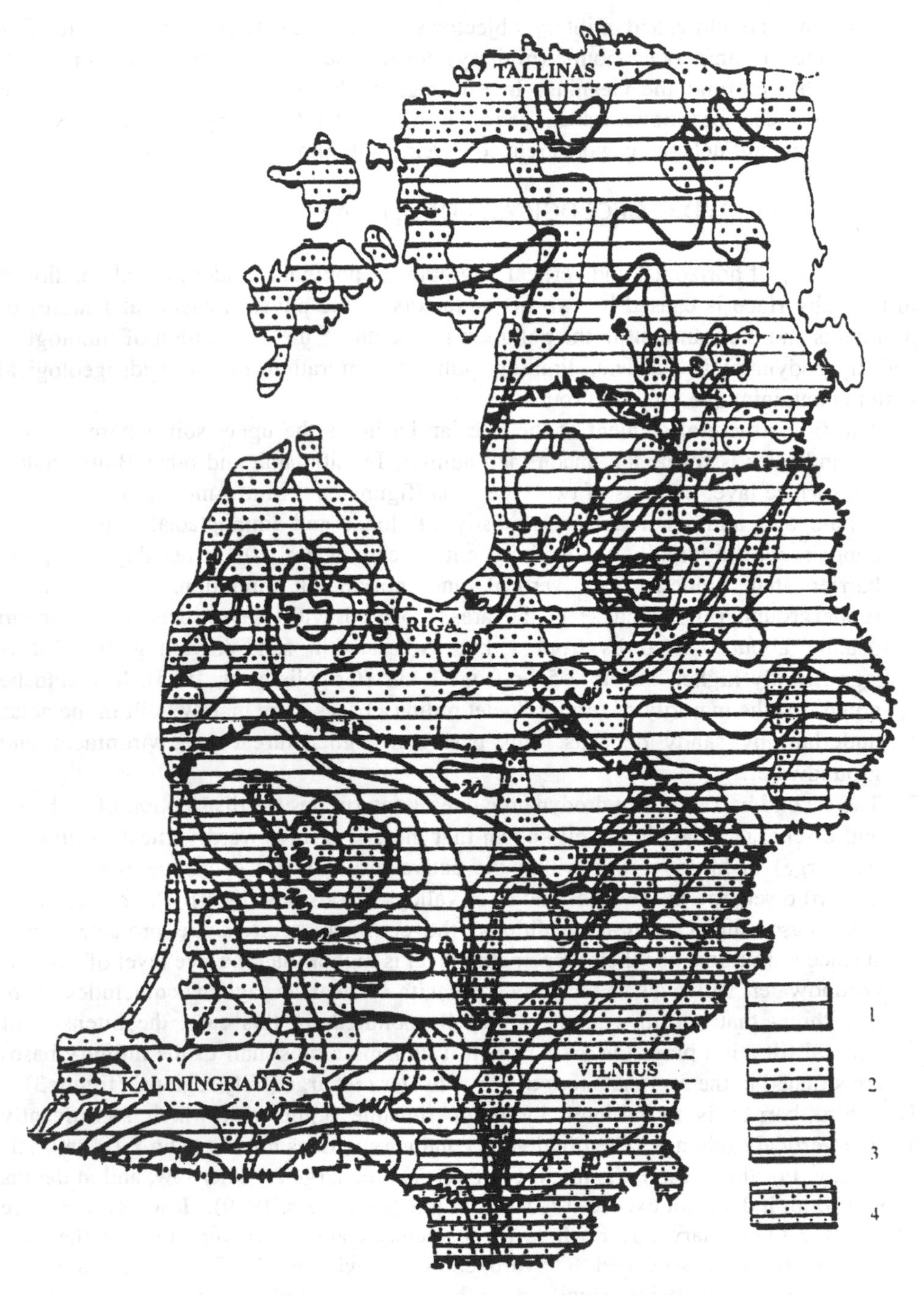

Figure 2. Types of the Quaternary cover in the Baltic States

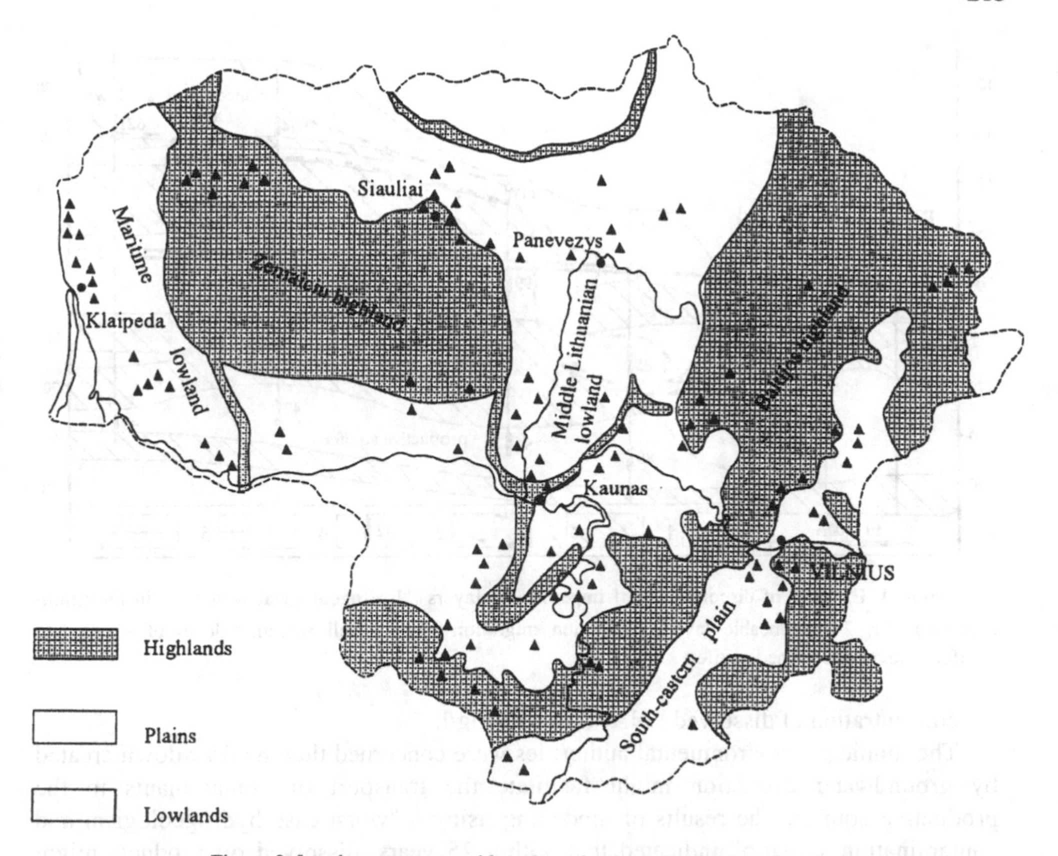

Figure 3. Landscape types and location of military bases in Lithuania

transports the accumulated pollutants (mainly oil products) from former military bases along with groundwater runoff to the rivers and other surface water bodies.

2.3. FORMATION OF GROUNDWATER RESOURCES AND WATER EXTRACTION

Groundwater streams formed under natural conditions do not possess high hydraulic gradients; and hence, rates of horizontal and vertical filtration are not high. However, once fresh groundwater begins to be extracted, the hydraulic gradient increases, filtration intensifies and, more importantly, levels of unconfined and confined aquifers are redistributed. For the same reason, the confined aquifers that were well-protected under natural conditions are now also affected (Grigelis, et al, 1985). Several case studies were made in Lithuania in order to assess the potential contamination of well fields by military activities. One such case was Zokniai military airport in north Lithuania, which is located at a distance of 2 km from the municipal well field. More that 1 million m^3 of shallow groundwater at the base is polluted by hydrocarbons. The thickness of free phase oil products above the groundwater table reaches 79 cm and the

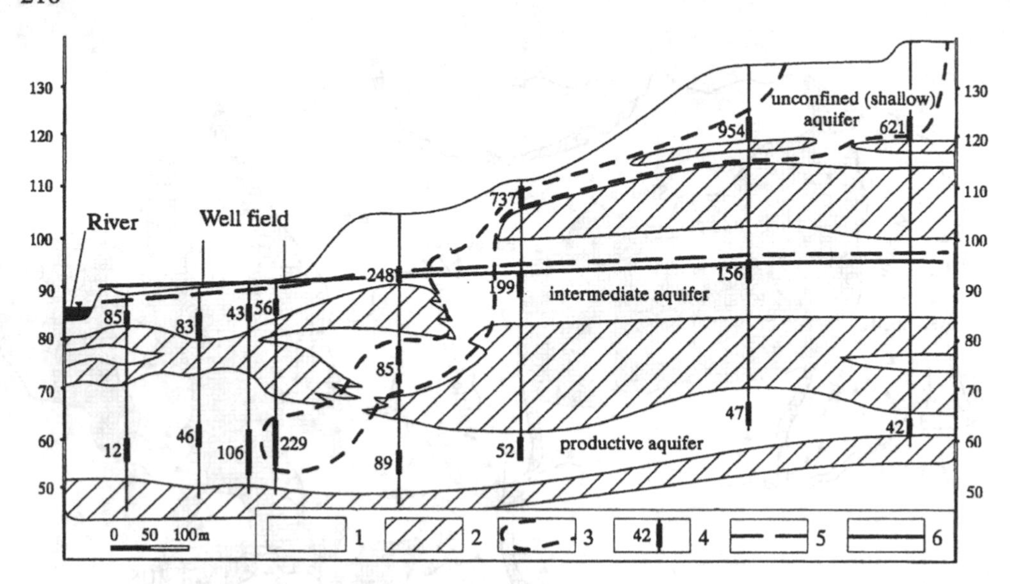

Figure 4. Example of discontinuity of impermeable layers ("hydrogeological window") in theVilnius area.1-aquifer, 2-impermeable layer, 3-pollution migration path, 4-well screen, 5-level of unconfined aquifer, 6-level of confined aquifer.

concentration of dissolved LNAPL is 10.5 mg/l.

The municipal environmental authorities were concerned that the drawdown created by groundwater extraction might facilitate the transport of contaminants to the productive aquifer. The results of modelling using a "worst-case hydrogeological and contamination scenario" indicated that within 25 years, dissolved oil products might reach a depth of 115 m and migrate 1650 m from the border of the air base towards the municipal well field (Paukštys and Gregorauskas, 1995). This suggests that the productive aquifer is safe for 25 years; however, well field operation will not be halted after the modelled period. What will happen with the contaminants after 25 years? The question has to be answered.

Groundwater extraction in most urbanised parts of Lithuania created regional, overlapping drawdowns. This is the case with Klaipëda and Ðiauliai where there is a possibility of transboundary overlapping drawdown development between Lithuania and Latvia. It is difficult to predict the fate and transport of contaminants in the areas where hydraulic relations between the aquifers are disturbed by human activity. In such areas even well-protected deep aquifers become vulnerable to pollution. Nevertheless, many military bases were constructed around major cities within the zone of influence of the well fields.

Evaluating military impact on potable water resources the regional and local aspects should be distinguished. Regional impact is being assessed using general information and local impact is based on more detailed hydrogeological data and results of mathematical modelling. Both regional and local aseessment of military contamination in Lithuania are discussed below.

3. Regional assessment of military impact

A large number of contaminated military bases may have a substantial regional impact on the environment. Investigations of soil and groundwater contamination at former military bases in Lithuania have been carried out by a number of national and international teams. However, due to the large number of military bases and limited financial resources, almost all previous contaminant studies were of a local character. Although the regional context is also of great importance, few references exist from such studies.

This and the following chapters are based on data from the report "Assessment Methodologies for Soil/Groundwater Contamination at Former Military Bases in Lithuania". The report is the final document of a cooperative effort between scientists from the Geological Surveys of Lithuania and Norway, the Canadian Department of National Defence and the Norwegian Defence Research Establishment. The co-operation project was supported by the NATO Linkage Grant.

For the regional environmental assessment of military impact, data from all previous studies carried out by private consultancies and government institutions were collated and summarised. Particularly useful information was obtained from the report by I. Kruger Consult in partnership with the Baltic Consulting Group (1995) entitled "Inventory of Damage and Cost Estimate of Remediation of Former Military Sites in Lithuania" (Kruger, 1995). Existing geological and hydrogeological information was also examined. The following procedure was adopted for the evaluation of regional contamination from military bases in Lithuania:

- all military bases were classified according to their perceived potential environmental impact;
- the natural geological-hydrogeological conditions existing at each of the bases were categorised according to their vulnerability and inherent potential for contamination;
- a groundwater-monitoring network was established at three selected military bases and water-level measurements and sampling were carried out at these bases for two years;
- contaminant transport modelling was carried out at a number of selected military bases;
- a groundwater run-off map for the region was compiled;
- a limited number of military bases assumed to be representative on a regional basis were identified;

- the contaminant discharge from each of these selected bases was estimated and used as a basis for extrapolation to estimate the total contaminant flux from all bases.

3.1. SHALLOW GROUNDWATER CONTAMINATION

The aquifer most vulnerable to impact by means of military contamination is shallow (unconfined) aquifer.

Shallow groundwater is found throughout much of Lithuania. It is utilised as drinking water by 1 million inhabitants extracting water from 350 000 dug wells, mostly in rural areas. Groundwater depth is closely related to landscape and drainage conditions. The groundwater level normally lies between 0.5-5.0 m below ground level (b.g.l.) while in upland areas it may reach 10 m below ground level. The vulnerability of shallow groundwater depends on both the lithology and thickness of the vadose (unsaturated) zone.

The amplitude of groundwater level fluctuation depends on the lithology of the water-bearing strata which is often the same as that of the vadose zone. The average annual water-level fluctuation in sandy soils is 0.2-0.7 m whilst in tills and clays the fluctuation amplitude reaches 1.5-2.0 m and more. Water level fluctuations create a "smear zone" in the soils contaminated by oil products. Large areas of Lithuania are composed of sandy soils. Shallow groundwater is therefore poorly protected from contamination on both a local and regional scale. A shallow aquifer acts as an initial barrier protecting deeper confined aquifers from pollution. Contaminants penetrating the vadose zone of a shallow aquifer can be transported by groundwater flow to surface streams. Figure 5 demonstrates that in more than 60 per cent of all cases, military bases are located on very vulnerable geological formations - sand, gravel or sandy loam.

3.2. CALCULATIONS OF CONTAMINANT FLUX

The military bases assessed have been categorised according to their groundwater run-off characteristics. Groundwater run-off has been quantified as an integrated index representing the rate of groundwater discharge into surface streams per unit area. This index reflects the surface geology of the area: the lithology and thickness of vadose zone, the lithology of water-bearing sediments and groundwater recharge conditions. Given the groundwater run-off and contaminant concentration, the flux of contaminants transported to surface streams by the groundwater flow can be estimated. This amount is used as the main index for the regional assessment of military pollution. Shallow groundwater resources and run-off in Lithuania have previously been evaluated in detail (Sakalauskiene 1983).

68 military bases are located on sandy soils where the groundwater run-off index is 1-5 $l/s.km^2$ (100-400 $m^3/day.km^2$). In this case the flux of contaminants discharged into surface streams depends on the contaminant concentration, the distance to the stream, and the rates of attenuation or degradation of contaminants in the aquifer. For the assessment of the regional impact of military pollution on the environment, the 160 bases on which oil contamination was detected were considered. These include 15 military sites where more detailed hydrogeological investigations were carried out and

contaminant concentration is known as well as 14 other military bases located less than 200 m from surface streams, with oil contaminated upper soils. The bases were superimposed on a digitised version of the topography/groundwater run-off map linked to global co-ordinates.

Given the concentration of pollutants at typical military bases and the groundwater run-off index, it is possible to estimate contaminant transport to the surface water bodies, and consequently assess the regional impact of military pollution on groundwater and the environment in general. In these estimates, attenuation of pollutants, e.g. by biodegradation, is not taken into account.

These estimates therefore represent an upper limit to the pollution potential. For the regional assessment of military pollution, typical cases were selected where identified oil contamination was located less than 200 m from surface streams.

These included the 15 bases where more detailed hydrogeological investigations were undertaken. The distance of 200 m was selected, as mathematical modelling of the Kazlu Ruda oil storage site indicates that in sandy soils oil products will typically not migrate more than 200 m from the pollution source in 10 years. Site selection was based on two considerations:

(i) that the dominant part of groundwater run-off discharges into surface waters (rivers and lakes);

(ii) that the site lies less than 200 m from the surface water. As stated above, previous research has indicated that hydrocarbons dissolved in shallow groundwater can be transported up to 200m by groundwater flow within a period of about 10 years. However, this rule-of-thumb can be easily destroyed by groundwater extraction. One such cases is the Điauliai (Zokniai) military base described above.

Groundwater run-off figures for the 15 military bases where hydrogeological investigations were undertaken have been calculated. Difficulties were encountered as contaminant concentrations are only known for these 15 bases. At other bases, contaminant concentrations could only be estimated by analogy with other sites with similar military activity and geological and hydrogeological conditions. A groundwater run-off map at the scale of 1:500 000 was used for the assessment of potential contaminant discharge. Given the groundwater discharge index and the area of contamination, the quantity of affected groundwater can be estimated from the following equation:

$$Q = M \times F,$$

where

M is the groundwater run-off index ($m^3/day/m^2$);

F is the size of the contaminated area (m^2).

Q = quantity of affected groundwater (m^3/day)

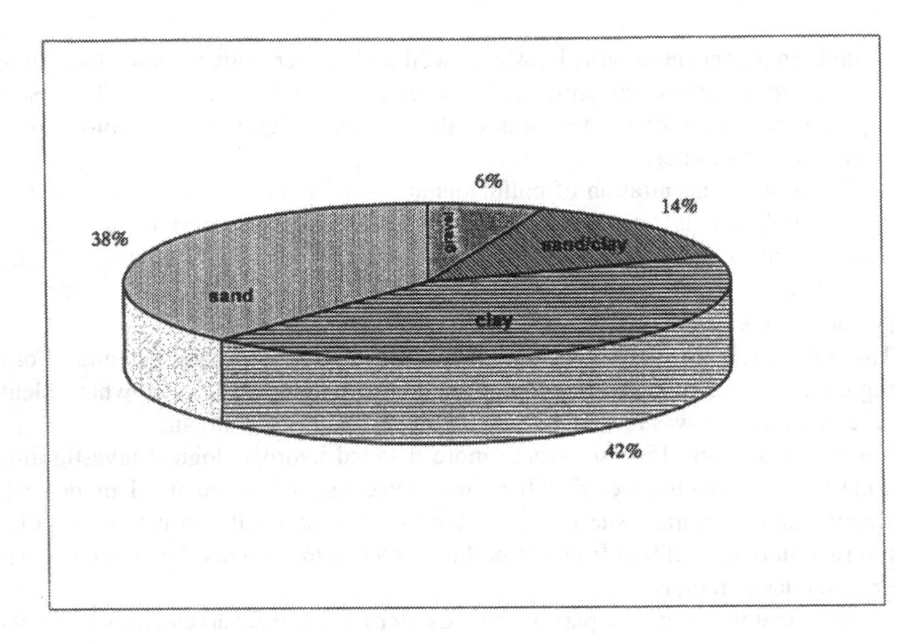

Figure 5. Soil types at contaminated sites

Investigations indicated that significant discharge of polluted groundwater only occurs into surface water bodies (i.e. rivers and lakes) and could be observed in groundwater discharge zones in sandy soils.

The following assumptions were made for the assessment:

- if a military base is located within 200 m of a river or lake, contaminants seeping into the groundwater will reach the river or lake;
- if the soil is polluted, contaminants will affect shallow groundwater (especially if the soil is sandy);
- oil contaminants do not significantly degrade biochemically during migration over a distance of 200 m.

Given these assumptions, the amount of oil contaminants reaching the shallow groundwater was calculated using the equation:

$$L = C \times Q,$$

where:
C = background oil product concentration (mg/l).
L = contaminant flux (mg/day)
Q = water flux through base (l/day)

Using these assumptions the contaminant load discharging with groundwater may be calculated. From the three bases where the most reliable results were obtained (Kazlø

Rûda, Pajuostë and Valèiûnai), 1155 g/day of dissolved oil products are modelled as entering surface streams. This is equivalent to 0.42 tonnes/yr.

From the remaining 12 bases, up to 25045 grams a day of hydrocarbons are polluting rivers and/or lakes, or 9.14 tonnes a year.

The total contaminant emission from these 15 bases is 26200 g/d or 9.6 tonnes per year. The biggest source of hydrocarbon pollutants (2100 g/day) is the Điauliai (Zokniai) air base.

For the 35 bases the approximate dissolved oil concentration was determined from single sample and it was calculated that 6056 g/day or 2.21 tonnes/year of oil products reach surface water bodies. It was also calculated that at the 14 bases located on sandy soils less than 200 m from a surface water body and where oil concentration is unknown, groundwater run-off is rather low at 22.45 m^3/day. Assuming that up to 1 mg/l of hydrocarbons is dissolved in the groundwater and that all this groundwater flows into surface streams, a total of 22 g/day or 0.008 tonnes/year is discharging into rivers.

Thus, given the available data it can be concluded that approximately 11.8 tonnes/year of dissolved oil products are discharging from groundwater into surface water bodies. Although very approximate, this figure indicates a significant adverse environmental impact of military pollution. It is known that 1 g of dissolved hydrocarbons can contaminate 2000 - 3000 litres of fresh water.

4. Methodology for local investigations of military contamination

As mentioned previously, the international team consisting of scientists from the Geological Survey of Lithuania, Geological Survey of Norway, Canadian Department of National Defence and Norwegian Defence Research Establishment were awarded the NATO Scientific Affairs Division Grant for the development of methodology for investigations of soil/groundwater contamination at former military sites in Lithuania.

Pajuostë military air base was selected as a pilot study area where site specific methodologies for soil/groundwater investigations were demonstrated and transferred to Lithuanian environmental institutions. A wide set of field methods has been used for investigations of military pollution, albeit due to the lack of time somewhat superficially, during the field work. Monitoring well drilling, geophysics (ground - penetrating radar and magnetic gradiometer), photoionisation survey and soil/groundwater sampling are the main methods to be mentioned.

Five monitoring wells were established, 38 water and 59 soil samples were collected and analyzed for 1800 analytes in laboratories in Canada, Norway and Lithuania.

4.1. GENERAL INFORMATION ON THE SITE

The Pajuostë airport is situated at the eastern side of Panevėþys town at the confluence of the Juosta and Nevėþis Rivers 2 km to the east of the town. Panevėþys has some 130 thousand inhabitants. The river Nevėþis is the western border of the airport and the Juosta River constitutes the northern boundary of the area (figure 6). The base is situated on a flat moraine plain which is cut by shallow valleys (3-6 m) of local streams. The surface run-off is directed towards the streams.

The area of the airport is 816 hectares. Military activities at Pajuostė began in the 1930s, and after 1945 the airportwas used by the Soviet Union as an air base for transport aircraft. In 1950, the airport was reconstructed and further enlarged to include about 54 aircraft, 300 transport vehicles and more than 140 buildings, providing an infrastructure to support up to 1,200 soldiers.

After the withdrawal of the Russian army in 1993, Pajuostė airport came under the control of the Ministry of Defence of Lithuania. It is planned to use the airport for military and civil transport aircraft.

Special installations with 7 m high soil walls were built at the base for the protection of aircraft. Thirty-two such installations were constructed, using 17 thousand m^3 of soil. Following the withdrawal of Russian troops the airport was left in very bad shape. Many buildings are ruined, much garbage has been left along the territory, and unidentified chemicals in open containers have been abandoned.

The airport is located within a heavily urbanised vicinity. Neighbouring villages use groundwater from dug wells for their drinking water supply. The main waterworks of the town Panevepys lies 1.5 - 2.0 km southwest of the airport with estimated groundwater resources of 65 thousand m^3/day.

The limited time available in Lithuania for the field work restricted the amount of work that could be achieved. The discussions above have indicated future activities that should be carried out.

Analyses of shallow groundwater included detection of main cations and anions, trace elements and oil products. As expected, several samples contained petroleum products. They were found in 7 of the 18 water samples and 15 of the 39 soil samples. Phenols were detected in 5 of the 18 water samples and total organic carbon (TOC) levels ranged from 4.0 to 154 mg/L, although the second highest concentration was only 43 mg/L. No organic priority pollutants were found in any of the six samples analysed; some phthalates were found, but these were attributed to the sampling containers. PCBs were not detected in any of the soil samples.

Elevated levels of metals were found in some samples, but generally contamination in the samples collected was limited. Elements that were found at elevated levels included cadmium, chromium, mercury, nickel and zinc. None of the seven anions were found in particularly high levels except in the landfill water sample.

Analysis of soil and groundwater samples shows that the most polluted area is the oil storage area. Concentration of oil products in soil differs from 1000 mg/kg (according to analyses conducted at Lithuanian laboratories) to 368 mg/kg (results obtained from Canadian laboratory) and 40 mg/kg (Norwegian laboratory). The total area of polluted soil constitutes about 30,000 m^2. In some oil storage areas, a concentration of oil products of 17,261 mg/kg was detected at a depth of 0.5 m.

According to general groundwater composition (main anions and cations) in the oil storage area, a concentration of permanganate oxidation (indirect index showing concentration of organic compounds) of 22-60 mgO_2/L was detected, with the total hardness of groundwater ranging from 10 to 37 mg-ekv/L. Lithuanian drinking water standards limit the mentioned values to 2.0-5.0 mg/L and 7 mg-ekv/l respectively.

Investigations of trace elements indicate that the groundwater at some location places is polluted by iron (3.4-7.5 mg/l. The maximum permitted concentration (MPC) for drinking water is 0.3 mg/l), lead (0.039 mg/l, MPC being 0.03 mg/l), chromium (0.096 mg/l, MPC 0.05 mg/l). The highest concentrations of oil products in shallow groundwater were detected in the oil storage area, reaching 886 mg/l (according to the data of Lithuanian laboratories).

The main findings from the field work are as follows:

- Analysis of a sample obtained from a barrel at the chemical storage area showed that it was mostly 1,2 dichloroethane. There were several barrels present in this storage area which should be sampled and analysed. Volatile hydrocarbons (1,2 dichloroethane in particular) should be analysed in future water samples from the site; the two maintenance areas and chemical storage area should be targeted.
- Cadmium was found in high concentrations in a drainage ditch at the chemical storage area along with petroleum products. The source and extent of the cadmium contamination should be ascertained and its impact assessed.
- Groundwater contamination by petroleum compounds was found near the fuel storage areas. The extent of this contamination should be investigated and its impact assessed.

Contamination has also been found at the two maintenance areas and the landfill. These places require further investigation to ascertain if there is significant environmental impact.

From the reconnaissance investigation and chemical sampling carried out at Pajuostë, the following conclusions can be drawn:

The petroleum product levels observed indicate that there is a potential for contaminated groundwater from this location to migrate to other areas and to resurface and contaminate surface water. Further monitoring wells need to be established in order to fully scope the problem and define the plume of contamination. Remediation of the area may be required if the plume of contamination is migrating into the river or threatens the town's drinking water. However, if this is not the case, other processes such as natural bioremediation, possibly assisted by nutrient addition, may suffice to decrease contamination to acceptable levels. An oil-absorbent boom could be placed across the stream leading from the area to contain surface-runoff of hydrocarbons, although protocols for the maintenance of this facility and the disposal of collected product would have to be made.

It was concluded after the field work at Pajuostë that no universal methodology for assessment of environmental contamination can be recommended which is specific to military bases. The following procedure was proposed that can be adopted at most contaminated sites (including civilian):

- Initial desk study incorporating collation of existing data on geology, hydrology, hydrogeology, site use and history. Environmental audit of past and present contaminant use, storage and disposal, working practices. Audits of chemical and fuel consumption and loss. Assessment of reliability of information sources.
- Reconnaissance investigation with coverage of large areas using rapid and relatively cheap techniques such as trial pitting, soil gas surveying (PID), geophysics and

augering. This, together with information from the desk study, should allow provisional identification of contaminant «hot-spots».

- Preliminary semi-quantitative risk-assessment of the existing contaminant situation. Identification and provisional ranking of contaminant sources, pathways and receptors.
- Detailed site investigations of contaminant hot-spots, usually using intrusive techniques such as drilling and borehole installation.
- Quantitative risk assessment, often involving contaminant transport modelling, to identify areas and/or contaminants which should be prioritised for remediation.
- Remediation-targeted site investigation, to establish the parameters necessary to design remedial options.

Risk assessment and cost/benefit analysis of remedial options. The risk assessment will identify risks to humans and the environment during remediation, and will quantify the risk reduction effected by the remediation. The cost/benefit analysis will assess the benefits (in terms of risk reduction and alleviation of legal liabilities) in the light of financial, and possible environmental, costs incurred during remediation.

Some problems will be specific to the investigation of military, bases, however. The most important of these is the secrecy and lack of information surrounding some military installations. It is probably true that environmental information from military bases is easier to access in many former «Warsaw Pact» countries than in many NATO countries today. Whether the comparatively free flow of environmental information in eastern Europe will continue into the future is difficult to assess. Chemicals and fuels are often known in military circles by (from the civilian point of view) somewhat obscure codes or abbreviations. This situation was encountered at Valèiûniai rocket fuel storage area where a Soviet propellant was identified only by a code name and its exact composition could not readily be determined by Lithuanian scientists. The situation was eventually clarified through contacts with military agencies in the UK having knowledge of the names used by Soviet forces for propellants.

5. Conclusions

Different approaches could be utilised to assess the regional and local impact of the military on the environment. Regional assessment is based on general information on anthropogenic load, geological-hydrogeological conditions of the area determining its vulnerability to pollution and also takes into account groundwater abstraction in the region. Some interpolation and extrapolation should be used to produce approximate figures of military impact to subsurface. There are natural protection barriers of potable groundwater resources in the form of soil properties and hydraulic properties of aquifers. However, low permeable soils are often defected by permeable zones or "hydrogeological windows" and hydraulic head is diminished by groundwater abstraction. Potable water consumption in the major cities of Lithuania have created the overlapping drawdowns. In areas such as these well protected aquifers become vulnerable to pollution. Former military bases were constructed without taking into

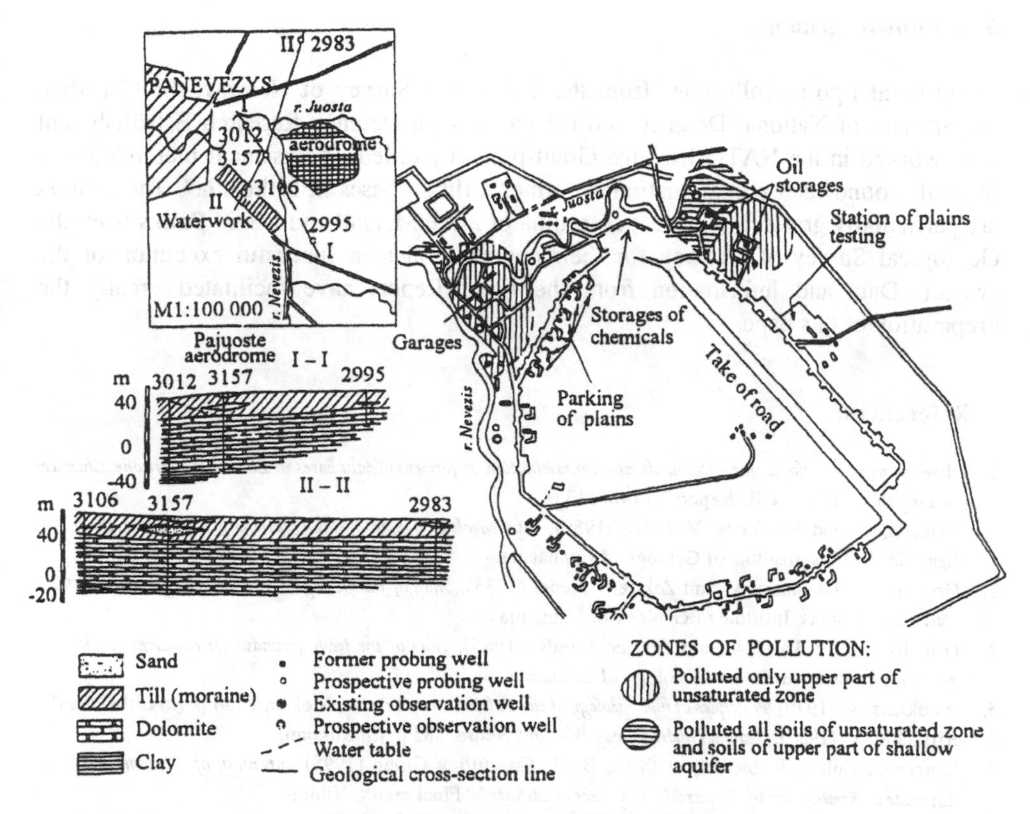

Figure 6. Situation scheme of Pajuostė air base

account the vulnerability of the site. Preliminary calculations of regional military contamination using sparse data indicated that 11.8 tonnes/year of dissolved hydrocarbos are discharging into surface water streams with contaminated groundwater.

The methodology for the detailed assessment of contamination was developed by an international team consisting of scientists from the Geological Surveys of Lithuania and Norway, the Canadian Department of National Defence and the Norwegian Defence Research Establishment. The demonstration study was supported by the NATO Linkage Grant. A set of field methods was used including intrusive techniques, geophysics and soil gas survey. The Linkage Grant project served as a good training programme and partnership between Cooperation Partner and NATO country scientists.

The economic situation in all post-communist countries does not permit the remediation of many polluted sites. In Lithuania for example, only one former fuel storage of Valèiûnai is being cleaned up at present. Pending financing for the remediation of contaminated sites, at least groundwater monitoring should be organised at the most polluted bases.

6. Acknowledgements

Many thanks go to colleagues from the Geological Survey of Norway, the Canadian Department of National Defence and the Norwegian Defence Research Establishment who worked in the NATO Linkage Grant project entitled "Assessment methodologies for soil/groundwater contamination at former military bases in Lithuania". The authors are particularly grateful to project co-ordinator Arve Misund and David Banks from the Geological Survey of Norway for their contribution to a succesful execution of the project. Data and information from the Final Report have facilitated greatly the preparation of this paper.

7. References:

1. *Assessment methodologies for oil/groundwater contamination at former military bases in Lithuania. Part one: Chemical investigation.* (1996) NGU Report 96.146. 113 p.
2. Grigelis, A. and Juodkazis, V. (eds.) (1980). *Hydrogeological map of the Quaternary deposits of the Baltic Republics.* Vilnius, Institute of Geology of Lithuaania.
3. Grigelis, A., Juodkazis, V. and Zektser, I. (eds) (1985). *Map of the fresh groundwater safe yield of the East Baltic area.* Vilnius, Institute of Geology of Lithuania.
4. Grigelis, A., Juodkazis, V. and Zektser, I. (eds) (1985). *Map of the fresh groundwater resources of the East Baltic area. Vilnius,* Institute of Geology of Lithuania.
5. Juodkazis, V. (1979) *Principles of hydrogeology of the Baltic States,* Vilnius, Mokslas, 144 p. [in Lithuanian].
6. Kiriuchin, V. (1987). *Regional hydrogeology. Moscow,* Nedra, 383 p. [in Russian].
7. Kruger Consult in Partnership with the Baltic Consulting Group (1995) *Inventory of Damage and Cost Estimate of Remediation of Former Military Sites in Lithuania,* Final report, Vilnius.
8. Paukštys, B. and Belickas V. (1996) Detection of groundwater contamination at former military bases in Lithuania. *Regional Approaches to Water Pollution in the Environment. (Eds. Rijtema, P. and Eliað, V.*). NATO ASI Series. 2.Environment-Vol.20. 71-91 pp.
9. Paukštys, B. and Gregorauskas, M. (1995) Assessment of the environmental impact from former military bases in Lithuania. *Achievements of geological science - for the environmental protection.* Vilnius, 46-49 pp. [in Lithuanian].
10. Sakalauskienė, D., et al.(1983) *Explanatory note for the mapping of fresh groundwater resources of the Baltic States, scale 1:500 000,* Vilnius, 55 p. [in Russian].

BIOASSAYS USED FOR DETECTION OF ECOTOXICITY AT CONTAMINATED AREAS

R. ROJÌÈKOVÁ[1], B. MARŠÁLEK[1], B. DUTKA[2], R. MCINNIS[2]
[1] *Institute of Botany, Academy of Sciences of the Czech Republic, Kvìtná 8, 603 65 Brno, Czech Republic*
[2] *National Water Research Institute, AEPB, PO Box 5050, Burlington, ON, Canada, L7R 4A6*

Abstract

Ten bioassays were used for the evaluation of acute, (sub)chronic toxicity and genotoxicity of an unknown soil from a military base and six model soils that were contaminated at former military bases with the major pollutants: kerosene, petroleum and used motor oil. The test species represented algae and higher plants, invertebrates and bacteria. The ability to detect toxicants in test soils and their water extracts was investigated and the sensitivity of each bioassay was compared with regard to the tested material (soil vs. water extract from soil) and other test species.

The most sensitive bacterial assays (Toxi-ChromoPad and ECHA Biocide Monitor) were able to detect acute toxicity in most samples. Another bacterial assay, MetPAD, responded positively to the presence of metals. The least sensitive test of the whole battery was the bacterial luminescence assay. The two phytotoxicity tests were similar in sensitivity. The crustacean test Thamnotoxkit F detected acute toxicity only in three samples. The maturation of nematodes was the most sensitive endpoint in nematode assay that was able to indicate potential genotoxicity in most extracts. The Muta-ChromoPlate identified the presence of mutagenic agents in all soil extracts, unlike the SOS-Chromotest Pad, which was positive for genotoxicity only for the most toxic soil with kerosene.

A minimised battery of bioassays for detecting toxicity in investigated soils could be composed of a bacterial assay (Toxi-ChromoPad or ECHA), a phytotoxicity assay (a seed germination and root elongation test) and an invertebrate test (nematode assay). Basic information about genotoxicity could be given by the Muta-ChromoPlate.

1. Introduction

The most effective information about contamination at polluted sites is given by a combination of chemical and biological analyses. To get a quick and cost-effective view of a problem, a battery of toxicity tests can be used to indicate potential sources of contamination, which will be further specified by chemical and other analyses.

F. Fonnum et al. (eds.),
Environmental Contamination and Remediation Practices at Former and Present Military Bases, 227–232.

Responses of test organisms to a toxic material are different, therefore more species are needed to cover all the toxicity scale of an evaluated sample which usually contains mixtures of organic and inorganic pollutants. There are no rules according to which individual assays should be incorporated into batteries, but it is generally thought that sensitive test organisms in a battery should represent different trophic levels and their sensitivity patterns should complement each other.

The toxicity assays applied in this study were selected to represent decomposers (bacteria), producers (algae, higher plants) and consumers (invertebrates) and to be able to assess toxicity both in soils and in soil extracts. The samples were prepared in such a way that they illustrated a commonly polluted soil in military areas. The potential acute/(sub)chronic/genotoxicity of these soils was assessed by the battery of 10 bioassays.

2. Material and Methods

500g of soil was mixed with 50 ml of a contaminant (kerosene, petroleum or used motor oil) and was occasionally mixed over a period of 24 h. Six prepared soil samples and 1 unknown sample from a military base (Tab. 1) were used for water extraction: 250g of soil and 250 ml of deionized water were mixed and continuously shaken for 2 hours, then centrifuged at 3000g for 15 min and filtered (paper filter N. 391, 0.07 mg, Niederschlag, Germany). Some of the samples were filtered through a membrane filter (0.45 mm pore size, Synpor, CZ) before use in an algal assay, a bacterial luminescence test and a Muta-ChromoPlate. The medium extract for nematode assay was prepared according to [1]. Analysis for eight metals was performed by atomic absorption spectrophotometry.

TABLE 1. Soil samples with model contaminants

Sample	Type of soil	Model contaminant (producer)
A	sand	kerosene (Severochema, Liberec)
B	clay	kerosene (Severochema, Liberec)
C	sand	petroleum
D	clay	petroleum
E	sand	used motor oil (Paramo, Pardubice)
F	clay	used motor oil (Paramo, Pardubice)
G	clay-loam	unknown soil from military base

Phytotoxicity assays. The 72-h growth of alga *Raphidocelis subcapitata* in microplates [2] was measured spectrophotometrically. The initial algal concentration was 40 000 cells/ml, and the medium was used according to standard [3]. The results were expressed as % growth inhibition in the highest test concentration, or EC50 (a concentration that inhibits growth in 50%) was calculated using regression analysis. The ability of watercress seeds (*Lepidium sativum*) to germinate, produce a root and a hypocotyl when placed into soil extract was observed after 72 hours [4]. The medium was used according to standard [3]. The results were expressed as % inhibition of germination or % inhibition of root/hypocotyl elongation in the highest test concentration. EC50 was calculated using regression analysis when possible.

Invertebrate assays. The 24-h mortality test with crustacean *Thamnocephalus platyurus* is based on hatching test organisms from cysts [5]. EC50 (a concentration that kills 50% of organisms) was calculated by the moving average method. By monitoring a controlled population of the nematode *Panagrellus redivivus* juveniles over their 96-h growth period, lethal (mortality) and sublethal (chronic toxicity = inhibition of growth or genotoxicity = inhibition of maturation) effects are assessed [1]. The results were expressed as % survival, growth and maturation.

*Bacterial acute toxicity assays.*Colorimetric bacterial assays (3 hours) are based on the ability of toxicants to inhibit *de novo* synthesis of enzyme β-galactosidase in mutant *Escherichia coli.* The MetPAD is more sensitive to metals [6]. The Toxi-ChromoPad assesses directly, without extraction, the presence of bioavailable toxicants in soils [1], and the Toxi-Chromotest is used for extracts [1]. The results were expressed as EC100 (a concentration that inhibits enzymatic activity in 100%) which was estimated visually. The ECHA Biocide Monitor is based on a small absorbent pad impregnated with bacteria and a growth indicator system to detect bacterial respiration. The colour reaction indicates the presence of toxicants [1]. *Photobacterium phosphoreum* was used in a 5-min assay to detect toxicants that inhibit bacterial luminescence [7].

Bacterial genotoxicity assays. The SOS-Chromotest Pad is a test for the presence of bioavailable genotoxicants in solid samples based on a colorimetric reaction of enzyme β-galactosidase in the bacterial strain *Escherichia coli.* The greater the amount of β-galactosidase produced, the greater the genotoxicant concentration in a sample [1]. The Muta-ChromoPlate is used for determining the presence of Ames mutagens in water samples (the test organism is *Salmonella typhimurium).* The test lasts 5 days. The results express the statistical significance of the difference between a treatment and a background mutation [8].

3. Results and Discussion

The results given in Table 2 show that the overall toxicity of the tested model soils was approximately the same, and that none of the samples was extremely toxic. At least half of the assays responded positively to toxic substances present in soils. The clay soil F with used motor oil, which can be assigned as the most toxic sample, caused an effect in each of the applied tests, and it was the only soil in which the SOS-ChromotestPad detected genotoxicity. F was followed closely by sand soil E, clay soil with kerosene B and both soils mixed with petroleum (C and D). Sand soil with kerosene (A) was the least toxic, with only five bioassays detecting toxicity in it. Soil G, which contained unknown contamination from a polluted military area, was less toxic than other model soils. Only four tests were able to show a positive result in this sample. The toxicity of G was partly comparable to sand soil contaminated with kerosene (A).

According to the Muta-ChromoPlate, all the tested soils contained highly mutagenic substances ($P<0.001$). However, the SOS-Chromotest showed genotoxicity only in the clay soil with motor oil (F). Similarly, inhibition of maturation in nematodes used as an indirect assay for detection of potential genotoxicants indicated 64% effect in sample F. Moreover, maturation was suppressed in 95% in clay soil with kerosene (B). The other

TABLE 2. Toxicity results for ten bioassays and seven soils

		Sample					
	A	***B***	***C***	***D***	***E***	***F***	***G***
algal growth inhibition	EC50=12.7%	20%	45%	33%	EC50=9.1%	EC50=28.9%	EC50=22.8%
seed germination inhibition	21.8%	27.3%	37.5%	NT	30.4%	NT	NT
root elongation inhibition	EC50=95.7%	EC50=114%	EC50=22.36%	27%	EC50=25.2%	EC50=87.8%	NT
hypocotyl elongation inhibition	NT	42.5%	EC50=54.%	30%	EC50=57%	25%	NT
Thamnotoxkit F	NT	EC50=14.4%	NT	NT	EC50=22.3%	EC50=34.3%	NT
nematode % survival: medium extract	95	100	94	47	77	84	95
water extract	93	96	91	97	91	93	93
nematode % growth: medium extract	100	91	97	51	58	44	100
water extract	92	86	102	83	94	59	83
nematode % maturation: medium extract	93	5	64	56	66	36	102
water extract	106	69	115	64	89	40	73
ECHA: soil	T	T	T	T	T	T	T
ECHA: water extract from soil	T	NT	NT	NT	NT	NT	± T
Toxi-ChromoPad: soil (EC100)	NT	12.5%	12.5%	3.125%	50%	50%	50-25%
Toxi-Chromotest: water extract from soil	NT	NT	NT	NT	NT	NT	NT
MetPAD (EC100)	100%	100%	100%	100%	50%	100%	NT
bacterial luminescence assay	NT	35%	29%	38%	31%	22%	NT
SOS-Chromotest Pad (CIP): cytotoxicity	NT	NT	NT	2	NT	NT	NT
genotoxicity	NT	NT	NT	NT	NT	6	NT
Muta-ChromoPlate	P<0.001 (5ml)	P<0.001 (5ml)	P<0.001 (2.5ml)	P<0.001 (2.5ml)	P<0.001 (2.5ml)	P<0.001 (15ml)	P<0.001 (2.5ml)

T = toxic, NT = not toxic, ± T = slightly toxic

inhibition of algal growth, seed germination, root and hypocotyl elongation = results given as % of inhibitory effect greater than 20% of control in 100% sample or EC50; % survival, growth and maturation of nematodes= in 100% medium extract (nutritive medium) and in 10% water extract

SOS-Chromotest Pad: CIP = Colour Index Profile (11-digit number representing the colour index from the lowest to the highest sample dilution), genotoxicity value = total sum of the net non-spiked CIP values which are > 2, cytotoxicity value = total sum of the net spiked CIP values which are > 2; Muta-ChromoPlate = level of statistical significance for a positive sample tested at a volume of x ml samples either did not show any genotoxic effect on *E. coli* and *P. redivivus* or the maturation of nematodes was inhibited in less than 50%.

The bacterial tests with the exception of the luminescence assay appeared to be the most sensitive assays in the applied battery. The high sensitivity of the Toxi-Chromo Pad and ECHA Biocide Monitor could be connected with the solid particles that were eliminated from the water extracts by centrifugation and filtration. This is supported by the results from the tests on water extracts when in most cases no toxicity was recorded by the same test species. The MetPAD test, which indicates mainly metal contamination, was also sensitive for all extracts of model soils. The chemical analyses showed that the soils did indeed contain several heavy metals that the bacterial strain in MetPAD is sensitive to (Tab. 3). On the other hand, the bacterial luminescence assay was the only test that did not detect inhibition higher than 38% in any sample. The reason for this observation may be that the highest concentration of water extracts tested was 45%.

TABLE 3. Heavy metal content in soils

Metals [mg/kg of dry weight]	Sand soil	Clay soil
Arsenic	18.8	33.0
Cadmium	< 0.2	< 0.2
Lead	16.8	20.5
Mercury	0.25	< 0.2
Copper	39.0	27.9
Zinc	284	165
Chromium	37.8	59.0
Nickel	18.8	26.2

The bacterial assays were followed in their sensitivity by the phytotoxicity tests. The root elongation of watercress was the most sensitive endpoint of the three used in this assay. The toxicity detection potential in algal and higher plant assay slightly differed for some samples (B and G). The application of the seed test may be more suitable for soil testing, as water extracts can be turbid and the spectrophotometric measurements in algal assay might interfere with the turbidity.

The crustacean *Thamnocephalus platyurus* was sensitive to water extracts of soils that were indicated above as the most toxic. Nematode survival and growth was not affected in more than 50%, with the exception of the most toxic sample F, which exhibited chronic toxicity to the test organism (56% inhibition of growth). When soil extracts were prepared in a nutritive medium (100% medium extract) and tested directly without any further dilution, the inhibitory effects detected were higher than in water extracts diluted 1:10 with the medium. Water samples and extracts of solid phase samples are usually concentrated 10x by flash evaporation to make up for the 1:10 dilution that occurs in the test protocol [9].

4. Conclusions

Although the toxicity of the soils in this study was assessed mostly with assays designed for testing aqueous samples, the presence of potential toxicants was indicated by most of them. The use of aqueous tests for evaluation of soils, sediments or sludges is quite frequent in ecotoxicology, as these tests are usually simpler and quicker than terrestrial test systems. As concluded in [10], if an effect is detected in aquatic ecotox tests, a deterioration in the habitat function may also be expected. However, if no effect is detected, a deterioration cannot be excluded. Nevertheless, some of the present tests can be used for quick and simple direct toxicity assessment of solid samples (Toxi-ChromoPad and ECHA). Thus, important information about toxicity action of the solid phase can be obtained before extraction procedures are initiated.

The extensive battery of bioassays used in this study offers an opportunity to select representative and sensitive tests that can detect toxicity in investigated soils in the most informative way. Such a miniaturized battery might contain:

- a bacterial test (Toxi-ChromoPad for direct soil evaluation),
- a phytotoxicity assay (a seed germination and root elongation test seems to be more suitable than an algal test if the soil extracts are turbid),
- an invertebrate test (a nematode assay can give more information about a sample than Thamnotoxkit, because chronic or genotoxic effects can be recorded).

With the increasing significance of the genotoxic potential of many substances, the incorporation of an genotoxic assay (Muta-ChromPlate) into an ecotoxicological battery can provide interesting additional information about the character of the sample under investigation.

Acknowledgements

This study was supported by NATO Linkage Grant No. 960 334. We would like to express special thanks to Dr. Anne Kahru for toxicity testing on luminescent bacteria.

5. References

[1] Dutka, B.J., McInnis, R., Jurkovic, A., and Liu, D. (1996) Water and sediment ecotoxicity studies in Temuco and Rapel river basin, Chile, *Environ. Toxicol. Wat. Qual.* **11**, 237-247.

[2] Blaise, C., Legault, R., Bermingham, N., Van Coillie R., and P. Vasseur (1986) A simple microplate algal assay technique for aquatic toxicity assessment, *Tox. Assess.* **1**, 261-281.

[3] ISO 8692 (1989) Water quality - Fresh water algal inhibition test with *Scenedesmus subspicatus* and *Selenastrum capricornutum*, ISO, Geneva.

[4] Máchová, J., Svobodová, Z., and Vykusová, B. (1994) Ecotoxicological evaluation of solid waste elutriates, Edition of methodologies, Research Inst. of Fish Culture and Hydrobiology. Vodòany [in Czech].

[5] Persoone, G. (1991) Cyst-based toxicity tests: I. - A promising new tool for rapid and cost effective toxicity screening of chemicals and effluents, *Germ. Jour. Appl. Zool.* **78**, 235-241.

[6] Bitton, G., Koopman, B., and Agami, O. (1992) MetPad ™: a bioassay for rapid assessment of heavy metal toxicity in wastewater, *Wat. Env. Res.* **64**, 834-836.

[7] Kahru, A. (1993) *In vitro* toxicity testing using marine luminescent bacteria (*Photobacterium phosphoreum*): the Biotox™ test, *ATLA* **21**, 210-215.

[8] Rao, S. S. and Lifshitz, R. (1995) The Muta-ChromoPlate Method for measuring mutagenicity of environmental samples and pure chemicals, *Environ. Toxicol. Wat. Qual.* **10**, 307-313.

[9] Dutka, B. J., Liu, D.L., Jurkovic, A., and McInnis, R. (1993) A comparison of four simple water extraction-concentration procedures to be used with the battery of bioassay tests approach, *Environ. Toxicol. Wat. Qual.* **8**, 397-407.

[10] Debus, R. and Hund, K. (1997) Development of analytical methods for the assessment of ecotoxicological relevant soil contamination. Part B - Ecotoxicological analysis on soil and soil extracts, *Chemosphere* **35**, 239-261.

COMPATIBILITY OF GROUNDWATER SAMPLING AND LABORATORY TECHNIQUES

Some practical aspects of investigations at military sites

S. JANULEVIÈIUS
Hydrogeological company "Grota"
Eiðiðkiø plentas 26,
LT - 2038, Vilnius, Lithuania

1. Introduction

Investigations of military bases in Lithuania began in 1992 when Russian army was leaving the country. Up till now five regional projects related with groundwater contamination were performed at military bases.

During the inventory and investigations it was mentioned that the main pollutants were oil products at the military sites. Soil contamination by them was identified at 70 % of all the bases. Performing the separate projects not enough attention was paid to interlaboratory calibration or intercalibration between sampling techniques, etc. Analysing the data of all the projects related with hydrogeological investigations at military sites, some essential differences of concentrations of oil products in groundwater are noticed. Sometimes differences between the maximum concentrations of dissolved hydrocarbons in groundwater reach thousands of times at the same places. Analysing the results of separate investigations it is evident that the problem is not so actual due to the identification of mostly hazardous sources of contamination and/or spreading of free phase oil plumes. But in cases of precise assessments, forecasts and simulations the need of correct values is undoubtedly necessary. So the causes of compatibility of groundwater sampling and laboratory techniques are on the discussion in Lithuania at present.

2. Some Data Under Discussion

Beginning from 1992 the following regional projects related with groundwater contamination at former Soviet military bases have been performed:

(a) Preliminary Ecological Investigations (Institute of Geography, EPM, 1992, 1993),

(b) Assessment of Soil and Groundwater Contamination at 10 Mostly Polluted Sites (Lithuanian Geological Survey, Hydrogeological Company "Artva", 1993, 1994),

F. Fonnum et al. (eds.),
Environmental Contamination and Remediation Practices at Former and Present Military Bases, 233–238.

(c) Inventory of Damage and Cost Estimate of Remediation of Former Military Sites in Lithuania (Kruger Consult in partnership with Baltic Consulting Group, 1995),

(d) Investigations of Groundwater State at Former Soviet Military Bases (Lithuanian Geological Survey, Hydrogeological Company "Grota", 1994 - 995),

(e) Assessment Methodologies for Soil/Groundwater Contamination at Former Military Bases in Lithuania (NATO, Geological Survey of Norway, Lithuanian Geological Survey, Canadian Department of National Defence, Norwegian Defence Research establishment, 1995 - 1997).

Carrying out all the above mentioned projects groundwater samples were taken from 15 mostly contaminated and hazardous military bases (there were 271 such bases). Oil products from 14 of them have been analysed (see table 1).

TABLE 1. Maximum Concentrations of Oil Products in Groundwater, 1993 - 1995

Military territory	Area of polluted ground water km^2	Maximum concentrations, mg/l		
		LGI (oil products)	EPM (oil products)	Grota (BTEX)
Kėdainiai airfield	0.4	6.06	26.0	-
Pagegiai fuel storage	0.14	0.71	13.0	-
Pajuoste airfield	0.13	36.5	-	0.07
Karmelava rocket base	0.008	-	3.1	-
Pilainiai engineering regiment	0.03	-	0.22	-
Alytus paratroops unit	0.001	51.5	-	-
Pabrade fuel base	0.021	15.0	-	-
Kazlu Ruda fuel base	0.03	22.5	-	0.9
Valciunai fuel base	0.1	>1000	38.0	99.0
Raudondvaris fuel base	0.01	0.64	-	-
Kairiai tank regiment	0.03	3.2	-	-
Zokniai aerodrome	1.0	>1000	0.7	-
Nemenchine civil defence centre	0.03	-	35.0	-
Vilnius military unit	0.014	-	1.3	-

Table 1 represents data obtained in different laboratories. Groundwater analysis was performed in laboratory of LGI (Lithuanian Geological Institute) during the project (b), while in laboratories of EPM (Environmental Protection Ministry) and Hydrogeological Company "Grota" - correspondingly during the projects (c) and (d). Besides, the samples were collected at different time and in some cases - at different points of pollution plume. So differences up to some limits could be objective. However, as it is shown below, we should reject concentrations higher than 1000 mg/l as uncorrect ones.

To improve the reliability of the data, carrying out the last investigations (e) one sampling was performed on 22 09 1994 at military airport Pajuoste in North Lithuania. Groundwater samples (for hydrocarbons) from the same wells were analysed simultaneously in laboratories of LGI, Norwegian Defence Establishment and Royal Military College of Kingston, Canada. Difference of results was not so essential as that shown in table 1, however they could be discussed as well (see table 2).

TABLE 2. Concentrations of Hydrocarbons in Groundwater, mg/l (Pajuoste Military Airport)

Well No.	Date	LGI (oil products)	Grota (BTEX)	Norwegian $(<C_{15})$	Canadian (light fraction)
1	1994 09 22	7.0	-	6.4	4.2
1	1994 11 25	36.5	-	-	-
1	1995 12 15	-	0.07	-	-
2	1994 09 22	7.4	-	63	3.4
2	1994 11 25	75.3	-	-	-
2	1995 12 15	-	2.02	-	-
3	1994 09 22	0.1	-	<0.04	<0.2
3	1994 11 25	0.2	-	-	
3	1995 12 15	-	0.0	-	-

Trying to avoid the unconformity of data it is necessary to review the factors which could influence the objectivity of such data, i.e. all procedure of collecting and analysing of the sample. Whole procedure includes drilling and installation of the wells, sampling, preparing samples for analysis and/or storing them and chemical analysis as well.

3. Drilling Techniques

At the first stage of investigations (a) only soil samples from the depths less than one meter were analysed. Samples were taken from mostly polluted areas of ground surface. Drilling was not used. Sampling of soil was performed using a special 1m length sampler which should be hammered into the soil. Groundwater samples were not taken at this phase.

During the latest investigations (b, c, d, e) groundwater level detection wells were drilled using hand augers. Drilling was conducted by folding hand augers with removable drill bits. Depending on the soil conditions, 70 mm or 100 mm diameter augers were used. At intervals of 0.3 m augers were removed from the borehole. The appearance of the smell of sample were used as indicators of soil contamination. Soil samples were collected from the intervals where pollution was suspected. The maximum depth of handdrilled wells was 6 m. In the areas investigated the shallow groundwater table is usually shallower than this.

For deeper subsurface investigations hollow stem auger and rotary mud drilling rigs were used. Four types of wells and boreholes were established during the site investigations: 1) test boreholes, 2) probe - test wells, 3) monitoring wells and 4) recovery wells.

Test boreholes were used for the general assessment of geological and hydrogeological conditions of the site. Only soil samples were taken.

Probe - test wells were drilled for the determination of groundwater quality, depth to groundwater level and the assessment of free phase oil thickness. Wells sunk to shallow groundwater were drilled by hand auger; the deeper aquifers were reached by using a drilling rig.

Monitoring wells were drilled for groundwater sampling, level evaluation and continuous observations. Monitoring wells of various diameters were installed following the delineation of a pollution plume by test boreholes. After the drilling of boreholes, stainless steel and PVC casing was installed together with appropriate screens.

In spite of the fact that all the above mentioned investigations were carried out by various companies, possibilities of choosing the drilling and well installation methods are very similar for each Lithuanian company and almost all of them used well known standard methods. So drilling techniques should not cause essential influence for the results.

4. Sampling and Laboratory Techniques

Various types of pumps were used for groundwater sampling. Type of pump was chosen depending on the yield of a well. Wells with very low yield were sampled using bailers or Waterra system. Analysis of sampling procedure showed that in several cases getting a free phase oil into the groundwater samples was not avoided. This occurred when they were taken from low yield wells particularly using bailers. During first investigations samples were placed into special bottles with carbon tetrachloride used for the extraction of oil products. So in some cases oil products were extracted not only from water but also from free phase oil. That is why extremely high concentrations of hydrocarbons in water samples (more than 1000 mg/l) appeared in some reports. Two similar cases are presented in table 1 (Valèiunai oil base and Zokniai airport). The latest investigations at these and analogicaly contaminated civil sites showed that even under the free phase oil layer concentrations of hydrocarbons in groundwater in most cases do not reach one hundred mg/l.

Carrying out the international project (e) the results for petroleum products obtained were similar except of one sample (see Table 2) the results of which 63 mg/l and 3.4 mg/l were found. This difference may well be due to aeration of one of the samples during collection with Waterra system. The procedure requires the oscillation of the plastic tube and unless this is done in a carefully controlled manner, aeration of the water samples can occur resulting in loss of volatile components; the contaminant in this case was the more volatile light fuel fraction.

Thus, it seems that the reasons of most significant errors in determination of hydrocarbons concentrations in groundwater were due to not very suitable procedure of sampling and preparing samples for chemical analysis.

The rest of reasons could be attributed to laboratory methods.

According to the International Commission of Unified Methods of Analysis "oil products" are realised as sum of aliphatic, alicyclic and aromatic (BTEX) hydrocarbons. They form 70 - 90 % of all compounds of oil and oil products. In other words it is the sum of polar and poorly polar compounds, which easily dissolve in hexane. Maximum allowable concentrations for these compounds are determined at present.

In the laboratories of Lithuanian Geological Institute and EPM concentrations of oil products were defined by IR (Infra Red) ray spectrophotometric method. It is based on light absorption in infrared range by methyl and methylen groups and comparison of recorded spectrograms. Using this method it is possible to define aliphatic and alicyclic hydrocarbons, which are weekly soluble in water. It means that they exist in amulgate state in a water sample. More soluble aromatic compounds are weekly absorbed in the IR part of ray spectrum, and that is why it is possible to define about 10 % of them. So if the main part of oil products consists of BTEX, very low concentrations will be seen using IR spectrometry method even in cases when groundwater sample has a strong smell of oil product.

At the laboratory of company "Grota" aromatic hydrocarbons (BTEX) in water were determinated by gas chromatograph DANI 86. 10 using capillary column and flame ionisation detector. Samples were injected by "Hadspace" autosampler.

Carrying out the international project (e) water samples were analysed in both laboratories (Norwegian and Canadian) by gas chromatography with flame ionisation detectors (GC/FID) after dichlormethane extraction.

5. Conclusions

During environmental investigations in Lithuania it was noticed that the data related with concentrations of oil products in groundwater are different when carrying out different projects.

Groundwater samples from the same sites were taken at the different time and analysed using various methods. So some differences concentrations are natural. Significant errors were made in sampling procedure and preparing of sample for analysis and these factors caused essential differences in data.

However, various laboratories have different possibilities of determining different organics in water so a set of analysis carried out at particular sites could be very useful as well as continuos groundwater quality monitoring at the selected sites. Such means could improve or hasten the selection of optimal sampling and analysis methods for the solution of concrete problem.

6. References

1. LGS, Artva. (1993) *Estimation of damage caused to groundwater by former Russian military objects. (Project)*., Vilnius, Lithuania, 1993 [in Lithuanian].
2. . Paukštys B.(Editor in Chief) (1997). *Assessment Methodologies for Soil/Groundwater Contamination at Former Military Bases in Lithuania* NGU, Norway.
3. Kruger Consult in partnership with Baltic Consulting Group. (1995) *Inventory of Damage and Cost Estimate of Remediation of Former Military Sites in Lithuania, Final report*, Vilnius.
4. Seirys N. (1997) Subsurface Remediation at a Former Military installation in Valciunai (Lithuania). *NATO ARW*

ADVANCED METHODS FOR THE IN-SITU REMEDIATION OF CONTAMINATED SOIL AND GROUNDWATER

JOSEPH A. PEZZULLO, P.E.
Vice President
Terra Vac Corporation
241 Norsam Drive
Langhorne, Pennsylvania 19047 USA

Abstract

Throughout the world, governments and their people continue to struggle with the complexities of hazardous and toxic wastes which already exist, while they also attempt to control and minimize the discharge of more waste. Much of the hazardous waste which is most serious is not readily visible; that being waste which exists in the form of contaminated soil and groundwater. Most commonly, subsurface contamination exists in the form of petroleum hydrocarbons, volatile organic compounds, and inorganic compounds. At many civilian and military facilities, petroleum hydrocarbons and volatile organic compounds in the subsurface. However, advances in applied remediation technologies have resulted in proven remedies to successfully remove the source of subsurface contamination while effectively controlling its migration in the subsurface. Soil and groundwater remediation is expensive, yet we may soon recognize a completely new definition for the designation of "waste as these innovative methods of remediation can transform the contamination into "bankable' commodities.

1. Introduction

The paradigm of "Out of sight, out of mind,' had long been the typical reaction of governments, citizens, and regulatory agencies when considering hazardous and toxic wastes. It was quite easy to adopt this type of attitude in the past, because discussions an the environment predominantly focused on the pollution which was readily visible to the naked eye, such as city smog and soot, industrial smoke stacks, wastewater discharges to rivers and water courses, mining operations, landfills and many others. Certainly these visible sources of pollution have been and continue to be major issues.

Only recently, that being in the past fifteen years or so, have we recognized a major shift in awareness toward forms of pollution which are not readily observed. People now realize that contamination in the subsurface soils and groundwater can be extremely harmful. A lady from the Newly Independent States (NIS) living on the outskirts of a

F. Fonnum et al. (eds.),
***Environmental Contamination and Remediation Practices at Former and Present Military Bases*, 239–256.**

former military base bails a bucket of groundwater from a well in her backyard. It smells likes an oil can. She proceeds to light a match and seconds later the bucket of "water" is on fire. This is the water her children drink and bathe in and she is furious. "What are they going to do? " she cries. "How and when will our water be cleaned'

Environmental restoration of subsurface contamination in the soils and groundwater has undergone considerable advances during the past decade, with in-situ remedies being the most favored technologies. *In-situ* refers to environmental cleaning methods which treat the contamination in the ground at the source without excavating the soil Excavation and dumping into a landfill is considered a last option, since the only thing accomplished by this method is moving the problem from one place to another. Furthermore, it has been observed that all landfills eventually leak, and the contamination once again leaches into the subsurface. Landfilling treats the symptom not the cause, and it poses unnecessary risks of exposure to workers and communities. Furthermore, contamination is sometimes found in places where excavation and treatment is simply not feasible, such as beneath buildings and airport runways. The preferred method, therefore, is to treat the contamination at its source, on-site and in-situ while also mitigating the migration of the contamination away from its source.

Certainly there are many publications covering the problems of subsurface contamination and the costs to remediate the properties. One such source of information can be found on the Internet under "North Atlantic Treaty Organization" (NATO)/ Base Closures. Here, the Bonn International Center for Conversion (BICC) contains a useful source of information on the political economic, and social ramifications of military base closure, redevelopment and environmental concerns (Cunningham, 1997). There are serious economic impacts to base closures and re-conversion to civilian use. Although communities may eventually profit from the land returning to civilian use, base conversions result in job losses and adverse cash flow to the local communities who used to thrive from the bases being in their communities. Add these acute economic impacts to the fact that these bases are typically abandoned in a severe state of environmental deterioration, and the results are at best socially severe. How can these situations be resolved ? Can these problems be turned into opportunities ? One answer is the "bankable" environmental project. By converting wastes into resources through innovative recovery programs it is possible to clean-up these sites with partially self-finding methods and at the same time create jobs and opportunities for the local communities.

This paper addresses certain issues pertaining to subsurface contamination by various contaminants, some innovative technologies to remediate such contamination, and potential economic solutions for funding clean-up projects. Presented herein are a few innovative technologies for on-site in-situ subsurface restoration, the general process of how they can be applied and one innovative case where these methods have been applied such that the project funded itself. Although the problems of nuclear contamination are hereby acknowledged and critically important, the attention herein is focused on subsurface contamination in the form of petroleum hydrocarbons and volatile organic compounds, which by far comprise the most widespread forms of contamination in the subsurface soils and groundwater at military base civilian airports, machine parts manufacturer chemical and industrial facilities, gas stations dry cleaners refineries pharmaceutical operations and many others.

2. Subsurface contamination

Typically, the source of subsurface contamination involving petroleum hydrocarbons or volatile organic compounds initiates from surface spills, leaking storage tanks both above and below ground, heap leaching fields, and deteriorated or ruptured pipelines. Some of the more common petroleum contaminants include gasoline, aviation fuel, diesel, kerosene, and fuel oil. Volatile Organic Compounds (VOCs), which are commonly utilized for degreasing agents both in the machine and electronics manufacturing industries and dry cleaning businesses, may include tetrachloroethylene (PCE), trichloroethylene (TCE), dichloroethylene (DCE), 1,1,1-trichloroethane and many others. Semi-Volatile Compounds (SVOCs) like phenols and naphthalenes are also quite common.

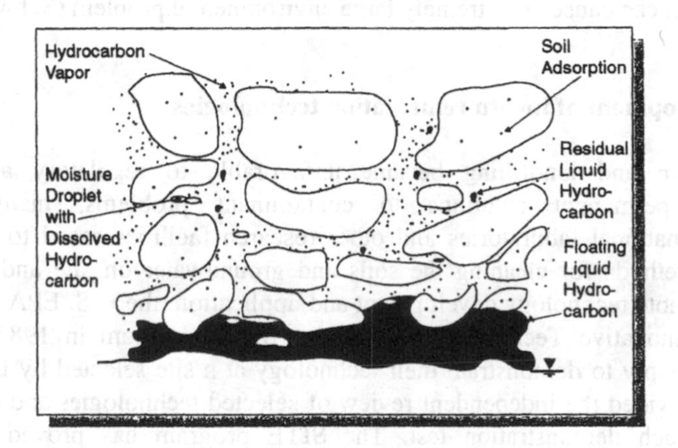

Figure 1 Hydrocarbon distribution in the subsurface

When petroleum hydrocarbons, VOC's and SVOCs are spilled into the soil, they may exist in several different phases as depicted in Figure 1. The aqueous phase is that which is dissolved in the soil's pore water content. If the contamination is volatile and the soil porosity is sufficient, the vapor phase fills the soils air porosity portion. To a lesser extent. some portion of the chemical is adsodoed to the soil matrix. Finally, if the concentration is high enough there may be free phase liquid product. If the density of the compound is lighter than water as in the case for petroleum hydrocarbons, the product floats on tbe groundwater table and is commonly referred to as Light Non-Aqueous Phase Liquid (LNAPL). Conversely, if the spilled compound is heavier than water, like chlorinated hydrocarbons or VOC's, then the product sinks through the groundwater and is typically called Dense Non-Aqueous Phase Liquid (DNAPL). This latter phase of contamination comprises a large portion of the contemporary groundwater problems, and it is one of the most challenging to locate and remediate.

To what extent do petroleum hydrocarbons, VOCs and SVOCs contribute to soil and groundwater contamination? In the United States for instance, such compounds are the major contaminants at more than 90% of the Superfund Sites. Another question one may ask is, "How much spillage can cause an environmental hazard?" Several factors must be evaluated to respond to this question. Soil type, grain size, porosity, permeability, annual rainfall data and depth to groundwater are some important considerations as well as apparent risk to human health and welfare. In other words, is groundwater utilized for drinking water in the area? What are the biological receptors for the compounds and what, if any, are the natural attenuation capabilities of the subsurface itself? Notwithstanding the technical evaluation of all these factors, an example would be to consider a groundwater plume which is 1000m long, 100m wide, and 20m in depth with an average concentration of 100 parts per billion (ppb), which is 20 times the United States Environmental Protection Agency (U.S. EPA) recommended limit for most solvents in drinking water, contains only 80 kg of chemicals, or about 56 liters of a chlorinated solvent such as TCE. Considering that a single 206 liter drum (55 gallon) drum of VOCs weighs about 300kg, it is not difficult to envision that a single drum of spilled solvent can cause an extremely large environmental problem (Schwille, 1988)

3. The development of in-situ remediation technologies

As excavation and landfilling became unfavorable to regulatory agencies when considering permanent solutions to contaminant problems, many companies, universities, national laboratories and other research facilities raced to develop more innovative methods for cleaning the soils and groundwater on site and in-situ. In an effort to promote technology development and application, the U.S. EPA introduced the Superfund Innovative Technology Evaluation (SITE) program in 1987 wherein any entity could apply to demonstrate their technology at a site selected by U.S. EPA. The U.S. EPA provided the independent review of selected technologies and also published reports on each demonstration test. The SITE program has proved successful in launching several innovative treatment technologies.

One of the first SITE demonstration applications accepted by the U.S. EPA in 1987 was for the demonstration of in-situ vacuum extraction to remove petroleum hydrocarbons and VOCs from the subsurface soil without excavation. At the site selected, an EPA contractor had been pumping groundwater for approximately 10 years, and they extracted approximately 90 kg. of TCE during this period. Over a 56 day demonstration, the in-situ vacuum extraction system removed nearly 1,300 kg TCE. In short, the in-situ SVE process removed more TCE in one day than groundwater pumping alone had removed in over two years. The SVE process also proved reliable in various soils, including silts and clays (U.S. EPA, 1989).

Since this demonstration, the vacuum extraction process has become one of the most widely acclaimed methods for source control and remediating soils contaminated by petroleum hydrocarbons and VOCs. Experts around the world agree that removal of the source of contamination in the soil is often the first step in the effective control of groundwater contamination, and there is widespread support for vacuum extraction as the preferred method for primary contaminant source removal. A variety of publications address the effectiveness and limitations of vacuum extraction, and the U.S. EPA has even gone so far as to publish a reference handbook detailing the design, operation, evaluation and costs associated with vacuum extraction treatment systems (U.S. EPA, 1991).

Vacuum extraction is attractive not only because of its proven efficiency as a standalone remediation process, but also because of its flexibility. It is easily adapted to a variety of subsurface problems, and can be combined with any number of additions and enhancements to address more complex cases of subsurface contamination.

4. The process of vacuum extraction

In general terms, the Soil Vacuum Extraction (SVE) process exerts a negative pressure gradient to the subsurface soils and thereby induces air flow toward the extraction well. A schematic drawing of the SVE process is depicted in Figure 2. As the air flows through the soil pore spaces, the contaminants are volatilized in place and migrate toward the extraction well where they are removed and treated at the surface, normally by activated carbon, catalytic oxidation or other vapor treatment methods. Figures 3 and 4 feature typical SVE process diagrams.

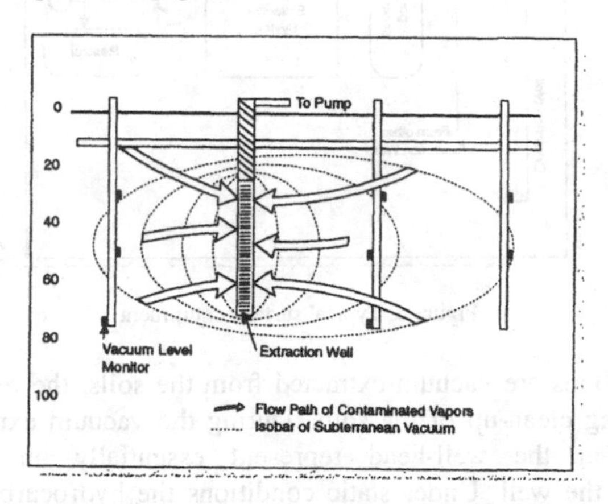

Figure 2 Soil vapor extraction (SVE) process

Under normal static conditions within the soil matrix, petroleum hydrocarbons and VOC's are partitioned between four possible phases: 1) vapor, 2) liquid, 3) dissolved in soil water, 4) adsorbed to solid particles. These four phases define the aggregate contaminant concentration in the soil. The vapor phase partitioning is a complex function of water content, organic content, solubility, temperature, and vapor pressure. It is not necessary to define the exact relationship between soil concentration and vapor concentration as a function of time in order to understand that reductions in extracted vapor concentrations are driven by continuous partitioning from the liquid phase into tbe vapor phase which corresponds to reductions in soil contaminant concentrations.

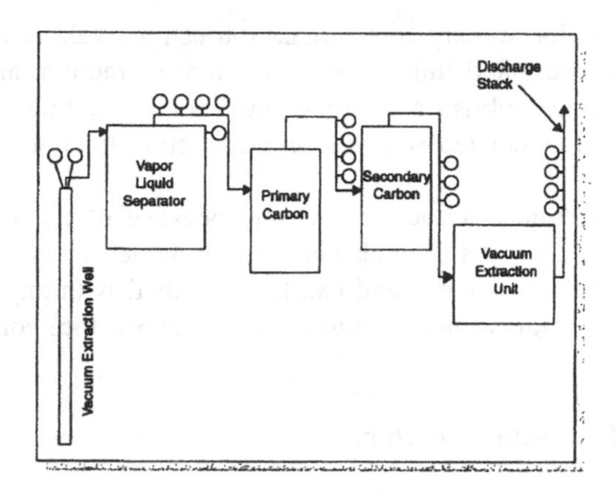

Figure 3 Typical SVE surface equipment

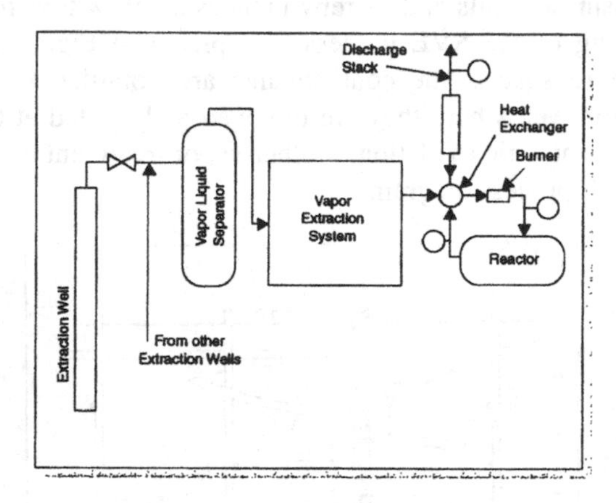

Figure 4 Typical surface equipment

When hydrocarbons are vacuum extracted from the soils, the removal rate declines with time, indicating clean-up of the soils. During the vacuum extraction process, the vapors extracted at the well-head represent essentially an aggregate soil-gas concentration near the well. Under static conditions the hydrocarbon concentration in the soil vapor is proportional to the aggregate contaminant concentration in the soil.

As vast volumes of soil vapor are removed by the vacuum extraction process, fresh air naturally recharges into the soils from the surface. In some cases, air may be injected, but generally air injection is not required. The clean recharge air moves through the contaminated zone and the hydrocarbons are partitioned from the soil matrix to the vapor phase and move to the extraction wells. With vacuum induced volatilization and air stripping of the soil matrix, clean-up occurs continuously.

While contaminant vapors are removed from the soils pore volume, the three other phases (liquid, dissolved, adsorbed) of hydrocarbon contamination vaporize in place, further reducing the aggregate soil concentration. Since hydrocarbons vaporize readily (depending on molecular weight, temperature, and vapor pressure), the vacuum extraction process continuously drives the contaminants within the soil matrix to the vapor stem, and the clean-up can be monitored by analyzing the extracted gas concentrations. Figure 5 shows a typical exponential decline of extracted soil gas concentration from an SVE system.

In certain cases, the contaminants consist of LNAPL and DNAPL, or the SVE process is in close proximity to the groundwater table. In such instances, the groundwater table may upwell as a result of the vacuum applied to the subsurface, as shown in Figure 6. Such upwelling reduces the area of open well screen, causing reduced subsurface air flows and lower contaminant extraction rates. To mitigate these effects, it is practical to simultaneously pump the groundwater while performing SVE, a process known as Dual Vacuum Extraction^{TM}.

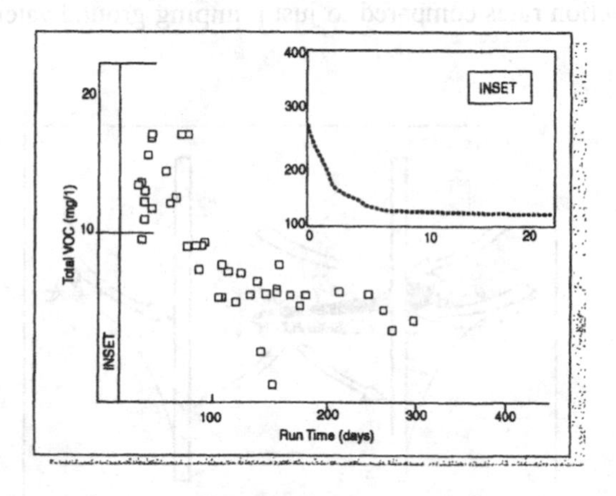

Figure 5 Typical concentration VS time

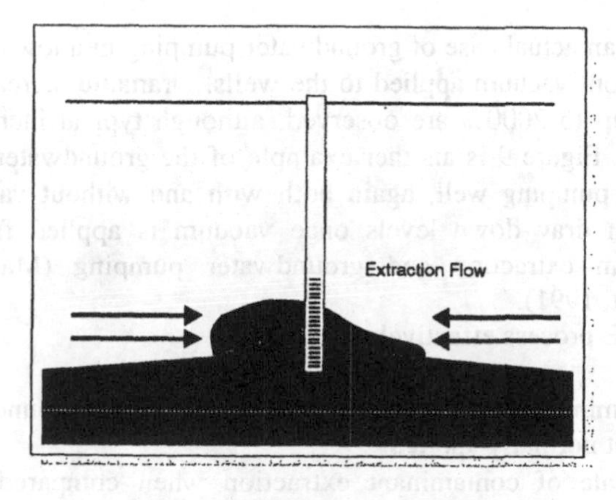

Figure 6 Groundwater up welling

5. Dual Vacuum Extraction™

Dual Vacuum Extraction™ (DYE) operates in essentially the same manner as SVE, except that DVE is a combined process whereby groundwater extraction is used to lower the groundwater table in the SVE well and thereby expose more soil volume that can be cleaned by the vacuum extraction process alone. Figure 7 represents the DVE process. Groundwater extraction can be accomplished by a variety of methods depending on the site specific conditions. Such methods may include submersible groundwater pumps, ejector or jet pumps, vacuum enhanced entrainment, or a combination of methods. Typically, DVE results in overall reduced costs because groundwater and soils are treated from the same extraction wells. Also, it has been noted in field experience that the negative pressure exerted on the recovery well can substantially increase groundwater extraction rates compared to just pumping groundwater without vacuum in the well.

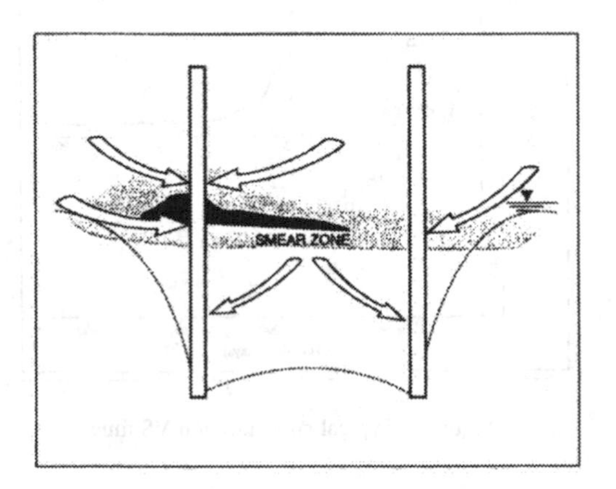

Figure 7 Dual vapor extraction ™ (DVE) process

Figure 8 shows an actual case of groundwater pumping in a low permeability aquifer both with and without vacuum applied to the wells. Dramatic increases in groundwater pumping rates of up to 2000% are observed, although typical increases from 10% to 100% are common. Figure 9 is another example of the groundwater draw-down versus time for an actual pumping well, again both with and without vacuum applied. The striking increase in draw-down levels once vacuum is applied further supports the synergy of vacuum extraction and groundwater pumping (Malot, et. al., 1990; Trowbridge and Ott, 1991).

Overall, the DVE process effectively:

- Recovers contaminants from above and below the static groundwater level, where SVE alone is not normally applied.
- Increases the rate of contaminant extraction when compared to the individual processes of SVE or groundwater pumping alone.

- Increases the rate of groundwater extraction, especially in lower permeability soils, and significantly increases the capture zone of a pumping well.
- Increases the magnitude and rate of groundwater level drawdowns and mitigates the migration of contaminants away from the source
- Increases the subsurface radius of influence of the SVE wells.
- Reduces the number of extraction wells required at a particular site.
- Reduces the overall cost of remediation.

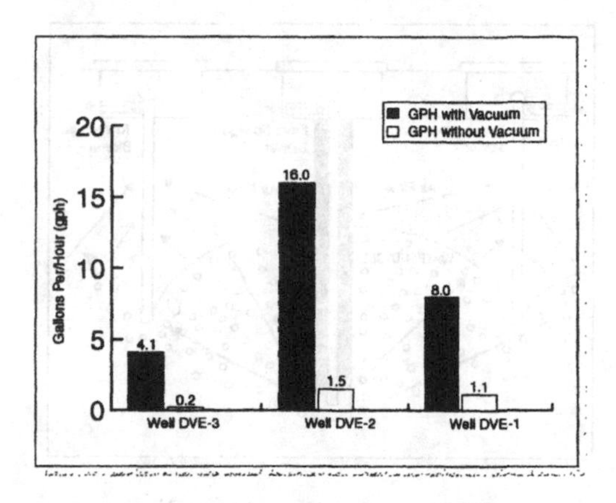

Figure 8 Groundwater extraction with/without vacuum applied

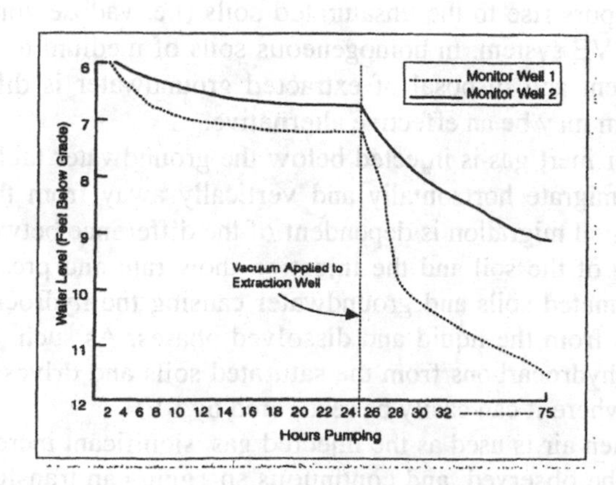

Figure 9 Groundwater depression with/without vacuum applied

6. Air Sparging (SpargeVAC™)

One of the more recent technological advances is the combination of air sparging and SVE or SpargeVAC . In contrast the DVE process described above, wherein groundwateris simultaneously extracted from tbe SVE welt the SpargeVAC™ process functions like an in-situ air stripper, and it is most favorable for applications in higher permeable settings. The orientation of SVE and air sparging wells may be vertical or horizontal, or any combination of the two, depending on the site specific conditions. Figure 10 is an example of one option for a SpargeVAC™ application.

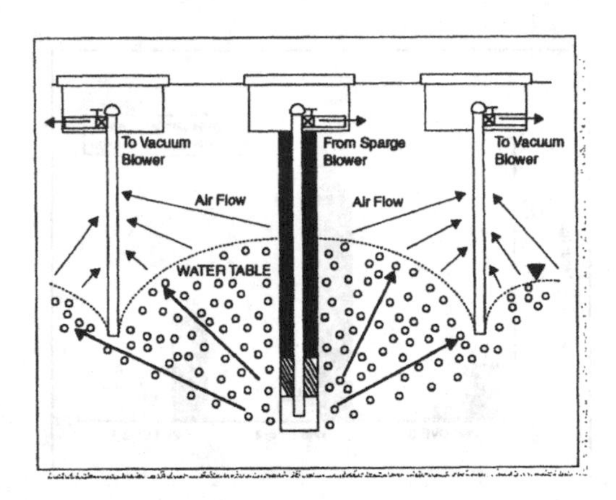

Figure 10 Air spargeing (spargeVAC™) process

Integrated SpargeVAC™ systems have been demonstrated to effectively and simultaneously clean both soils and groundwater in-situ, without groundwater pumping. SpargeVAC™ involves the injection of clean air or other gas (i.e. nitrogen) into the saturated zone to effect partitioning of adsorbed and dissolved hydrocarbons to the vapor phase. The vapors rise to the unsaturated soils (i.e. vadose zone) where they are recovered with the SVE system. In homogeneous soils of medium to high permeability, or where the treatment and disposal of extracted groundwater is difficult or costly, a SpargeVAC™ system may be an effective alternative.

When air or other inert gas is injected below the groundwater table, the injected gas forms bubbles that migrate horizontally and vertically away from the injection point. The extent of horizontal migration is dependent of the difference between horizontal and vertical permeability of the soil and the injection thow rate and pressure. The injected gas contacts contaminated soils and groundwater causing the hydrocarbons to partition into the vapor phase from the liquid and dissolved phases. As such, the SpargeVAC™ process desorbs the hydrocarbons from the saturated soils and drives the contamination into the vapor state where it can easily be recovered by SVE.

Additionally, when air is used as the injected gas, significant increases in subsurface oxygen content can be observed, and continuous sparging can transform the essentially anaerobic(no oxygen) conditions below the groundwater table into aerobic (oxygen rich) conditions. Often SpargeVAC™ raises the soil and dissolved oxygen content to significantly enhance the aerobic biodegradation of hydrocarbons.

In one example of a SpargeVAC™ application, a 21-fold increase in extracted toluene concentrations and a 300% increase in TCE and PCE concentrations was observed (Piniewski et. Al., 1992). In this project groundwater pumping alone had previously been utilized for five years, pumping 200 m^3 per day. Over the five year period, groundwater concentrations were reduced from 19 mg/1 (ppm) to 3 mg/1. Using the combined SpargeVAC™ process, the groundwater contamination was reduced further by 93% to less than 150 ug/1 (ppb) in four months, and virtually eliminated any above ground treatment of the groundwater.

In another example at a major international airport in New York, a SpargeVAC™ remedy removed more than 45,000 kg of aviation fuel in less than one year with no groundwater treatment system. The system consisted of more than 4000 feet of horizontal air sparge and SVE wells which were installed beneath the jetliner taxi ways. Utilizing the horizontal drilling methods, there was no interruption to the busy daily air traffic at the airport.

7. Bioremediation (BIOVAC®)

Further advances for in-situ remediation were made almost without any intended technological innovation, but essentially by intuition and empirical recognition at variousoperating sites. Since the first discovery, BIOVAC® or bioventing has become another highly acclaimed processes for remediation of the full range of hydrocarbon compounds. The successful biodegradation of petroleum hydrocarbons is well documented in the literature. Biodegradation in the environmental sense refers to the breakdown of harzardous substances by micro-organisms or their byproducts. It occurs naturally in all sorts of environments, an example being the decomposition of organic materials in the soils of a forest.

Biodegradation falls into two major categories, aerobic degradation, which requires oxygen, and aerobic degradation, which does not. For instance, the biodegradation of petroleum hydrocarbons such as gasoline, aviation and diesel fuel is essentially aerobic, while the biodegradation of halogenated hydrocarbons such as trichloroethylene (TCE) and tetrachloroethylene (PCE) is essentially anaerobic.

Hydrocarbon biodegradation is fundamentally an oxidation-reduction or electron transfer process. Biological energy is obtained through the oxidation of reduced materials and microbial enzymes catalyze the electron transfer. Essentially the hydrocarbons are oxidized (i.e. donate electrons) and an electron acceptor (e.g. oxygen) is reduced (i.e. accepts electrons). The electrons are removed from organic substrates to capture the energy that is available through the oxidation process. A large proportion of the microbial population in the soil depends upon oxygen as the terminal electron acceptor for metabolism. Loss of oxygen induces a change in the activity and composition of dhe soil microbial population. When oxygen is absent, organisms may switch to available nitrates (NO_3-) and sulfates (SO_4-) and perhaps iron oxides [e.g. $Fe(OH)_3$], manganese (Mn_3^+, Mn_4^+), or bicarbonate(HCO_3-) as the election acceptors if the microbes have the appropriate enzyme schemes to be able to utilize such compounds. However, oxygen is the most preferred electron acceptor because microorganisms gain more energy from aerobic reactions. As a result' aerobic biodegradation of hydrocarbons is generally must faster than aerobic degradation.

Though each petroleum hydrocarbon has a unique stoichiometric equation for biological oxidation, the mass of hydrocarbons biologically oxidized can generally be estimated from the following equation for hexane:

$$C_6H_{14} + 9.5\ O_2 = 6\ CO_2 + 7\ H_2O$$

From this equation if follows that roughly 1 kg of hydrocarbons is degraded for every 3 kg of carbon dioxide (CO_2) produced. In practice a 50% yield factor is assumed as CO_2 is also utilized by microorganisms for the biological synthesis of cellular mass.

The goal of the BIOVAC® process is to use aerobic degradation. Generally it is the oxygen deficiency in the soils which are contaminated with fuel products that limits the rate of biodegradation. One very effective method to bring oxygen to the subsurface soils is by moving air through the soil, and the SVE, DVE, and SpargeVAC™ processes are ideally suited for this task. As the vacuum extraction system operates, it withdraws vast amounts of contaminant loaded air from the subsurface and fresh air containing 200,000 ppm oxygen recharges into the soils to reoccupy the pore space. In the case of SpargeVAC, the oxygen-rich air is injected below the groundwater table to provide a large zone of aeration wherein there is active bioremediation. With oxygen introduced, biodegradation occurs as a natural reaction, and the rate of degradation can be evaluated by monitoring CO, in the extracted vapors. Normally the populations of indigenous microorganisms is sufficient and it is required to introduce any special microorganisms. However, in some cases where the population of indigenous microorganisms is too low or they have been substantially reduced by fatal toxicity to some compound, then it may be an alternative to use a specially formulated or laboratory grown microorganism.

Other important factors for a successful BIOVAC® operation are the careful monitoring and regulation of nutrients (i.e. nitrogen and phosphorous), temperature, soil moisture and pH. Although site specific tests must be performed to determine the optimal operating condition, as a rule of thumb one might target a carbon-nitrogen-phosphorous ratio of about 100-5-1, temperature of 10C to 30 C, soil moisture of 20% to 50% and a pH range of 6-8.

For design and operational guidelines, it is customary to obtain soil samples and perform a series of studies to determine optimum site specific conditions for enhanced bioremediation. Typically these studies might include a carbon dioxide respiration study which indicates the microorganism population and a metabolism study which consists of experimental and control groups to evaluate optimal nutrient combinations.

8. Integrated vacuum extraction and pneumatic soil fracturing (FRACVAC™)

In most applications of SVE, the pressure differential (i.e. vacuum) which is propagated in the subsurface guides the contaminants toward the extraction well through advective transport At sites with low permeability soils, the contaminant extraction rates by vacuum alone may not be sufficient to meet soil clean-up objectives within the desired time, as the flow of contaminants through the low permeability subsurface is controlled more by diffusion than by advection.

Figure 11 depicts the difference between advective and diffusive flow. Advection is a result of the differential pressure in the subsurface while diffusion is controlled by differences in the contaminant concentrations. Advective flow is orders of magnitude more rapid than diffusion. Increasing the number of subsurface paths for advective flow results in increased contaminant mass removal and accelerated remediation.

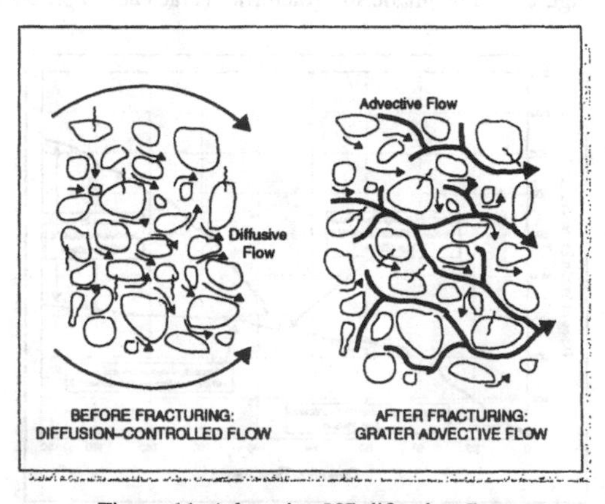

Figure 11 Advective VS difussive flow

One technique to enhance advective flow in the subsurface is pneumatic soil fracturing (ie. FracVac™). Here, the process injects pressurized air into the subsurface, creating micro fractures for the SVE system to withdraw the contaminants. The process is similar to hydraulic fracturing, a technique used for years in the petroleum exploration and production industry to increase the yield of hydrocarbons from low permeability rode formations. Figure 12 provides a schematic drawing of the pneumatic fracturing process.

For successful pneumatic fracturing, one must overcome the forces within the soil matrix. These forces include the overburden pressure gradient, matrix tensile strength, fluid pore pressure, and other matrix stresses. The fracture geometry is highly dependent on the matrix stress conditions near the injection point, and fractures will propagate preferentially along existing and faults and fissures.

In one particular case pneumatic fracturing was tested in a glacial till consisting of a firm clay with an estimated hydraulic conductivity of 10^{-7}cm/sec. The results were a three-fold increase in contaminant extraction rates. Figure 13 shows the results of the fracturing program compared to the theoretical extrapolation of the production curve without pneumatic fracturing.

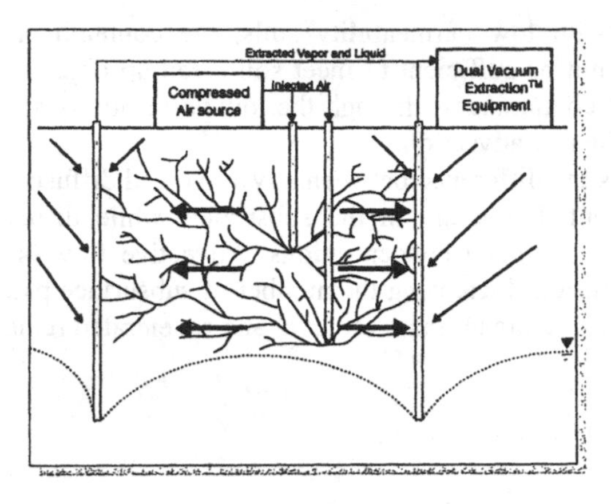

Figure 12 Pneumatic soil fracturing (FracVac^{TM}) process

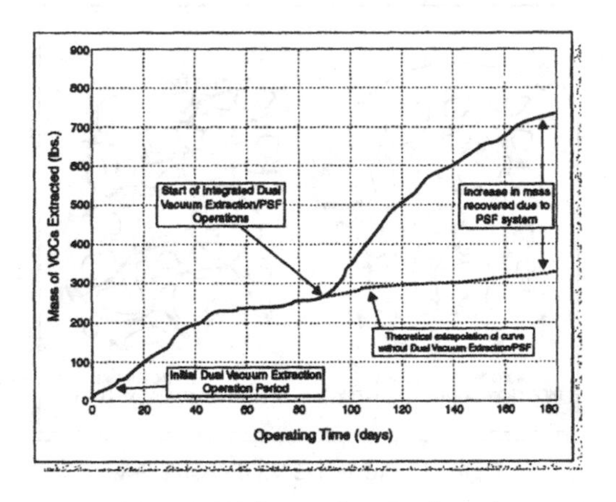

Figure 13 VOC extraction after fracturing

9. Soil heating

Much research and development is currently focused on methods to enhance contaminant removal by in-situ soil heating. SVE works faster on contaminants with a higher vapor pressure, and the higher the vapor pressure the more quickly the contaminants can be removed.

Although SVE has successfully removed heavier compounds and SVOCs, substantial savings in clean-up time and cost are recognized with increased temperature of tbe soils. Heating the soil offers several advantages in remediation by SVE. As a rule, each increase of 25 C doubles the contaminant vapor pressure and the Henry 's Law Constant (i.e. vapor pressure divided by water solubility). For any given contaminant concentration in the soil, heating the formation to about 70 C will increase the amount of contaminant vapor concentration by a factor of sixteen. Average ambient temperatures in the subsurface soils are about 10 C. In general, the features of soil heating include:

- Increased contaminant volatility
- Faster soil desiccation
- Enhancement of soil microbial population

Several methods exist to increase the subsurface soil temperature. Some of these include:

- Hot air injection
- Steam Injection
- Electric soil heating (6 phase soil heating)
- Radio frequency heating
- Oxidant injection

Each of these methods have unique advantages and disadvantages which must be evaluated on a site-specific basis. For instance, steam has a higher heat capacity and overall more efficient subsurface heating capability than does hot air; however, the steam rapidly cools and condenses in the subsurface causing higher formation water saturation, lower relative air permeability and reduced vapor flow. Six phase soil heating is a very promising technology which involves splitting three phase power and heating the soils with electrodes while extracting the vapors with SVE. Resultant subsurface temperatures have been measured greater than 120 C, and contaminant extraction rates multiplied tenfold over extraction rates under ambient conditions, yet the capital costs for equipment and operating costs can be high.

A promising new advance in technology has been the use of hydrogen superoxide (H_2O_2) to oxidize the hydrocarbon in-situ, and the method has also proven useful to de-chlorinate VOCs.

10. In-situ oxidation (OxyVAC™)

Ihe OxyVAC process has gained wide acceptance as a proven in-situ remedial technology to address a variety of subsurface contamination problems. Because of its oxidation potential, hydrogen peroxide (H_2O_2), and more specifically, the hydroxyl radical is known to be an effective treatment method for removal of contaminants from soils and waste streams. The oxidation potential of H_2O_2 is a well known phenomenon which has been studied since the turn of the century. The fundamental reaction is the oxidation of an organic molecule, such as hydrocarbon, phenol or chlorinated compound to form a variety of oxidized products. In a complete oxidation reaction, one by-product might include CO_2, or if the oxidation is incomplete, then by-products might include alcohols, aldehydes or carboxylic acids, all of which are very biodegradable. Fundamentally, the oxidation by H_2O_2 generates no toxic by-products, is environmentally benign, and the H_2O_2 itself may be degraded by subsurface microbial enzymes. A typical OxyVAC™ process is depicted in Figure 14.

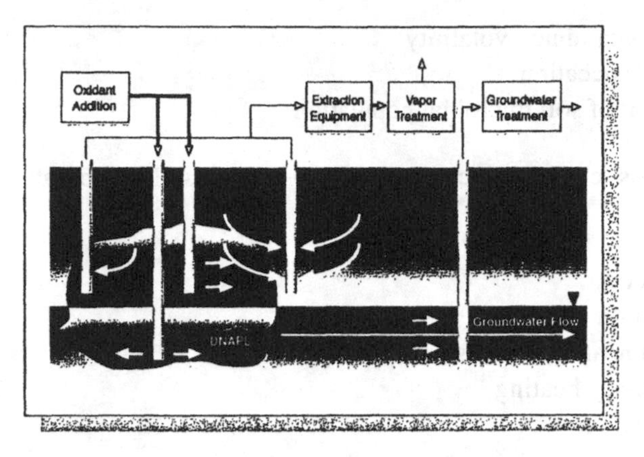

Figure 14 OxyVAC™ process

The catalyst of the oxidation is iron (Fe^{2+}) which when combined with H_2O_2 produces the hydroxyl radical (OH•) in a reaction commonly referred to as Fentons Reaction, named after the individual who first discovered the reaction in 1876. This reaction is written as:

$$H_2O_2 + Fe^{2+} \rightarrow OH\bullet + OH\text{-} + Fe^{3+}$$

This type of reaction is catalyzed when hydrogen peroxide contacts naturally occuring iron contained in soil and rock The heat produced by the reaction causes volatilization of the contaminants, which are then withdrawn from the subsurface by means of SVE. (Land, et. al. 1997)

In the case of halogenated organic compounds, H_2,O_2 has proven effective to dechlorinate the contaminant, and in-situ applications of OxyVAC™ for the treatment of PCE and TCE are proving very successful. TCE, for instance is reduced to carbon dioxide and water according to the following reaction (Gates, et. al 1995):

$$C_2HCl_3 + 4.5\ H_2O_2 \rightarrow 2CO_2 + 5\ H_2O + 3\ Cl$$

The introduction of hydrogen peroxide to the subsurface may be accomplished by a variety of methods, including direct injection to wells, surface injection, or by high pressure subsurface injection to open bore holes by methods similar to pneumatic fracturing or jetgrouting.

11. Funding soil and groundwater remediation

The "Bankable" Environmental Project

The most difficult part of nearly all remediation projects is the cost. With treatment cost running anywhere from $10 to $100 per ton for tbe in-situ metbods, and sometimes up to $1,000 per ton for excavation, incineration, and landfilling, the cost of digging and moving the dirt or "cleaning the dirt" in situ is painful in all instances, even though the economic bite from the in-situ options are typically less painful than those associated with excavation and treatment.

One proven method to mitigate the costs associated with remediating contaminated property is resource recovery, very similar to the common recycling programs for aluminum cans, plastics and paper. Likewise, with the innovative application of in-situ treatment technologies it is possible to recover and recycle certain products to offset some of the costs of remediation and in some instances, provide a favorable business platform.

The most promising prospects for resource recovery are found to be aging civilian and military airfields. Most of these air-bases have large aviation fuel storage facilities, and over the years the tanks and pipelines have leaked and plane loads of fuel have been dumped. It is not uncommon to find several meters of weathered aviation fuel floating on the groundwater table over several hectares. While this fuel can not again be utilized for aviation, it is generally suitable for use as fuel oil kerosene and diesel for domestic and agricultural purposes.

Costs of fuel in places like the Newly Independent States (NIS) have doubled, tripled or more over the past five years, and supplies are scarce in many locals. At one former Soviet air base there is an estimated 140,000 tons of aviation fuel in the subsurface distributed over four fuel storage facilities, and this is a small site compared to other air bases in the region. By using SVE and its sister technological advances in an innovative treatment train, approximately 70% of the hydrocarbon will be recovered for recycling, whilst the remaining portion is vaporized or biodegraded. Even if only 50% is recycled, at $200 (U.S.D.) per ton (approximate value in the region), the 70,000 tons of fuel represents a value of 14 million dollars. Significant in-situ soil and groundwater cleaning can be accomplished by SVE for this sum of money. Other sources of funding to bolster project financing may include the recycling or sale of unnecessary military equipment.

Innovative reinvestment schemes can be established whereby the remediation contractors are given full ownership to any revenues received from recycling or resale. The contractor in turn agrees to reinvest a certain percentage of the profits at each site to ensure that the remediation systems operate sufficiently long enough to meet certain risk-based environmental standards. It is proven possible for these remediation projects to pay for themselves, and in some instances provide sufficient additional funds to help clean other properties where resource recovery is not economically feasible.

Furthermore, local economies have the advantage of having contractors working in their towns, helping to support the local economies with their hiring of local labor and general living expenses. In the end, it is a win-win-win situation for all parties.

12. Conclusion

Subsurface contamination problems represent some of the most sensitive and complex environmental situations everywhere in the world, and it only takes a small amount of spilled gasoline or solvent to cause a large problem. Complex problems require innovative solutions, and fortunately there are well tested and proven effective technologies to replace excavation and landfilling as a remedial option. Of the methods to clean properties at the source of contamination and to mitigate the migration of contamination away from its source, the vacuum extraction technology along with its sister enhancements, Dual Extraction™, air sparging, bioremediation, pneumatic fracturing, soil heating, and in-situ oxidation have a long list of successes in instances where petroleum hydrocarbons, VOCs and SVOCs are concerned It has been proven time and time again that an aggressive approach to remediate the source of contamination is nearly always the most cost effective and permanent environmental solution. Furthermore, when the source of contamination contains an economically feasible amount of recoverable product, a well planned program of recycling allows these projects to pay for themselves.

13. References

1. Cunningham, Keith B. (1997) *Base Closure and Redevelopment in Central and Eastern Europe,* Bonn International Center for Conversion, Report No: 11.
2. Schwille, F. (1988) *Dense Chlorinated Solvents in Porous and Fractured Media,* Lewis Publishers, Chelsea, Michigan, U.S.A.
3 U.S. Environmental Protection Agency (1989) *SITE Program Demonstration Test: Terra Vac In-Situ Vacuum Extraction System,* Publication No: EPA/540/5-89/003 a,b, Risk Reduction Engineering Laboratory, Office of Research and Development, Cincinnati, Ohio, U.S.A.
4. U.S. Environmental Protection Agency (1991) *Sod Vapor Extraction Technology Reference Handbook* Publication No: EPA/540/2-91/003, Risk Reduction Engineering Laboratory, Office of Research and Development, Cincinnati, Ohio, U.S.A.
5. Malot, J.J. and Piniewski, R.J. (1990) *Innovative Technology for Simultaneous In-Situ Remediation of Soil and Groundwater,* Hazardous Materials Controls Research Institute, Cleveland, Ohio, U.S.A.
6. Trowbridge, B.E. and Ott, D.E. (1991) *The Use of In-Situ Dual Vacuum Extraction for remediation of Soil and Groundwater,* National Groundwater Association, Dublin, Ohio, U.S.A.
7. Land, C.A., et. al (1997) *Process for Soil Decontamination by Oxidation and Vacuum Extraction,* United States Patent No: 5~615,974; PCT/US94/01315, Washington, D.C., U.S.A.
8. Gates, D.D., et al. (1995) *Chemical Oxidation of Volatile and Semi-Volatile Organic Compounds in Soil,* Air and Waste Management Association, 88th. Annual Meeting, San Antonio, Texas, U.S.A.
9. Dual Vacuum Extraction™, SpargeVAC™, BIOVAC®, FracVac™, and OxyVAC™ are registered United States trademarks of the Terra Vac Corporation.

GROUNDWATER CONTAMINATION CLEAN UP - REALITY OR ILLUSION?

BJARNE MADSEN
Department of Hydrology and Glaciology
Geological Survey of Denmark and Greenland
Thoravej 8 DK-2400 Copenhagen NV
Denmark

Abstract

In most cases the data and the background information provided as the basis for remediation activities do not describe in sufficient detail the variability of the flow and transport mechanisms controlling the clean-up processes. Thus the efficiency of the remediation techniques are often overestimated, and excessive trust is placed in our ability to clean up contaminant plumes and restore the groundwater quality. In light of various unsuccessful projects implemented in Denmark as well as abroad, there is an obvious need for a critical review of the reliability and efficiency of the remediation techniques and their justification in relation to restoration of contaminated groundwater reservoirs. This paper examines some of the key problems associated with clean-up techniques illustrated by an example from an abandoned chemical waste dump in Denmark.

1. Introduction

Remediation and clean-up of contaminated groundwater by "pump-and-treat" methods is a well-developed technique which has been widely used in Denmark as well as in other countries around the world. A variety of methods have been developed in order to remove contaminated groundwater by pumping, accompanied by on-site advanced water treatment. The efficiency of the methods are further developed to also include systems of recirculation and *in-situ* remediation using additives and microorganisms.

Apparently this procedure has been quite successful when compared to the billions of dollars spent on groundwater restoration in the western world during the past decades. Lately the developed methods have been heavily marketed in the Eastern European countries for remediation of numerous sites with extensive groundwater contamination.

The efficiency and accordingly, the success of the methods has generally been evaluated on the basis of two main characteristics: First, the ability to control the movement of contaminant plumes and prevent it from further propagation (hydraulic fixation), and second, the ability to remove contaminants from the groundwater zone. While the first property by careful design can be effectively applied, the benefits of the second

F. Fonnum et al. (eds.),
Environmental Contamination and Remediation Practices at Former and Present Military Bases, 257–265.

property are far more debatable than is generally considered. The reason for this originates in the different requirements for accuracy and thereby the scale to be considered. Flow control (volumes) can be managed adequately on average potential and gradient assessments which do not depend very much of the scale, whereas the processes related to contaminant transport and remediation are velocity controlled. In this case, succesful analyses are therefore highly dependent on the scale and the possibility for obtaining sufficiently detailed knowledge on the variability of hydraulic conductivity, sorption, and degradation characteristics.

The excessive confidence in the clean-up efficiency of "pump-and-treat" methods may be explained by the criterion for success that is generally applied, which in many cases is too much based on average considerations, and only based on the concentration of polluting substances in the pumped water as opposed to the real changes within the aquifer. The concentration of the pumped water will generally decrease with time during continuous pumping, and after sufficient time will be below the acceptable limits, thus confirming the modelling predictions. However, this procedure gives only a very coarse measure for the entire content of contaminants in the groundwater reservoir describing some average situation. The impact from heterogeneities and the variability of the geologic and hydraulic conditions and thereby the distribution of the residual concentrations will be reflected by looking only at the pumped water. Much more detailed monitoring is required at least if any improved control of the remediation process is needed.

Lately a number of North American researchers have been emphasising the problems with the large investments in pump and treat systems, which in their opinion in many cases cannot be justified, (Nyer [1], Bredehoeft [2] and MacDonald *et al.*) [6]. The methods have also in our opinion limitations which need to be paid great attention in order not to waste money on fruitless efforts which to some degree might be solved by natures` own ability to control the processes.

2. The Hydrogeological Reality

Assuming ideal conditions where the groundwater flow can be characterized by complete uniformity (homogeneous and isotropic conditions), and where the contaminating substances do not adsorb to the aquifer material, it may from a practical point of view be possible to remove any undesirable contaminants to a sufficiently low level within a limited period of time. Unfortunately such ideal conditions will not be represented by any of the geological environments in the real world. Even major sandy outwash deposits, which traditionally have been viewed as uniform sandboxes will generally turn out to be rather heterogeneous with regard to the scale required for describing contaminant transport.

Apparently secondary variations in the structure of the sedimentary units have often significant impact on the variation in hydraulic conductivity. In lower permeable zones the migration of the solutes will be significantly slower than in high permeability zones, primarily due to low velocities, and secondarily due to adsorption. This will often result in considerable variations on the detailed scale in flow velocities and transport patterns and be observed on a larger scale as dispersion.

During a clean up situation the pumping will increase the flow velocity, especially within the zones of preferred flow. These zones will often represent a relatively limited part of the entire cross section. Assuming a uniformly distributed concentration in the pumped aquifer, the amount of contaminating substances in these high producing zones will consequently be small in relation to the total quantity of substances, and will be relatively quickly flushed out. Subsequently the lower permeable zones slowly liberate material to the high yielding zones by diffusion. This will altogether result in a decline in the concentration of contaminants in the pumped water, but considerably less than for a homogeneous situation as can be seen from figure 1. If, on the other hand, the effect of pumping is measured *in situ* within the low permeable units, the concentrations will only change very little as compared to this!

The difference between the apparent aquifer concentration represented by the pumped water, and the true aquifer concentration (and its distribution) will be much more significant when the heterogeneity of the geology increases as is the case in real layered aquifers, not to mention deposits with fracture and matrix interaction (dual porosity) like in limestone and clayey till, Fredericia *et al.* [3]. In those situations one will typically observe a significant increase in the concentration of the pumped water if the pumping is stopped and restarted after some time. If the criterion for success had been based on *in situ* measurements in the contaminant plume, the result would have been quite different and exposed the true clean up process. The situation will be illustrated by the following example from the Skrydstrup Landfill.

3. The Skrydstrup Waste Disposal Site

In 1984, investigations of the waste disposal site at Skrydstrup in the county of South Jutland were initiated. The hazardous waste buried in the former gravel pit was identified to include mainly chlorinated solvents which a local refrigerator factory had deposed over a seven years period. The waste material was excavated, and the residues which could not be removed by other means were disposed of at a special reserve.

Parallel to the removal of the contaminant, source mapping of the extent of the contaminant plume was initiated. It was estimated that approximately 30 tonnes of chlorinated solvents had been disposed of, and an unknown fraction had over the years leaked to the groundwater zone. Based on this information, a "pump-and-treat" system including well abstraction in central parts of the plume was established in order to remove contaminated groundwater and prevent further migration of contaminants towards Jernhyt Stream and the Vojens water works only some few kilometres downstream. The pumped water was to be cleaned in an on-site purification plant including aeration and carbon filtration processes, and reinfiltrated within the former waste disposal area. The provisional investigations showed that the aquifer would be restored after 16 -20 years of continuous pumping.

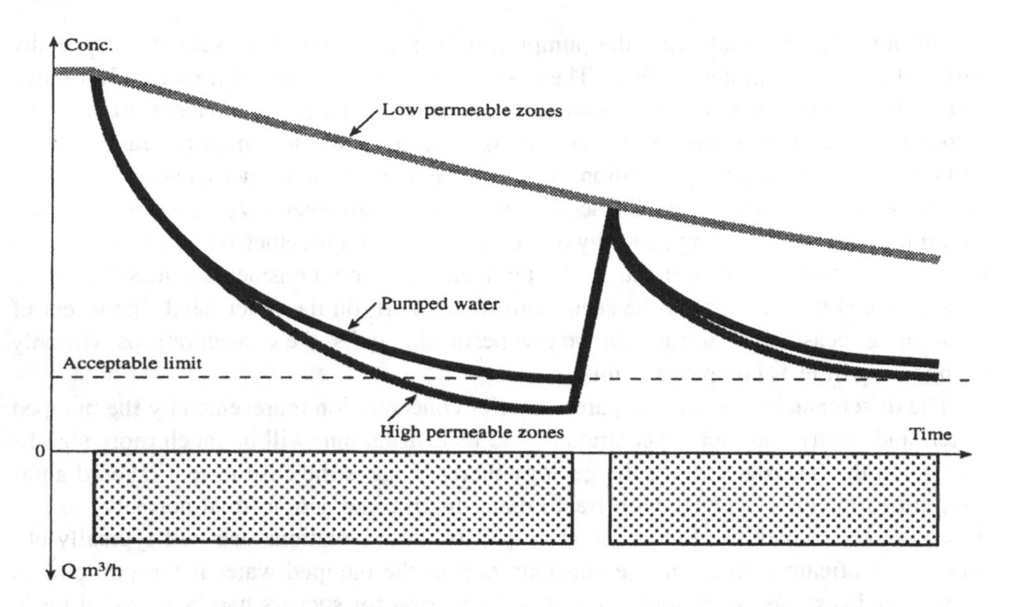

Figure 1. Decline in concentration as a function of pumping periods and observation point .

The mass of contaminants extracted during the first few years of operation corresponded well to the assumptions and predictions made for the operation of the system. Therefore, it was thought that the "pump-and-treat" system was operating according to schedule. As planned, the "pump-and-treat" system was to be evaluated in order to deter mine if changing the well field schedule would improve the efficiency of the "pump-and-treat" system. However, based on supplementary field investigations of the contaminant distribution, which were supported by solute transport modelling, it became obvious that the remediation was not having the required effect [4]. Polluted groundwater was measured in close proximity to the recipients, a condition which was supported by the most recent modelling results. This also indicated that the contaminated groundwater was already discharging to Jernhyt Stream.

A subsequent attempt to optimize the original pumping strategy on the basis of detailed model simulations was not very promising, as the following problems were identified. In the first place, the leakage of contaminants to Jernhyt Stream had already started and would be increased during the coming years no matter which remediation strategy was selected. Even a tripling of the existing pumping rate corresponding to approximately 110 m^3/h would not be sufficient to prevent further spreading of contaminants.

Secondly, of the original estimated 24 tonnes of contaminants in the aquifer, only five tonnes had been removed by pumping. Of the remaining contaminants only an additional five tonnes may in practice be extracted from the aquifer by pumping during a 30 year period, assuming effective and realistic pumping strategies. Five tonnes would still remain in the lower permeable parts of the aquifer and approximately 10 tonnes of pollutants would discharge into the Jernhyt Stream riparian area.

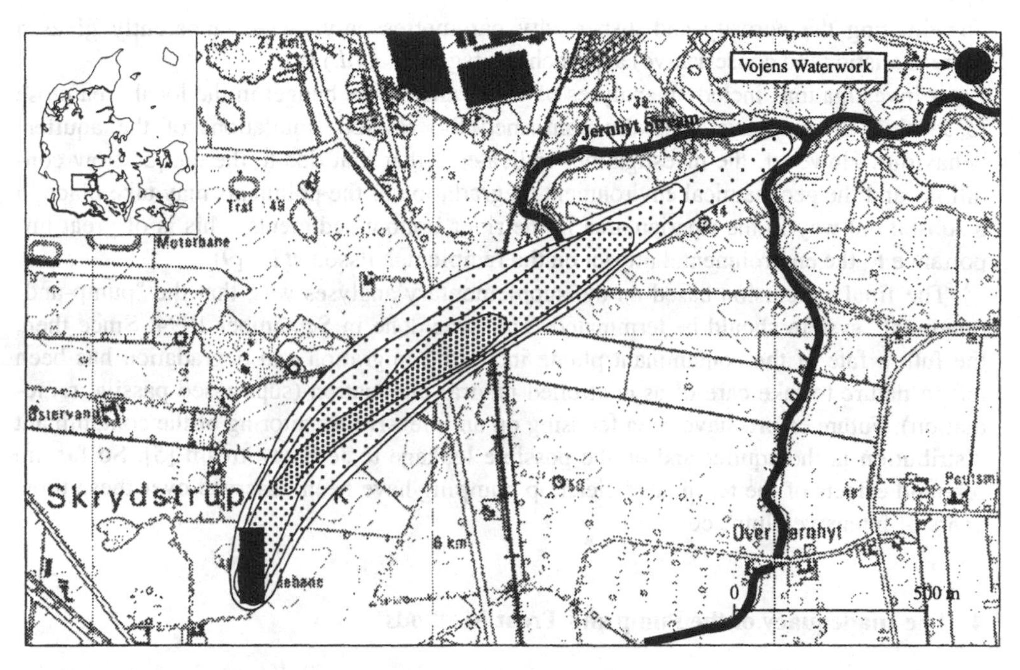

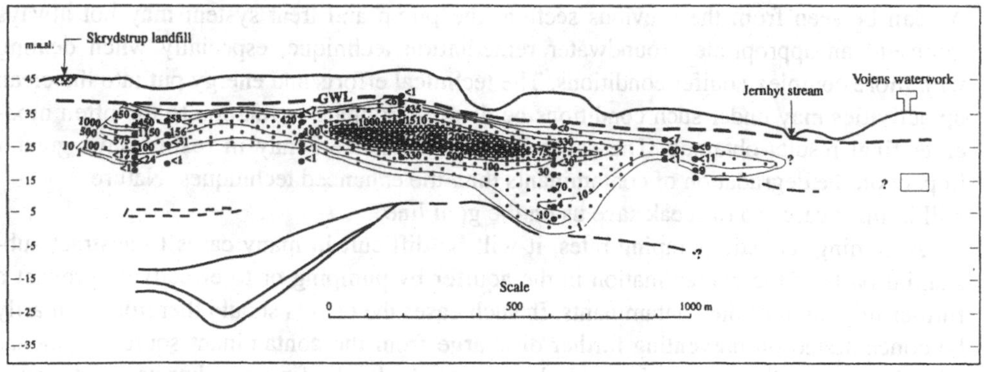

Figure 2. Map and cross section.
The area of the Skrydstrup Waste Disposal Site showing the extent of the contamination plume in 1992.

The original predicted migration of the contaminant plume on which the remediation strategy has been based was found inaccurate as additional results were obtained from field observations. The significant corrections of the preliminary conclusions made are closely related to the assumptions and basic conditions for the model setup. During the initial simulations, it was assumed that the pollutants would be degraded in the aquifer in accordance with some fixed rate of decay based on general experience. However, detailed field investigations have shown that the occurrence of such degradation cannot in general be justified at this site, and that retardation by sorption is the only noticeable process involved. Consequently, the existing transport model calibration has been based

on adjusting the porosity and dispersivity parametres only, which apparently gives a more realistic (and conservative) approach (Refsgaard *et al.*) [8].
Later investigations including analyses of water and solute budget in the local area close to Jernhyt Stream have generally confirmed the revised simulations of the aquifer-behaviour. However, the investigations indicate further that due to the unique flow conditions and the geochemical environment, degradation of the pollutants may take place in a substantial way in the riparian area in the river bottom sediments. This is of great importance to the environmental impact on the recipient, Nilsson *et al.* [9].

The final conclusion based on the supplementary analyses was, that the "pump-and-treatment" system should be terminated. This was done in September 1994. Since then, the future fate of the contaminant plume including migration and degradation has been left to nature to take care of as controlled natural attenuation (supervised passive remediation). Future efforts have been focusing on an intensive monitoring of the contaminant distribution in the aquifer and on the possible leakage to Jernhyt Stream [5]. So far, no rebound effects of the terminated clean up pumping have been observed over these three years of intensive plume control.

4. The Inadequacy of the Pump and Treat Methods

As can be seen from the previous section, the pump and treat system may not always represent an appropriate groundwater remediation technique, especially when dealing with more complex aquifer conditions. The technical efforts and energy put into the clean up activities may under such conditions be disproportionate in relation to the often moderate final results obtained. The capability of nature itself may in reality have greater impact on the degradation of contaminants than the enhanced techniques. Nature will in most cases so to speak save us on the goal line.

Assuming realistic pumping rates, it will be difficult in many cases to abstract substantial parts of the contamination in the aquifer by pumping or to effectively prevent a further migration of the contaminants. In such cases the efforts should therefore primarily be concentrated on preventing further discharge from the contaminant source by means of well tested techniques, such as simple capping, hydraulic fixing by horizontal drainage systems and/or pumping wells, construction of vertical flow barriers such as reactive walls or funnel and gate instumentation, etc. In close connection with these techniques a controlled "passive remediation" of the existing contamination is applied, which relies on natures` own degradation capacity accompanied by intensive monitoring.

In order to put the things into a wider perspective, in the oil business a long and laborious struggle has taken place with the purpose of extracting as much oil as possible from the subsurface. Over the years gigantic amounts of money have been spent on squeezing a few extra drops of oil out of the underground by enhanced recovery techniques including hydraulic fracturing of the oil bearing formations, and by injection of gas and water containing detergents, chemicals etc.

In spite of this, it has only been possible to recover approximately 15 - 20 % of the total volume of oil in a number of the fields in the North Sea and up to 20 - 30 % at other places where the conditions are more favourable. Even though the situation from a purely physical point of view is not directly comparable with the groundwater contamination problem, these low success rates of oil recovery are in themselves suggestive and make it difficult hard to believe that it should be possible in the field of environmental management to obtain an approximate cleaning up of aquifers contaminated with hazardous solutes by applying traditional pumping techniques. It requires at least carefully prepared methodologies based on detailed information on geological structure and flow and transport mechanisms on small scale level to establish the necessary background as a minimum for expecting a satisfactory rate of success of pump and treat systems on a pragmatic operational level.

The complete success which can be defined as total removal of contaminants from the aquifer can never be obtained by pumping only. Degradation processes have to be included. The basic reason behind this is very closely connected to the fundamental physical concept of irreversibility, which characterizes the behaviour of complex systems like flow in a non-homogeneous geological environment. The precondition for having the possibility to operate complicated systems as a reversible process is knowing all structures down to the details where they can be regarded as homogeneous. In geological terms, this means that theoretically it is necessary to operate on a grain-size scale (in practice on lamina scale). Sufficient description on this level may be possible in laboratory experiments, but is obviously impossible for large-scale systems like contaminated groundwater reservoirs. The problem can theoretically be analysed by making comparison to the second law of the theory of thermodynamics which states that the entropy of processes in closed systems will either increase or remain constant. In practice this means that there will be a heat loss in every prosses, or in groundwater a loss of contaminants which cannot be recovered. So, it is not exceptional that one cannot restore aquifers completely by pumping. There is a basic physical explanation for this!

5. Clean Drinking Water in the Future

Recognizing the weakness of "pump-and-treat" methods as a tool for cleaning up contaminated groundwater, this point should be reflected in the discussions of future planning on how the limited resources for safeguarding groundwater quality can be prioritized. However, the "pump-and-treat" method will still be one of the tools to be considered for implementation of protection strategies. It is therefore of great importance to further develop and adjust the techniques. It can especially be recommended to design any pumping system on the basis of very conservative considerations with regard to the flow conditions in the aquifers, as well as to the geochemical environment and the chemical characteristics of the solutes.

The most immediate consequence of a more critical review of the "pump-and-treat" method would probably be that a number of groundwater contaminated sites cannot be cleaned up within manageable limits, either in relation to time or to economy. As a logical consequence it has so far yet to be generally accepted that there will be areas where the groundwater is not clean. It will therefore be of great importance to identify and protect areas of significant drinking water interest which in future are going to form the backbone of the Danish water supply.

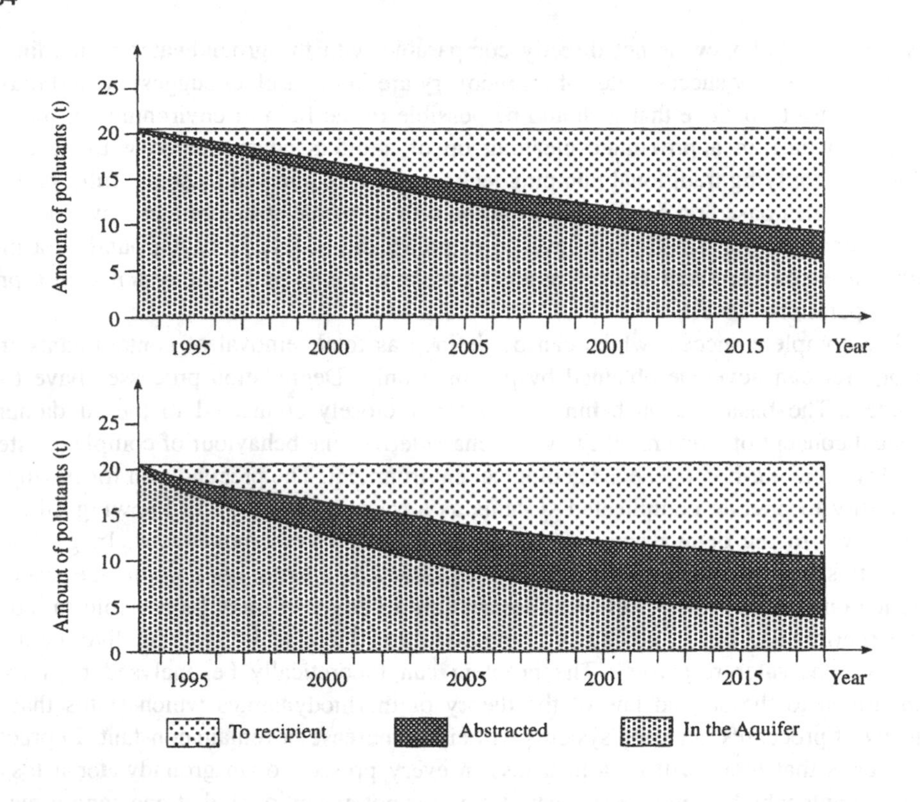

Figure 3. Calculated division of future quantities of contamination for two alternative remediation strategies.

a) Continuation of the existing remediation strategy including pumping of 35 m^3/h from the same wells as previously.

b) Increase of the pumping rate to 55 m^3/h from optimum located abstraction wells. Application of the separation pumping technique may be discussed as the contaminant plume is of limited vertical extent.

The experience gained during recent years has fully substantiated that also in groundwater management prevention is preferable to cure. Therefore the resources spent on preserving good quality groundwater should primarily be concentrated on long term solutions and preventative measures including sustainable exploitation instead of provisional remediation and short-sighted technical clean up solutions. Because, once deteoriated the groundwater will need a very long period to recover - with or without remediation. The possible alternatives of today's protection measures including changed land use practice, forestation, nature conservation, etc. should be part of a future integrated planning of the groundwater resources and to the widest possible extent be utilized for the protection of future water supplies.

This paper is to a wide extent based on an article published at the Geological Survey of Ireland's 150th anniversary symposium on environmental geology in 1995, Madsen [7].

6. References

1. Nyer, E. K.,(1993) Aquifer Restoration: Pump and Treat and the alternatives. *Groundwater Monitoring Review*, winter 1993, 89 - 92.
2. Bredehoeft, J. (1992) Much contaminated groundwater can't be cleaned up. Editorial, *Groundwater*, Vol. **30**, No. 6, Nov.-Dec.
3. Fredericia, J., Larsen, F. and Madsen, B. (1992) Groundwater pollution in clayey till areas (in Danish). *Vand & Miljø* **3**, 74-77.
4. The County of Southern Jutland (1993) 3D-modelling of the contaminant migration from the Skrydstrup Waste Disposal Site. Reassessment of the existing pump and treat system (in Danish). Customer report no. 15. DGU & DHI.
5. The County of Southern Jutland (1996) Monitoring Report of the Skrydstrup Waste Disposal Site (1995 - 1996). (in Danish). GEUS No. 107.
6. MacDonald, J. A. and Kavanaugh, M. C. (1994) Restoring Contaminated Groundwater: An Achievable Goal? *Environ. Sci. Technol.*, Vol. **28**, No. 8, 1994.
7. Madsen, B.(1995) Clean-up of contaminated aquifers - a well preserved illusion? GSI 150th anniversary environmental geology symposium. Dublin, October 1995: The role of geology and hydrogeology in environmental protection. Geological Survey of Ireland.
8. Refsgaard, A., Nilsson, B. and Flyvbjerg, J. (1995) Skrydstrup Waste Disposal Site - a Case Study. Proceedings of the 68th international conference, WEFTEC'95, Vol 2: Residuals & biosolids / Remediation of soil & groundwater, Miami Beach, Florida, Oct. 21 - 25, 1995.
9. Nilsson, B., Madsen, B. Aamand, J. and Refsgaard, A. (1997) Natural Attenuation of Dissolved Chlorinated Solvents at the Skrydstrup Site, Denmark. Symposia S10: The Anatomy and Attenuation of Chlorinated Solvent Plumes in Granular Aquifers. GSA Annual Meeting, Salt Lake City, Utah, October 20-23, 1997.

REMEDIATION PRACTICE AT A FORMER MISSILE BASE IN LATVIA

H. WHITTAKER, D. COOPER
Emergencies Engineering Division, Environmental Technology Centre, Environment Canada
Ottawa, Ontario, Canada, K1A OH3
R. WHITTAKER
Health Protection Branch, Health Canada
Tunney's Pasture, Ottawa, Ontario, Canada, K1A OL2
J. AVOTINS, V. KOKARS, J. MALERS, J. MILLERS, S. VALTERE
Riga Technical University
Kalku iela. 1, Riga, Latvia, LV 1058
DZ. ZARINA
University of Latvia
Raina bulv. 19, Riga, Latvia, LV 1050

1. Introduction

Severe environmental pollution of soil and groundwater exists in Latvia from activities that took place at former Soviet military bases. Since 1994, the former owners of these territories have retaken possession of their lands and have started to use their properties for agriculture, exploitation of forests and recreation. These activities have increased the exposure of populations to toxic substances, with subsequent effects on human health.

Considering that one of the two most acute environmental problems present in Latvia involve the wastes left behind after the territories were abandoned by the Russian army, a policy goal based on the National Environmental Policy Plan for Latvia [1] states that one of the priorities is:

"to achieve significant improvement of the environmental quality and ecosystem stability in areas where it presents increased risks to human health, while at the same time preventing deterioration of environmental quality in the rest of the territory".

A joint project under the auspices of the Canadian International Development Agency (CIDA) and Environment Canada sought to create an Environmental Centre of Excellence at Riga Technical University. This objective was accomplished while demonstrating techniques for assessment and remediation at the former missile base located at Barta, Latvia. These activities were carried out in conjunction with the

F. Fonnum et al. (eds.),
Environmental Contamination and Remediation Practices at Former and Present Military Bases, 267–276.

Canadian environmental consulting company Gartner-Lee International and Riga Technical University.

The specific goal of the project was to instruct Latvian specialists and supply them with technologies in order to assess and remediate highly polluted areas that might be detrimental to both the environment and human health.

In addition to CIDA and Environment Canada, the worldwide Latvian community, Riga Technical University and Health Canada have contributed to the project budget of $1,5 million Cdn.

The project was organized in three phases. The first phase involved the training four specialists from Riga Technical University in Canada for periods varying from six months to one year. During the second phase, four more specialists were trained in Canada and the initial assessments of the former military bases at Tazi and Barta were undertaken. The third phase, which began in June, 1996 involving the demonstration of Canadian remediation equipment was accomplished with the participation of Latvian and Canadian specialists. Specialized equipment used in the assessment and remediation phases is retained by Riga Technical University.

2. Assessment

During the second phase of the project, a visual assessment was performed at the *SAMIN* spill site in Barta and at the nitric acid spill site in Tazi. This was followed by chemical analyses of surface samples. These results were used to delimit the perimeter of the polluted area. Further evaluation of the site required the drilling of wells to obtain samples for the analysis of soil and groundwater. These were performed at the field laboratory in Cimdenieki, Riga Technical University and the Environment Canada laboratories [2].

SAMIN, which is a mixture of 2,4- and 2,5-xylidine isomers is toxic at low concentrations (0.5 ppm through inhalation, skin exposure or by ingestion of food or water). According to ACGIH (American Conference of Government Industrial Hygienists) it relates to an environmental air pollution level of 2.5 mg/m^3. This corresponds to the maximum workplace concentration during an eight hour period. Xylidines belong to a group of compounds that have been found to be potentially mutagenic and carcinogenic. These effects are similar to those found for aniline compounds. Xylidines have manifested nephritic, pulmonary and general pathogenic effects. There are cases reported in the scientific literature where lethal effects have been reported by cutaneous absorption. Delayed health effects have been observed, as well as cumulative effects. Health and safety considerations were paramount for the whole duration of the project, with all field workers being outfitted with protective clothing including Saranex® coveralls, respirators, rubber and latex gloves, and rubber boots.

The 1995 assessment phase results indicated that:

1. There was an area of 15m x 5m contaminated with xylidine.
2. The concentration of xylidines in soil reached 20,000 ppm (2%)

3. The dissolved xylidine concentration in groundwater was as high as 500 ppm
4. The nitric acid spill was delineated and found to be localized and not posing an immediate danger to human health; for these reasons, it was deemed that the best approach would be for natural degradation processes to take their course.

A further assessment of the xylidine spill was carried out in August 1997. The results indicated that the "hot spot" was still present but that, instead of being at a depth of 5-10 centimetres, it was now primarily at a depth of 50 centimetres and that it had spread somewhat horizontally.

3. Bench-scale Testing of Remediation Procedures

Initial research, undertaken at Environment Canada in preparation for the field demonstration, using contaminated soil and water samples from Barta, Latvia as well as spiked samples allowed the effect of operating parameters to be investigated.

Laboratory analyses of xylidines were performed using Gas Chromatography with a flame ionization detector. Additionally, a rapid analytical method for the determination of spiked samples was developed. This UV/VIS spectrophotometric method was based on the reaction of xylidine with para-dimethyl-aminobenzaldehyde (DMBA) and trisodium pentacianoamino ferroate (TPF) to allow for a shift of the absorbance signal from the UV region towards the visible zone of the spectrum.

In bench scale testing, soil washing achieved removals of up to 88.6% of xylidines from soil, but these occurred only after long residence times involving mixing for 24 hours with high water to soil ratios, in the range of 8 by weight. It was also observed that Barta soil contained a high volume of very small solids which were difficult to remove from the washing solution. The option of washing with an acid solution gave better results in a much shorter time period.

Solvent extraction was investigated as a possible option. The use of methanol as a solvent allowed for xylidine recovery in the range of 90%. However, this was a costly process which would often require additional treatment for the removal of the solvent. For these reasons, other options were pursued.

The boiling point of xylidines is in the region of 213 °C and 228 °C. The research undertaken at Environment Canada indicated that effective and relatively quick removal of xylidines might occur when soil was heated to temperatures in the neighbourhood of 200 °C. The rate of removal was in the range of 99% and this could be accomplished within a fifteen minute time-frame. At these removal efficiencies, bioremediation would still be useful as a polishing step. In order to optimize the recovery of xylidine using the low temperature thermal desorption (LTTD) process, it was found that the moisture content should be near 10%. Steam coming off the soil acts as a carrier gas which serves to enhance the xylidine desorption while lessening its destruction and oxidation by reducing the oxygen concentration in the desorption unit. The gaseous emissions from the LTTD process can be condensed to obtain relatively concentrated xylidine and water solution/emulsion which does not carry over the finely dispersed solids from the soil.

The ground water remediation research centered on the process of steam stripping which is analyzed in [3]. Good results in concentrating xylidines were obtained by using this process. Optimal xylidine recovery was achieved using high initial xylidine concentrations and relatively high feed/steam ratio values. Additionally, the use of a pH adjustment step before the steam-stripping process was beneficial for the maximization of the recovery of xylidine in the column tops stream. It was possible to achieve concentration factors approaching 8 during optimized runs. The residual xylidine in the column bottoms could be destroyed using Advanced Oxidation Process.

Advanced oxidation research was conducted on spiked 2,4-xylidine and 2,5-xylidine using UV photolysis alone, UV/H_2O_2, photo Fenton's reaction and Fenton's reaction. The results of the photolysis research indicated that the using only photo-oxidation with 2,4-xylidine and 2,5-xylidine required an unacceptably long residence time to achieve non-detection levels. The results of UV/hydrogen peroxide experiments indicated that H_2O_2 concentration had a pronounced affect on UV exposures times needed to achieve non-detection levels of xylidines. Experiments with Fenton's reagent indicated that enhancing Fenton's reagent with UV light resulted in faster oxidation of both 2,4- and 2,5-xylidine.

The process train which was developed for processing of the contaminated soil, solids from the ground water filtration in the LTTD unit and the groundwater can be seen in Figure 1. The product stream from the LTTD and steam stripping processes is a concentrated xylidine solution. Xylidine is ultimately recovered in the form of xylidine salts. An advanced oxidation process is used to eliminate residual xylidine and oil products from both the steam stripping column bottoms flow stream and the water resulting from chemical treatment. Bioremediation can be used as a polishing step for the contaminated soil after low temperature thermal desorption treatment.

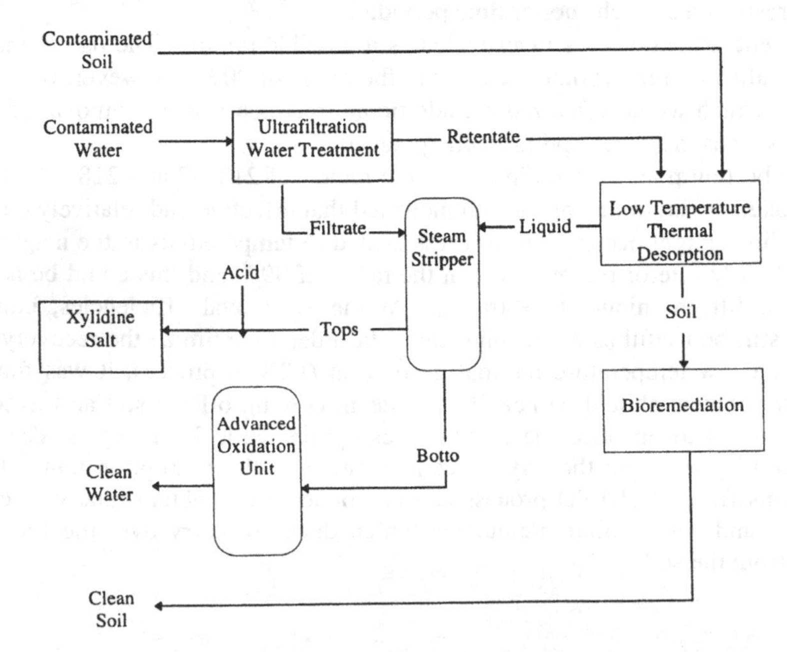

4. Groundwater Remediation

During the first phase of the remediation field work which occurred in August 1996, an emphasis was centered on the optimization of the process train for the decontamination of the xylidine contaminated groundwater. The *SAMIN* fuel spill had occurred at a missile fuel loading point located near a swampy area which had previously been filled with gravel and covered with concrete and asphalt. Consequently, the groundwater in the area is high in humic acids. It would have been difficult to treat this water by steam stripping or advanced oxidation processes because of the tendency of fouling to occur in the case of steam stripping while the low transparency might impede advanced oxidation techniques. The use of ultrafiltration allowed for the conversion of the dark brown opaque groundwater into a clear yellow-brown solution. Subsequently, steam stripping and advanced oxidation process were used on the groundwater both as it was pumped from the holding tank and after ultrafiltration. Descriptions of the processes are provided in below.

4.1. ULTRAFILTRATION

Ultrafiltration was carried out on several hundred litres of groundwater. The results indicate that ultrafiltration is a useful preliminary process for the removal of xylidine contamination from groundwater. In the process train, 90% of the groundwater was used as feed solution for the steam stripper and/or advanced oxidation process unit, while the remained 10% was utilized for wetting the soil for processing the low temperature thermal desorption (LTTD).

4.2. STEAM STRIPPING

The pilot plant steam stripping unit was capable of processing up to 1.3 litres per minute of contaminated feed water. Multiple runs were carried out, each of a duration of 1-1/2 hours or more. Raw feed, ultrafiltrate and condensed solutions (tops) from previous runs were used. The results achieved were quite satisfactory, with removal efficiencies of 98%. It was observed that runs which used feed from tops from a previous run resulted in the separation of pure product in the form of golden droplets suspended in the concentrated stream. The TABLE 1 contains the results from the 1996 work [4]. The results from the limited number of runs carried out in 1997 are similar.

4.3. ADVANCED OXIDATION PROCESS (AOP)

A 1 kW bench-scale unit was used for this work. Thirty litres of feed were used for multiple runs. The experimental period was increased to one hour to permit better modeling and obtain more detailed data. As sampling occurred at regular time intervals, it was possible to calculate the degradation of xylidine over time. Many runs were performed using the bottoms of the steam stripper as a polishing step, with the addition of peroxide. Other runs were performed using contaminated groundwater with and without ultrafiltration to determine the effectiveness of destruction of xylidines using this particular process.

TABLE 1. Steam Stripping Results

Run Number	Concentration (ppm)			Feed Rate (ml/min.)
	Feed	Tops	Bottoms	
1	~200	763	3	700
2	~200	789	3	700
3	~200	760	ND	700
4	161	971	2	1000
5	202	1083	4	1000
6	~200	1556	3	1000
7	1054	4239	8	700
8	180	910	3	1000
9	3	54	1	1200
10	111	712	6	1200
11	133	796	3	1000
12	219	1765	4	1000
13	204	1041	6	1000
14	985	4630	28	1000
15	-	-	-	-
16	14	95	ND	1200
17	235	1414	3	1300
18	206	1226	5	1300
19	193	1294	2	1300
20	17	109	ND	1200
21	911	5554	68	1200
22	~200	1915	15	1200
23	145	1374	8	1200
24	-			
25	150	1096	1	1000

5. Soils Remediation

Three technologies were chosen for remediation of the soils in the vicinity of the highly contaminated "hot spot". These included:

a. Soils Washing
b. Low Temperature Thermal Desorption (LTTD)
c. Bioremediation.

5.1. SOILS WASHING

In soil washing, the contaminated soil is contacted with an aqueous solution which transfers the contaminant from the soil to that solution. The active ingredient in the aqueous solution may be an acid, a base, a water soluble polymer or an adsorbent. In the testing carried out in 1997, a simple concrete mixer was used to provide contact between a nitric acid solution and the xylidine contaminated soil. The initial pH of the acid solution was 1.0 to 2.0. The soil to liquid ratio was 1:2 with 15 kilograms of soil and 30 kilograms of acid solution being used. Each trial used 2 batches of acid solution and one rinse with clean groundwater with contact times of 15 minutes for each stage.

The rejects from the screening of the soil for the LTTD were used as feed for the soil washing. As can be seen from the results in the table below, the testing indicated that soils washing was successful in removing the xylidines from the Barta base soils but that a polishing step such as bioremediation would be required. The aqueous solution from the process was treated by ultrafiltration, then used as feed for the steam stripper.

TABLE 2. Soil Washing Results

Run Number	Concentration (ppm)				
	Feed Soil	First Wash Effluent	Second Wash Effluent	Rinsate	Treated Soil
1	19568	2688	703	63	31
2	-	777	129	6	-
3	-	1202	212	37	-
4	-	1114	196	23	-

5.2. LOW TEMPERATURE THERMAL DESORPTION (LTTD)

Low temperature thermal desorption is a process in which a soil is contacted, either directly or indirectly, with hot gases. This results in the volatilization of volatile and semi-volatile compounds for re-use or destruction. In the work carried out at the Barta site in 1997, indirect heating of the soil was used to minimize the production of potentially more toxic by-products. The work was carried out in an LTTD unit manufactured in Canada.

There were initial difficulties with the unit. These included:

a. jamming of the soil in the auger tube
b. difficulty in controlling the temperature

These were overcome by ensuring the feed soil contained at least 10% water (the retentate from the ultrafiltration of the soil washing aqueous stream which contained xylidines was used) and by modifying the propane gas flow. As can be seen in the Table 3, the LTTD unit, is capable of removing xylidines from soil. Operating parameters had to be closely monitored and adjusted in order to successfully remove and concentrate the xylidine in a free phase liquid form. The concentrated liquid stream produced from the LTTD unit was used as feed for the steam stripper as indicated in the process flow diagram.

TABLE 3. LTTD Results

Run	Concentration (ppm)		
number	Feed Soil	Condensate	Treated Soil
1	234	2790	UDL
2	3374	226	358
3	1977	1202	UDL
4	2347	3824	100
5	-	3764	67

5.3. BIOREMEDIATION

There are two general ways bioremediation can be employed, in-situ and ex-situ. In both, the contaminated soil and groundwater are contacted with microorganisms which use the contaminant as a food source. These organisms may be naturally occurring or produced and augmented in the laboratory. In this project, bioremediation studies were carried out in the laboratory and in the field. All were ex-situ.

In 1995, work in Canada indicated that bioremediation was not possible on the highly contaminated soil (approximately 2% xylidines) from the "hot spot". This is in accordance with the results obtained in Estonia.

In 1996, testing was carried out at Barta using manure and straw for bioaugmentation of the screened, mixed soils from the "hot spot" and surrounding less contaminated soil. Mixing was carried out using a small rototiller. Although the initial results were not promising, these improved later and xylidine degrading microorganisms were isolated from the soil. These were augmented in the laboratory during the winter of 1996/97 for introduction into the Barta soil in August of 1997.

This summer, the contaminated soil was excavated and 2 biopiles were made. 15 litres of a solution containing the augmented microorganisms were added to one of the biopiles while the other was left to degrade, if possible, under natural conditions. Although no definite results will be available for at least several months, it is hoped that the use of the augmented bacteria will be successful in cleaning the contaminated soils to an acceptable residual level.

6. Current State of the Site

With the creation of the 2 biopiles, the bulk of the contaminated soil has been removed and placed on the asphalt surface at the former missile fuel loading point. At this location, the bulk of the contamination has been isolated and prevented from spreading further into the environment. There remains some contaminated groundwater under the asphalt and under the location from which the soil has been removed. Field assessments indicate that the contamination penetrates approximately 1.5 metres under the edge of the paved surface. This groundwater will move at the speed of approximately 1 metre per year and will very gradually spread. If bioremediation does not continue to provide acceptable results within a reasonable time-frame, then this contamination will have to be addressed using some of the advanced technologies.

7. Conclusions

In general, the objectives of the project have been achieved. Specifically, the following results were achieved over the past 3 years.

1. Four professors and four recent graduates of Riga Technical University have received advanced instruction and experience in contaminated site and hazardous chemical spill assessment and remediation.
2. Equipment has been provided for assessment, remediation and analysis activities.
3. The contamination at the Barta site has been mapped, partially treated and contained.
4. A process train for the cleanup of xylidine contaminated soil has been developed, field tested and patented.
5. The use of low temperature thermal desorption, soils washing and bioremediation for the cleaning of xylidine contaminated soil has been demonstrated.
6. The use of ultrafiltration, advanced oxidation and steam stripping for the cleaning of xylidine contaminated groundwater has been demonstrated.
7. 17 papers have been presented in 6 countries.
8. Two Masters Degrees have been awarded.
9. To ensure the sustainability of this work, a consulting firm has been formed and is submitting proposals on several environmental projects in the region.

8. References

1 *National Environmental Policy Plan for Latvia*, (1995), Gandrs, Riga.

Ladanowski, C., Whittaker, H., Sungaila, M., Bonneville, S., Whittaker, R., Malers, J., Millers,
2 Environmental (1996) Site Assessment in Latvia, in *Proceedings of 13th Technical Seminar c*
Chemical Spills, Calgary, Alberta, Canada, pp.377-398.

3 Punt, M., Whittaker, H., (1991) A Comparison of Steam stripping and Air Stripping for tl
Removal of Volatile Organic Compounds from Water, in *Proceedings of the Eighth Technic*
Seminar on Chemical Spills, Vancouver, B.C., Canada, pp. 117-132.

Avotins, J., Kokars, V., Ouellette, L., Whittaker, H. (1996) Evaluation of Steam Stripping f
4 Recovery of Xylidine at Former Missile Sites, in *Proceedings of 13th Technical Seminar c*
Chemical Spills, Calgary, Alberta, Canada, pp. 1-10

IMPLICATIONS OF SELECTED BACTERIAL STRAINS FOR CONTAMINATION OF OIL POLLUTANTS IN SOIL AND WATER

S. GRIGIŠKIS, V. CIPINYTE
Biocentras
Graiciuno g. 10, Vilnius, Lithuania

1. Introduction

The extraction of petroleum, either land-based or offshore, transport of oil overland or by sea, transfer and processing in refineries along with loading and discharging activities frequently cause considerable deterioration of the environment. This deterioration may take the form of leakage or accidents with the ensuing sporadic or continuous pollution of marine or terrestrial regions by petroleum hydrocarbons.

Many areas in Lithuania have been polluted by oil or oil products, with the worst cases being concentrated at the former military bases of the Soviet Army, the railway, or at hydrocarbon storage facilities. In 1988, the Lithuanian company *Biocentras* was the first organisation to apply biological technology in decontaminating the environment from oil pollutants. Following successful work in this field, *Biocentras* was appointed by the Lithuanian government in 1993 to spearhead the effort to solve problems in the treatment of hazardous waste originating from oil production.

Using the bacterial agent "Degradoilas" consisting of selected and cultivated oil-degrading bacteria (patented both in the Republic of Lithuania in 1992, IP No. LT311 IB and No. LT3112B, and in the United States on February 27, 1996, Patent No. 5494580), *Biocentras* cleans up oil-contaminated soil and water bodies, designs, constructs and operates off-site soil storage and bioremediation depots and bioreactors, purifies oil waste waters, modernises existing designs and installs new facilities for the treatment of oily waste.

2. Degradoilas - a hydrocarbon-degrading bacterial agent

2.1. COMPOSITION

"Degradoilas" may include from I to several strains of bacteria depending on the composition of the oil pollutant and the natural conditions in which the pollutant in found.

F. Fonnum et al. (eds.),
***Environmental Contamination and Remediation Practices at Former and Present Military Bases**, 277–281.*

All the pertinent strains of bacteria used in "Degradoilas" occur in nature; they are not gene-engineered and they have not been subjected to any genetic manipulation. They have been chosen from *Biocentras* repository of more than 300 strains of oil-degrading microorganisms collected from all over the world.

The relevant strains are as follows:

- *Azotobacter vinelandii 21,* isolated on Porto-Santo island, Portugal;
- *Acinetobacter calcoaceticus 23,* isolated on Porto-Santo island, Portugal;
- Pseudomonas sp. 9, isolated in Oslo fiord, near Horten, Norway;
- *Arthrobacter sp.* NY9, isolated from oil-polluted Siberian marshes near Nefteyugansk, Russia;
- *Pseudomonas sp. 31,* isolated from a railway bank in Vaidotai, Lithuania.

The strains differ in maximal activity with respect to various groups of hydrocarbons (aliphatics, aromatics, asphaltenes) and in their capacity to degrade hydrocarbons in various natural conditions (soil, fresh or sea water).

TABLE 1. Preferred substrate and natural conditions for selected hacterial strains

Bacterial strain	Chemical nature of the hydrocarbons	Characteristics of polluted area
Azotobacter vinelandii 21	aromatics	soil, water
Psedomonas sp. 9	aliphatics	water
Arthrobactersp. NY9	aliphatics	soil
Pseudomonas sp. 31	asphaltenes	soil,fresh water
Acinetobacter calcoaceticus 23	asphaltenes	sea water

Depending on the type of pollution, the strains may be combined for better degradation of hydrocarbons.

2.2. SAFETY OF "DEGRADOILAS"

The toxicity of the bacterial strains has been investigated on mice and rats at the Institute of Biochemistry of the Lithuanian Academ~v of Science. According to the findings of the Institute, neither of the five strains demonstrates any toxic properties.

The ecotoxicity of strains has been investigated at the Joint Research Centre of the Ministry of Environmental Protection with the ensuing finding that none of the five strains displays any ecotoxic properties.

2.3. CONDITIONS FOR DEGRADATION OF HYDROCARBONS BY "DEGRADOILAS"

"Degradoilas" works effectively amidst a broad range of physical-chemical factors and is not sensitive to various bacteriocidic and bacteriostatic compounds.

Temperature: minimal 8°C, optimal (30+3)°C, maximal 40°C.
pH: optimal 6,5-7,8.
Humidity: 10-100%.

Oxygen: the access to air is essential for "Degradoilas" action. The oxidation of one carbon atom of hydrocarbon requires approximately 3 atoms of oxygen. Heavy metals: The following table lists the concentrations of heavy metals' salts which cause a decrease in the rate of biodegradation by as much as 50%:
Ag^{+}- 5 mg/1; Hg^{+}- 15 mg/1; Pb^{2+}, Ni^{2+}, Co^{2+} - 25 mg/1; Cu^{2+} ,Cd^{2+}- 60 mg/1; Zn^{2+}, Mn^{2+}, Fe^{2+} 600 mg/1.

Salinity: An 8% concentration of NaCI decreases the biodegradation rate by no more than 50 %.
Fuel bacterial ihibitions: In concentrations of 50 mg/l, the common inhibitor in fuel quaternary ammonia salt cetrimide decreases the rate of biodegradation up to 50%.

2.4. DOSING OF MATERIALS FOR BIODEGRADATION.

The degradation of one ton of oil pollutants requires either 25 - 100 g of paste type frozen bacterial biomass or 200- 500 g of dry bacterial cell powder with the addition of common mineral fertilisers as follows: approximately 10 kg containing nitrogen and I kg containing phosphorus. The prepared substance is sprinkled on at a ratio of I - 5 1 of liquid on I m2 of polluted surface.

Biocentras participated in the establishment of two sites for the biological decontamination of oil pollutants. Near Kaunas, part of an asphalt concrete production plant was transferred to a site where polluted soil could be delivered and treated with "Degradoilas" and fertilisers. The plant's operational capacity is 800 m3/year.

Due to the reconstruction of an oil terminal enterprise, a large closed-loop facility for the decontamination of oil-polluted soil was built acording to *Biocentras* design and specifications near Klaipeda. The operational capacity of the Kiskenai site is 5000 m3 of polluted soil a year.

TABLE 2. The main large scale water purification projects performed by "Biocentras"

Location	Quantity of contaminant purified and/or area of clean up	Additional information
Lake Onega, USSR	100 t of fuel oil 6 km coastline	Oil spill resulting from collision of two tankers; service completed in July 1988.
Kuibyshev Water Storage Facility, USSR	18 km of coastline 120 t of fuel oil	Spill resulting from crack in hull of oil tanker. Service completed in May 1989
Artificial leather Plunge plant, Lithuania	60 t of fuel oil in the River Babrungas	Accident in heat power station. Service completed in April - July 1990.
Mazeikiai oil refinery, Lithuania	110 t of crude oil in the pond	Accident in heat power station. Service completed in September 1990.
Kirdeikiai collective farm, Lithuanian National Park	The Lake Pakasas coastline, 36 t of fuel oil	Accident in heat power station. Service completed in June 1991.
Mozyr oil refinery, Byelorussia	Approx. 5000 m3 of oil sludge, 6 ha of plant territory	Accident in waste water purification plant station. Service completed in July 1991.
Kaisiadorys glue factory, Lithuania	The River Lomena, approx: 110 t of fuel oil	Accident in the oil storage tank of heat power station. Service completed in July 1991.
Nemunas river, Lithuania	50 km of coast line, 500 t of fuel oil	Accident in the oil storage tank. Service completed in December 1992
Mazeikiai oil refinery, Lithuania	450 t crude oil spilled on 1,2 ha territory of refinery	Accident in oil pumping station. Service completed in September 1994.
West Siberian oil fields at Poikovo, Russia	13 ha of water bodies, marshes and soil polluted by crude oil spill	Accidents in pipelines. Service completed in September 1994.
West Siberian oil fields at Poikovo, Russia	20 ha of crude oil polluted water bodies, marshes and soil polluted by crude oil spill	Accidents in pipelines. Service completed in August 1995.
West Siberian oil fields at Nefteyugansk, Russia	51 ha of crude oil polluted water bodies, marshes and soil polluted by crude oil spill	Accidents in pipelines. Service completed in August 1996.
West Siberian oil fields at Nefteyugansk, Russia	50 ha of crude oil polluted water bodies, marshes and soil polluted by crude oil spill	Accidents in pipelines. Service completed in July 1997.

TABLE 3. The main large scale soil clean up projects performed by "Biocentras"

Location	Quantity of contaminant purified and/or area of clean up	Additional information
Juknaiciai, Lithuania	3 km along railroad; 140 t of fuel oil	Soil pollution resulting from train accident. Service completed in July 1989
Svente railway station, Latvia	360 t of fuel oil; 2,5 ha along	Railway accident railroad involving 6 tanks. Service completed in July 1991.
Ventspils railway station, Latvia	180 t of diesel oil spilled over 2,5 km of railroad	Railway accident involving 3 tanks. Service completed in June 1992.
Panostis railway station at oil terminal, Lithuania	0,46 km^2 of area polluted by fuel oil	Continually polluted area at oil pumping station. Service completed in August, 1992
Rykantai railway station, Lithuania	614 t of various oil products on 0,25 ha of forest	Railway accident involving 16 tanks. Service completed in October of 1993.
Mazeikiai oil refinery, Lithuania	About 150 t of oil products and oily waste deposits on 3 ha of territory	Continually polluted area, service completed

TABLE 4. Effectiveness of "Degradoilas" in decontamination of oil pollutants.

Polluted object	Pollutant	Concentration	
		Initial starting	Final
Soil	diesel oil	5,8 g/kg	33 mg/kg
		57,6 g/kg	122 mg/kg
	fuel oil	4,36 g/kg	68 mg/kg
		42,4 g/kg	320 mg/kg
	crude oil	11,3 g/kg	480 mg/kg
		91,0 g/kg	1750 mg/kg
Water	diesel oil	150 mg/kg	0,32 mg/1
	fuel oil	150 mg/kg	3 mg/1
	crude oil	450 mg/kg	2,5 mail
Waste water - Mazeikiai oil refinery	Oil products	13,5-31 mg/l	0,9-1,7 mg/l

REMEDIATION OF SOIL AND GROUNDWATER AT AIR BASES POLLUTED WITH OIL, CHLORINATED HYDROCARBONS AND COMBAT CHEMICALS

DR JAN ŠVOMA
AQUATEST SG, a.s.
Geologická 4, Praha 5, 151 00 Czech Republic
DR F HERÈÍK
KAP
Skokanská 80, Praha 6, 160 00 Czech Republic

1. Introduction

Groundwater around air fields and air bases in former Czechoslovakia was heavily contaminated by oil products, largely aircraft kerosene, in the past. The safeguarding of aquifers located near military and civilian airfields was started early enough. Three large, formerly Soviet air bases contribute significantly to the contamination of both surface and groundwater.

The threat to the quality of groundwater in the vicinity of airports is due to a combination of several environmentally unfriendly factors. Airports and airfields are usually built on plateaux, near groundwater divides, or in wide valleys. In the former instance, polluting oil products may migrate radially through the aquifers, and may pose threat to the water resources that are found at lower altitudes. In the latter case, the airport would be situated in sand-gravel sediments on valley terraces, which usually hold considerable amounts of groundwater.

Another characteristic feature of airport/field operation is the continuous handling of large volumes of fuels over long term, including their transport and storage. The air is constantly polluted with aircraft engine exhaust fumes.

Due to their unique role and function, airports and airfields cannot be relocated just because of environmental problems of ground or mineral water protection, even when the existing sites are very sensitive from a hydrogeological perspective.

The only feasible option is therefore to protect the geological enviroment as effectively as possible.

The results of the investigation and clean-up of groundwater and soil validate the theoretical assumption that massive groundwater contamination may be caused by oil spills extending over prolonged periods of time rather than infiltration of emissions from aircraft into the soil.

The contamination of rocks and soils at the former Soviet military air bases is all the more dangerous because of a long absence of environmental control, and the lack of opportunities to start the clean up process; the consequences can now be seen in civilian

F. Fonnum et al. (eds.),
***Environmental Contamination and Remediation Practices at Former and Present Military Bases*, 283–298.**

areas right in the hot spots of contamination, where ammunition is still present and additional contamination with chlorinated solvents and combat chemicals can be traced.

2. The Spreading of Aircraft Fuels and Chlorinated Solvents in Soil, Rocks and Groundwater

Our most up-to-date knowledge suggests that oil products spilled on the ground will move vertically through the permeable rocks' intergranular aeration zone to a less permeable layer, or else to the capillary fringe of groundwater. There the product tends to accumulate and spread horizontally. It floats on the groundwater table and is carried in the direction of its flow.

The level of the groundwater table moves, and an oil product retained in a rock formation for longer periods of time therefore tends to be separated into several zones in which it occurs in the gaseous phase or in the liquid phase (adsorbed to rock grains or completely saturating the rock, or possibly sealed under water); it is finally dissolved or emulsified in water. The migration of oil substances in each of the zones is governed by specific laws.

The formation of a gaseous envelope in the aeration zone of the rock structure around the oil substance under the ground is important particularly in the case of aircraft fuel. Gaseous oil hydrocarbons diffuse in all directions, also upwards to the surface. This can be put to use when mapping the polluted area.

When an oil product migrates through an as yet unpolluted medium, sorptive or capillary forces will retain a variable amount of it in the rock; the amount retained depends on the type of the rock and its water content, and on the properties of the oil product, and may vary between 5 and 30 l/cu.m. The migration is faster through polluted media, that is, in the case of recurring spills. On the other hand, a higher water content slows down further migration. In the capillary fringe zone, in which water and the phases of the oil substance migrate together, the migration is limited to the saturation range of 15 to 85 per cent. The oil product settled on the groundwater table will, in a homogeneous medium, finally take up an ellipsoidal shape elongated in the direction of groundwater flow, relatively close to the point of the spill (in the order of tens of meters). Over the extensive interface between the groundwater and the oil substance - which is formed at the boundary between the aerated and water-saturated zones of the rock medium - oil hydrocarbons are leached at a rate that depends on their solubility, and migrate through the aquifer. Oil hydrocarbons dissolved in groundwater are transported over much longer distances than the liquid phase of the oil substance itself (over hundreds and thousands of metres). Oil hydrocarbons, when dissolved, do not spread uniformly This is due to the "more permeable channels" that are formed and that contain higher concentrations of oil hydrocarbons in water. The oil hydrocarbons in solution are degraded by micro-organisms and adsorbed to rock particles; these two effects account for their loss over time, and for the gradual shifting of the contamination boundary. As a rule, a generally decreasing extent of pollution is observed over time, unless further contamination of the aquifer occurs. The short-term changes in the extent of pollution are related to the natural fluctuation of the groundwater table and to the rinsing of oil substances absorbed in the aeration zone.

The movement of water is considerably affected, or even determined by the network of crevices and tectonic faults in diagenetically consolidated sediments, igneous rocks, and metamorphites. Fractured parts act as zones of preferential groundwater flow, and thus simultaneously serve for the propagation of free oil products. As a result of this, and of the very low or zero residual saturation, the oil product may pass over distances of hundreds of metres within relatively short periods of time [14]. The kerosene leaked at the Prague Airport re-emerged again in the fissure springs of Cenomanian sandstones at a distance of 700 metres from the point of spill, in five months of the initial leak.

The physical and chemical properties of aircraft fuel pose limitations on the behaviour in an aquifer from the perspective of the spreading of pollution and its intensity. The two other important properties are surface tension and adsorption power. Density and viscosity at 10 °C are properties that affect transport of oil substances in the unsaturated rock medium. Jet fuels and aircraft kerosene, because of the values of these indices, penetrate faster than water through soils that have a water content of up to 15 per cent. The easiest and fastest penetration is exhibited by aircraft petrol, followed by aircraft kerosene.

Chlorinated solvents are represented at former Soviet air bases by the following chlorinated hydrocarbons (CHC): DCE, TCE and PCE. Due to their specific physical and chemical properties, CHC behave differently under the ground to other oil hydrocarbons.

The vertical movement of the CHC phase only starts, as with oil hydrocarbons, when the residual saturation threshold is overstepped. These critical values are 3 to 30 l/cu.m for dry soils, and 5 to 50 l/cu.m for water saturated soils [7]. These values are applicable only immediately after the accident, for CHC evaporate quickly and thereupon migrate in the gaseous phase. However, the physical and chemical properties, see Table 1, cause a substantial change in the mechanism of propagation. Since CHC are heavier than water, they penetrate the capillary fringe in the radial direction and further down to under the groundwater table. Employing inhomogeneities, CHC may migrate across the aquifer to the impermeable bedrock in which they accumulate in crevices and hollows. The high tension of the vapours causes a swelling of gaseous CHC under the arrested liquid phase in the aerated zone. The gradient makes the vapours rise to the surface of the terrain; if they find a way of venting into the air, the transition from the liquid to the gaseous phase is welcome for natural decontamination. Otherwise, an undesirable accumulation of vapours takes place. The vapours are subsequently dissolved by infiltrating precipitation and brought back into the aquifer.

Chlorinated hydrocarbons spread in groundwater mainly by convection, like oil products, and also by longitudinal and cross dispersion. Their much higher solubility generally causes higher CHC concentrations in groundwater in comparison with oil hydrocarbons. As a result of the phase's migration to as far as the impermeable bedrock, there are several zones of higher concentrations in the vertical direction. On the impermeable bedrock, the product spreads primarily in the direction of its tilt, and also accumulates in the local depressions in the aquifer's bottom. Near such pools, secondary zones of increased CHC concentrations in groundwater are formed.

Table 1. Physical and chemical properties of volatile chlorinated hydrocarbons [7]

Compound	*Density at 20°*	*Viscosity ($mm^2.s^{-1}$)*	*Solubility at 20° ($g.l^{-1}$)*	*Vapour pressure at 20° (mbar)*	*Water-air split factor*
Trichloroethylene (TCE)	1.5	0.4	1.1	77	2.74
Tetrachloroethylene (PCE)	1.6	0.53	0.16	19	1.22
Trichloroetane (TCA)	1.3	0.65	1.3	133	0.71
Dichloromethane	1.3	0.32	20.0	473	8.1
Trichloromethane (chloroform)	1.6	-	7.8	195	0.25
Water	1.0	1.0	-	23	-

3. Impacts of Emissions and Spills from Aircraft on the Quality of Atmospheric Precipitation, Soil and Groundwater

With a view to evaluate the effect of aircraft operation on the contents of oil hydrocarbons in soil and groundwater, snow, soil and groundwater were sampled in the 1980s at two civilian and several military airports and airfields in what then was Czechoslovakia. The samples were, as a rule, taken from three sites perpendicular to the axis of landing (take-off) of aircraft, at different distances from the fringe of the runway. Oil hydrocarbon contents in soil amounted usually to units, rarely tens of mg/100 g dry matter in the surface-layer soil; they were in the tenths of mg/l in fresh snow. Repeated sampling in 72 hours revealed up to a fivefold increase of oil hydrocarbons in the snow.

Zone sampling indicated a rapid drop in oil hydrocarbon content in the direction from the terrain's surface.

Table 2. Oil hydrocarbon content in soil and snow, Karlovy Vary Airport [10]

Distance of sampling line from runway, m	*Oil hydrocarbon content (mg /100 g dry matter)*	*Average oil hydrocarbon content (mg/100 g dry matter)*	*Note*
20	21.6 - 4.55* - 5.12	10.42	soil
100	3.34 - 3.01* - 3.19	3.18	soil
200	2.20 - 4.30* - 4.30	3.60	soil
200	(0.76) - (0.32)* - (0.37)	(0.48)	snow (mg/l)

* the middle sample was taken from the axis of the landing corridor

Oil hydrocarbon concentrations in the uppermost layer of soil showed a fan-like decrease from the flight axis along which they climbed to the maximum values at a distance of 50 to 500 from the verge of the runway.

Table 3. Oil hydrocarbon content in fresh snow, Prague Ruzynì Airport in 1978 [3]

Distance of sampling line from runway (m)	*Oil hydrocarbon content in snow (mg/l)*	*Average oil hydrocarbon content in snow (mg/l)*	*Maximum oil HC content in snow in 72 hrs (mg/l)*
150	0.22 - 0.60* - 0.41	0.41	3.3
450	0.38 - 0.78* - 0.27	0.60	0.48
600	- 0.49* -	(0.49)	0.65
750	- 0.30* -	(0.30)	0.63

* the middle sample was taken from the axis of the landing corridor

Table 4. Deep contamination of soil with oil hydrocarbons, Karlovy Vary Airport

Sampling site	HV 2	HJ 6	HJ 7	HJ 9	HJ 11
Sampling depth (m)	Oil hydrocarbons mg/100 g dry matter				
0.00 - 0.2		7.43			2.64
0.5	4.77	1.02	0.22	0.26	0.63
1.0	3.79	1.96	0.21	0.24	0.54
1.5				0.22	
2.0			0.45	0.29	0.62
3.0			0.36	0.32	

In January 1980 at the Línì military airfield [6], oil hydrocarbon levels in fresh snow along the runway were 0.11; 0.17; 0.22; 0.23; 0.26; 0.24 mg/l (0.20 mg/l on the average). At the other site, 100 metres away, the figures were: 0.12; 0.18; 0.19; 0.14; 0.11; 0.15 mg/l (0.15 mg/l on the average).

The average contents of oil hydrocarbons in the groundwater of granites and granite eluvium around the Karlovy Vary airport ranged from 0.05 to 0.24 mg/l, peaking at 0.10 to 0.70 mg/l. The last mentioned value was found on a tectonic fault that crossed the site of old oil HC spills from an underground tank.

Along the runway of the Èáslav military airfield [5], in the groundwater of chalk marlstones, low oil HC levels were found, having the following annual averages:

1983: 0.06 1988: 0.03 1989: 0.03 1991: 0.05 1992: 0.02 mg/l.

These values even closely approximate the "natural" background of oil hydrocarbons in the Czech Republic's groundwater, determined at 0.035 mg/l by Houzim [4].

Soil and atmospheric precipitation contamination with oil hydrocarbons infiltrated with emissions may cause the oil hydrocarbons to be present in groundwater in the order of a maximum of tenths of mg/l.

In the growing season, soil and groundwater contamination is reduced by natural biodegradation, and increased by the greater intensity of infiltrated precipitation. The maximum levels are therefore usually observed in early spring.

Concentrations of oil hydrocarbons in groundwater over 1 mg/l are due to other causes than aircraft emissions.

4. Strategy of Soil and Groundwater Protection around Airfields

The effort to protect soil and groundwater is focused on the following issues:

- Preventative protection - investigation of contamination and initial monitoring of water quality;
- Safeguarding measures to arrest the migration of the contaminant plume;
- Clean up, or decontamination of soil, rock and groundwater;
- Subsequent monitoring of groundwater quality;

Initially, oil spillages that posed threat or damaged the quality of water in rivers and wells (the Línì military airfield, the Praha Ruzynì international airport) were cleaned up in former Czechoslovakia. The clean up process was based merely on safeguard pumping from wells drilled into the hot spot of the contamination and on the downstream border of the contamination plume. A system of dual pumps was used as a rule where the cleaning of the water from the upper pump, with suction at the level of the lowered water table, was sufficient. The clean up gear consisted of gravity separators and filters filled with activated carbon or less expensive Vapex. The latter is produced by hydrophobisation of expanded perlite. The water was sometimes re-infiltrated above the contaminated area with the help of sprinklers whereby additional cleaning took place by aeration. Biodegradation of spilled oil using allochtonous bacteria that utilise hydrocarbons, with the addition of nutrients and gaseous oxygen in the aquifer, was successfully launched at the beginning of the 1980s.

For the purposes of cleaning up groundwater in gravel-sands in valleys, safeguard pumping was complemented by the erection of impermeable vertical barriers, e.g. at the Pøerov military airfield.

Under a governmental resolution, systematic surveys were started in the second half of the 1970s near expected large sources of contamination, i.e. around airfields and military bases under a programme of preventative groundwater protection. Water quality monitoring systems were set up; when major contamination of an aquifer was detected, emergency safeguard pumping was started. For economic reasons, but also due to the limited capacities of specialised companies, such projects were started as finances and time allowed, depending upon the importance of the particular site in terms of water management and the aquifer's vulnerability. The highest priority for starting preventative or remedial protection was ascribed to civilian and military airports located near curative mineral springs, or in hydrogeological structures important for the distribution of drinking water supplies. Emergency clean-up only involved decontamination of soil after new accidents when it was obvious that the contaminant was posing a threat to groundwater quality. This approach was in line with the legislation in place at that time because water protection was regulated by Water Act of 1973 while it was not until the early 1990s that legislation was passed for the protection of soil against contamination [13].

Table 5. Clean up at selected bases in former Czechoslovakia

Site	Clean up time	Quantity separated free kerosene (m^3)	Note
Praha Ruzynì	1973 - 1991	354.5	civilian airport
Praha Kbely	1992 - 1995	29.7	Czech military airfield
Èáslav	1983 -1995	535.3	Czech military airfield
Línì	1971 - 1984	528.0	Czech military airfield
atec	1988 - 1995	2055.1	Czech military airfield
Milovice - Bo í Dar	1983 - 1995	33.8	former Soviet air base
Mimoò - Hradèany	1990 - 1995	260.0	former Soviet air base
Sliaè Vlkanová	1981 - 1995	549.0	Former Soviet air base in Slovakia
Total		4345.4	

5. Peculiarities of Cleaning up Former Soviet Air Bases in Former Czechoslovakia

In 1968, the Soviet Army appropriated a number of military bases, including four major air bases: Milovice Bo í Dar, Hradèany u Mimonì, Olomouc, and Sliaè Vlkanová near Banská Bystrica in Slovakia.

The greatest obstacle to investigation and clean up was the strict ban on entering the bases for Czechoslovak supervisory agencies and environmentalists. Not only the absence of any rules, but also the Soviet troops' insensitivity to nature conservation in general resulted in a very careless handling of fuels, and even their intentional disposal in sewage canals, rivers, soil and even wells. Unlike Czechoslovak military and civilian airports, oil contamination at Soviet air bases is accompanied by chlorinated hydrocarbons. Another difficulty is the unexploded ammunition, which makes investigation and clean up procedures even more complicated. Although thorough pyrotechnic search was and is conducted before starting any excavation or drilling work, it is never risk-free. At the Milovice Mladá training base, 25 million pieces of infantry ammunition, plus fifty thousand artillery grenades and land mines were found on an area of 60 sq.km, and later disposed of. Some of the military material is older than the Soviet occupation; it dates back to World War II while some pieces come from the time of the Austrian-Hungarian Empire when a military camp was set up there in 1904.

6. Investigation and Clean up of Contamination: Case Histories

6.1 CZECH AIR BASES

Some degree of soil and groundwater contamination with oil hydrocarbons occurred at all of the twelve air bases in the Czech Republic in the past. The atec airfield in northern Bohemia; the Línì airfield near Plzeò in western Bohemia; Praha Kbely; Èáslav (central Bohemia); Pardubice (eastern Bohemia); and Pøerov and Olomouc in Moravia experienced the heaviest contamination. Investigation and clean up of groundwater, rocks and soil were carried out on all bases. In the above localities, the work is still going on. The clean up of the bases that are being transferred to the civilian sector is contracted through public tenders, see the map enclosed.

6.1.1 Línì

The base is situated in western Bohemia in the Plzeò Carboniferous Basin. Groundwater circulates in fissure permeable arcoses and sandstones. Their generally low hydraulic conductivity ($m.s^{-1}$) increases by one to three orders of magnitude in tectonic zones. The aquifer is drained by the Radbuza River that flows to the south and west of the field.

Oil contamination was detected in 1969 in domestic wells in neighbouring villages. Investigation was followed by the drilling of clean up wells, with the help of which emergency safeguard pumping was started in 1971 in the hot spots of contamination. These were four large-capacity underground tanks in concrete bunkers, two railway fuel filling stations and, above all, corroded pipes buried in the ground between the tanks and the stations. Furthermore, three car parking lots, two central fuel depots, a radar station and abandoned woods where water free of mud but containing "residual" oil products was disposed, contributed to the contamination. Aircraft kerosene predominated (90 per cent),while the other pollutants included car petrol and diesel oil (8 per cent) and lube oils (2 per cent). The water pumped out of the boreholes was treated in gravity separators and Vapex filters - the latter debuted here.

At the beginning of the clean up process, a 1 to 10-cm layer of kerosene covered 3 hectares of groundwater table. Dissolved oil hydrocarbons exceeded concentrations of 1 mg/l on an area of 77 hectares and 0.1 mg/l on 10.5 sq.km.

The clean-up effort had helped to reduce the free oil phase to merely point occurrences by 1981, while the area with dissolved oil HC over 0.1 mg/l had shrunk to 5.5 sq.km. Before that, in 1970, oil hydrocarbon contents over 1 mg/l were found in the Dobøany well that supplied drinking water at a rate of 30 l/s. The safeguard pumping carried out at the airfield along the tectonic line that connected the centre of the contamination and the Dobøany well, helped to include the well in water supplies again. By the end of 1982, a total of 528.8 cu.m of oil products, mainly kerosene, had been separated from the groundwater table thanks to the clean-up process.

6.2 FORMER SOVIET AIR BASES IN THE CZECH REPUBLIC

6.2.1 Milovice Bo í Dar

The Mladá training base, which has been used by the military since 1904, is located 40 km north-east of Prague, and covers an area of 60 sq.km, including the Milovice Bo í Dar airfield. The peripheral zones of the base have contact with the protection zones of the Káraný water project that supplies Prague and its satellite settlements at a rate of 1,900 l/s of high quality drinking water.

The air base includes the following facilities:

a) Bo í Dar airfield, with its above-ground and underground fuel tanks and technical support services, e.g. a laundry room discovered later, which had caused contamination with chlorinated hydrocarbons; it was detected only recently, and the area is now subject to atmogeochemical investigation;

b) A back-up supply system at Všejany, consisting of two large-capacity, partly buried tanks and a railway fuel filling station. These three sites are connected by a buried pipe, 3.5 km long, and three pumping stations. The pipe forks out at the airfield, one branch feeding the underground tanks and one supplying the filling station;

c) A part of the kerosene, and perhaps some rocket fuel - although not conclusively proven yet - were stored in a large central depot in the Milovice camp.

Table 6. Capacity of fuel tanks at the Mladá Milovice base

Site	Volume	Contaminant
Bozí Dar	11,000 m^3	90% kerosene, 5% petrol, 5% diesel oil
Všejany lower tank	1,000 m^3	kerosene
Všejany upper tank	2,000 m^3	kerosene
Milovice camp large-capacity storage	16,000 m^3	60% car petrol, 30% diesel oil, 8% kerosene, 2% gearbox oils

The groundwater is associated with isolated Quaternary gravel-sands whose hydraulic conductivity is n.10^{-5} m/s and, generally, with fissure permeable Turonian marlstones and limy sandstones whose hydraulic conductivity varies considerably in relation to tectonic exposure from n.10^{-6} m/s to n.10^{-3} m/s; the effective yield of the wells ranges from hundredths of l/s to ten l/s.

Oil hydrocarbon pollution. The great differences in permeability influence the migration of the free phase and solutions of oil products. The layer of the oil products on the groundwater table is discontinuous and irregular, and overall occurrences of the two components of the oil contamination do not coincide in terms of either time or space. In 1993, the groundwater table was covered by a layer of oil HC on an area of 1.7 sq.km. Groundwater containing over one mg/l of dissolved HC covered an area of another 500 hectares, an additional 650 hectares showing levels over 0.10 mg/l. Some 41,000 cu.m of contaminated soil containing 1 g HC/kg dry matter and more was earmarked for decontamination.

The kerosene layer on the groundwater table was tens of centimetres high. The greatest thickness was registered in the vicinity of the bottom tank at Všejany - 2.6 m in 1991. Over the six months that preceded the clean up work, the layer had subsided to 0.5 m which signalled the danger of the contaminant continuously migrating until a protective depression was created by pumping in 1992.

The notion that the base's grounds were heavily contaminated was proved as late as 1980 when a kerosene layer turned up in a house well in Zbo íèko, some 500 m downstream from the centre of the contamination. The time of the first occurrence of heavy contamination of two of the military's six water supply wells with oil and chlorinated (TCE, PCE) hydrocarbons, which exceeded the permissible C limit, is not known; the two wells are now out of operation.

It was only as late as 1982 that safeguard pumping was permitted, but then it was only permitted to drill the clean up wells on the outside of the fence around the base. Before 1988, when clean up was allowed at last within the perimeter of the airfield, although only on declassified spots, 5,660 l of kerosene had been separated from the hydraulic barrier in Zbozíèko.

In the meantime, signs of yet another oil accident appeared in 1987 when kerosene started to spill from a broken buried pipe into a stream above Všejany. Clean-up pumping commenced that same year, and was continued intermittently until 1995.

Systematic investigation and clean up of the base were only allowed after the Soviet Army's pull out in June 1991; nevertheless but, still, the work was only started after the outcome of a public tender was known in the autumn of 1994. Before that, emergency safeguard pumping was permitted at the base, however, only to the extent that the hot spots of contamination were identified, and subsequently tapped by boreholes. The clean up pumping at the airfield took place from 20 boreholes with a total yield of 10 to 32 l/s. The water was treated in gravity separators with Vapex and Fibroil filters. When high PCB levels were detected (hundreds of nanograms per litre of water), active carbon filters were employed. From the vicinity of the bottom tank in the Všejany forest, 4 to 10 l/s were pumped from 3 to 10 boreholes. This water was passed through Kutex filters and re-infiltrated upstream from the point of contamination. Pumping and similar treatment was carried out at the former railway filling station: about one l/s was, and continues to be, pumped from one borehole and a captation sump. At times, depending on the sudden appearance of oil film on the groundwater table, clean up pumping is operated from two to three boreholes next to the upper tank, with a total yield of 1 to 3 l/s and the water undergoing the same treatment.

The efficiency of groundwater clean-up improves on all the sites by applying in situ bio-reclamation when allochtonous bacteria and their nutrients are injected into boreholes and excavations. This, moreover, supports the development of autochtonous oil consuming micro-organisms. Near the bottom tank in the Všejany forest venting of soil gas was started successfully in 1995, both through shallow vertical boreholes and in 30-metre deep boreholes inclined at an angle of 45° that reach to under the bottom of the lower tank. The venting boreholes were activated by blasting.

Over the clean-up period, 34,700 l, or 29,500 kg, of oil products, largely kerosene, had been separated from all the various sites on the base by 31 December 1996.

The following table indicates that evaluation based on merely the separated free phase recovered from the groundwater table underestimates the total amount of the contaminant removed from under the ground. This applies in particular to the final stages of clean up when the graph of separated oil phase vs. time is asymptotic and the oil hydrocarbon layer drops to below 1 cm, or even turns into oil film.

Table 7. Total quantities of oil products removed (Všejany)

Period	Free phase separated (kg)	Filtration of dissolved oil HC	Venting (SVE) (kg)	Total (kg)
1995	786	132	330	1,248
1996	426	474	658	1,558
Total	1,212	606	988	2,806
%	43	22	35	100

The absolute increase in the quantity of oil hydrocarbons removed in 1996 was caused by opening in mid-1996, four new hydrogeological boreholes and three venting boreholes for the purposes of clean up.

The amount of oil hydrocarbons removed from the aquifer by means of in situ bioreclamation, calculated on the difference between the initial and final concentrations in water, and the volume of the groundwater in question, was 4,789 kg (sic!), which would represent 180 per cent of the oil hydrocarbons removed hydraulically and by venting [12].

The soil polluted with oil hydrocarbons was excavated at the time of removing the underground tanks, and was subjected to ex situ biodegradation; in situ biodegradation was only applied to the bedrock under buildings and roads, and in their vicinity. A combination of the two techniques was used on two sites along the route of a Všejany - Bo í Dar pipeline. So far, 570 cu.m of earth containing on the average 4,335 mg oil HC/kg dry matter has been removed here and taken for bioreclamation. Through mechanical homogenisation and aeration of earth, adding biological substrates and nutrients, the average oil HC content was reduced to 390 mg/kg dry matter over the year (the limit value prescribed was under 1,000 mg oil HC/kg dry matter). During in situ bioreclamation, the nutrients, biological substrate and cleaned water were absorbed in infiltration trenches and boreholes above the line of the pumped clean up boreholes. The substrate used, GEM 100, is based on the activity of Pseudomonas spec and Actinobacter spec. To enhance the efficiency of soil clean up, an environmentally safe surfactant based on higher alcohols was added to the substances to be absorbed. In situ bioreclamation took at least double the time of the ex situ soil clean up.

Between October 1995 and June 1997, as much as 32,760 kg of kerosene was removed from an allegedly emptied buried pipeline between Lipník and Bo í Dar through vacuuming, rinsing and pumping.

Chlorinated hydrocarbon pollution. In 1990, CHC contamination of groundwater was identified in the Jiøice 19 camp situated some 5 km north-east of the Bo í Dar airfield. PCE and TCE predominated in the hot spot of the contaminated area, while its fringes contained distinct levels of DCE and DCA degradation congeners. Groundwater clean up was started in 1992, the pumped out volumes gradually built up from the initial 2 l/sec to 11 l/sec in 1995. Pumping from two to seven boreholes was operated and water was cleaned in three stripping columns with CHC intercepted on active carbon filters. Cleaned water was sprayed by sprinklers in the upstream direction on the terrain's surface to infiltrate; deep infiltration was effected through injection into one borehole. To date, the clean-up effort has helped to remove 240 kg CHC from the groundwater; between 1994 and June 1997, a marked decline in CHC levels was observed in the main clean up borehole HJ 2: 482 - 387 - 177 - 117. At the periphery, CHC concentration in three boreholes dropped to under the clean-up limit value, and pumping was therefore discontinued there. Overall, 1,129,466 cu.m of water has been pumped out and cleaned up.

The treatment of contamination of the unsaturated zone has not been as successful. Atmogeochemical investigation revealed CHC (TCE only) contamination on an area of 600 x 200 m, with the maximum concentration amounting to about 100 mg/cu.m. In wells drilled to a depth of 20 to 40 m in atmogeochemical anomalies, CHC levels ranging from 60 to 110 mg/cu.m were observed during static sampling in line with the previous results; however, their mass flux was almost zero during suction. Locally, in places where calcareous siltstones and clastic limestones only have fracture permeability and the groundwater table lies between 30 and 40 m, venting had to be optimised as follows:

- venting boreholes were situated so as to coincide with atmogeochemical anomalies and tectonic zones (identified with the help of detailed geophysical investigation on the surface),
- in cases of low permeability, borehole blasting was used,
- flux intensity markedly increased when groundwater table was lowered by pumping.

Following this optimisation, CHC in inlets to vacuum pumps increased up to 300 - 500 mg/cu.m, with the pressure in the borehole mouth ranging from 8 to 18 kPa. In the Ji 1702 borehole (at the centre of soil air contamination), PCE predominated over TCE, in the other boreholes the two compounds were almost balanced. DCE was also detected in the exhausted soil air (up to 54 mg/cu.m). Two to four Siemens vacuum pumps with a capacity of 50 to 80 l/s were used for exhausting gaseous CHC hydrocarbons, to which four to 12 venting and combined boreholes were connected at four sites, depending on the prevailing CHC hydrocarbon concentrations. Downstream from the vacuum pumps, gravity separators and active carbon filters were mounted. Active carbon was regenerated on site. Between 1995 and June 1997, the following quantities were exhausted every year: 92; 571; 258; or a total of 921 kg of CHC; and 2,653,606; 5,606,276; 2,597,847; or a total of 10,857,729 cu.m of soil air. Sustained decline in CHC concentrations was only observed in more shallow venting fields with boreholes about 20 m deep. Most gaseous CHC are exhausted from three combined boreholes from which pumping goes on at the same time. The Ji 1702 borehole, which has the highest yield, gives about 40 kg CHC a month, without any marked tendency to decline.

As the clean-up process continues, groundwater pumping is increasingly motivated by optimisation of venting conditions rather than groundwater decontamination itself. In fact, groundwater containing CHC is pumped out of Ji 1702 from deep below the C limit of the Dutch list.

Combat chemical pollution. At the Soviet Army's ammunition depot found in a forest 1.5 km north of the Bo í Dar airfield, tear gas and increased beta radioactivity were identified in three underground tanks, at levels of 0.3 mg/l and 100-160 Bq/l, respectively. Chemical analysis revealed natural radioactivity of potassium whose concentration was determined at 3,000 - 5,150 mg/l (1 g of potassium has an activity level of 28.2 Bq/l). The potassium, or KOH, was obviously used by the military to decompose toxic chloroacetophenone to less dangerous acetophenone. Contaminated water was decontaminated at a plant in Rez u Prahy, some 1,320 kg. Also, 1,860 kg of gas masks and chemical-proof wear were taken to thermal decontamination. The soil excavated during tank removal was subjected to on site bioreclamation. In one month's time, no chloroacetophenone or its indications through cholinesterase inhibition were detected in this soil.

6.2.2 Hradèany

The Ralsko military training base, erstwhile grounds used by the Czechoslovak army, is located some 80 km north of Prague. The area covers 250 sq.km, including the Hradèany air base. It is built up of formations of excellently permeable Turonian sandstones with a thickness of 70 metres, which are not covered by any impermeable layer. It constitutes the most important groundwater reservoir in Bohemia, and some parts of it are already being tapped to supply water mains.

The aquifer's permeability is of the fissure-porous type, hydraulic conductivity averages at 8 m/day, reaching up to 20 m/day in the upper part of the aquifer.

Oil hydrocarbons are concentrated in soils, in a layer of 0 to 1.3 m above the groundwater level. The volume of contaminated soil was estimated at 202,000 cu.m and the total quantity of oil hydrocarbons in soil amounted to 2,789 tonnes. The contaminated area covers about 29 hectares. [1]

Groundwater remediation was started in 1989 by recovering the free phase through 24 boreholes. At that time, the layer had a maximum thickness of up to 6 m. To prevent the contaminant from spreading in the groundwater a hydraulic barrier was designed. The contaminated groundwater collected through clean up wells and on the hydraulic barriers (discharge 25 l/s) was treated in the existing sewage and oil water treatment plant.

The pollutants include jet fuel, diesel oil, petrol, fuel oil, and chlorinated hydrocarbons. There was leakage of various fuels and detergents when used in operations, which mixed naturally, dissolved, sorbed to soil and biodegraded. The outcome of those processes is a mixed, stratified contamination concentrated in the soil above the groundwater level. The hydrocarbon mixture penetrates the Plouènice River which drains the Middle Turonian aquifer.

Soil is most heavily contaminated at former railway filling stations.There c Contaminants form a layer 0 to 1.3 m thick above the groundwater table, and their levels range from an average of 7,040 mg/kg soil to a maximum of 34,000 mg/kg. After three years of the groundwater clean up effort, the free phase is now 0.1 to 3 cm thick on the groundwater table in the most contaminated areas. HC concentrations in groundwater amount to 120 to 140 mg/l near the water table. At a depth of 2 m, this parameter ranges from 1 to 20 mg/l.

The most hazardous HC pollutants include benzene, toluene and xylene. Their highest concentrations in the groundwater are 105 mg/l, while the total amount of contaminated groundwater is estimated at 217,000 cu.m.

Contamination with chlorinated hydrocarbons was tested by soil sampling. The area of contaminated soil gas covers 1.35 hectares, and the volume of contaminated soil is estimated at 54,000 cu.m. CHC pollutants are concentrated in two spots near the north-west perimeter of the air base.

The northern stretches, near the Boreèek water supply facility, contain groundwater contaminated with effluent from a rendering plant. PCE had been used in its operation for twenty years, some 130 tonnes a year.

The southern area is contaminated by TCE that was used at the air base. CHC concentrations in the groundwater of both areas average at 10,000 µg/l, to reach a maximum of 140,000 µg/l on the river's left bank.

The groundwater contaminated with chlorinated hydrocarbons is cleaned up by pumping from four boreholes and its passage through stripping towers. The monthly output is 40 to 50 kg of chlorinated hydrocarbons.

Remediation is based on venting the oil that contains jet fuel over an area of 2,500 sq.m through 14 boreholes. The monthly output dropped from 560 to 240 kg in seven months.

The quality of the clean-up has been proposed to meet the B to C limits of the Dutch list for toxic components. In respect to "non-toxic" hydrocarbons we assume that the C limit of the Dutch list will meet the environmentalists' expectations.

Since 1990, some 200 tonnes of kerosene have been separated from the groundwater table, while 80 and 320 tonnes of oil hydrocarbons have been removed by venting and bio-venting, respectively.

Since 1993, some 1,500 kg of CHC have been removed through pumping and stripping, and 200 kg of CHC by soil vapour extraction.

7. Conclusions

- Air bases pose a threat to and damage the quality of surface and groundwater. Particularly dangerous are the former Soviet bases, both because of the specific process of contamination and the very late start of remediation. Under the clean-up programme undertaken in the Czech Republic in the early 1970s, more than 5,000 cu.m of oil products have been removed from groundwater table and hundreds of tonnes of emulsified and dissolved oil and chlorinated hydrocarbons have been eliminated from aquifers by means of filtration and stripping. Tens of tonnes of oil and chlorinated hydrocarbons have been removed with the help of *in situ* venting and bio-venting.

The following are the most frequent causes of oil product leaks into the ground:

- Inadequate technical condition of storage and distribution facilities and of the respective safety equipment, inadequate safety measures. Especially dangerous are leaks from buried tanks, pipelines and fittings, the direct checking of which is impracticable. Even small leaks may release large quantities of oil products into the ground: the theoretical amount of a liquid with a viscosity equal to a unit, released from a hole 2 mm in size under a pressure of 0.15 MPa, is 900 cu.m over one year [9];
- Inadequate equipment and negligence in the handling of fuels. This result in spills, dripping, and overflowing in the process of filling and draining storage tanks. Overfilling is frequently due to the absence or malfunctioning of safety devices and level gauges.
- Unexpected emergencies such as tanker truck accidents or air crashes. Even though such instances are rare, they should be borne in mind in view of the intensive air traffic and the volume of the fuel handled.

The time consuming and cost intensive long-term remediation of aquifers, based only on the "pump and treat" method can be made more effective by

- detailed investigation of the sites on which uncontrolled contaminant migration has occurred, by means of photographic, geobotanical, atmo-chemical and hydrochemical detection of the pollutant, using mathematical modelling and GIS to forecast the pollutant migration, and applying risk analysis;
- combining pumping with rock rinsing and *in situ* water bioreclamation simultaneously with decontaminating the unsaturated zone by *in situ* venting and bio-venting.

The most effective and least expensive approach is preventative protection of aquifers based on the monitoring of groundwater quality. This is essential in hydrogeologically sensitive structures.

8. References

1. Èerný, J., Herèík, F., Soukup, L. (1994): *In situ* remediation of soil and groundwater polluted by jet fuel in Northern Bohemia, 5 p., Symp. Environmental Contamination in Central and Eastern Europe, Budapest, Hungary.

2. Chvojka, R., Komar, A., Švoma J. (1994): Addressing environmental damages caused by the Soviet Army, NATO CCMS pilot study on environmental aspects of reusing former military lands. MS, Workshop, 5 p., Ober Ammergau, FRG, 24 - 26 May 1994

3. Houzim, V. (1980): Groundwater protection against pollution, Research Report, 265 p. 04, MS, Stavební geologie Praha, Czech Republic (in Czech)

4. Houzim, V. (1986): Delimitation of protection areas with regard to permissible concentrations of organic compounds. Symposium on Groundwater Protection Areas, *19th Congress of the International Association of Hydrogeologists*, p. 424-432, Karlovy Vary, 8 -15 September 1986, Czech Republic

5. Jelínek, J. (1994 b): The Èáslav airfield - Report on Clean up. MS, AQUATEST, Stavební geologie, Praha (in Czech)

6. Karlín, F., Mašín P. (1982): The Línì airfield - Report on Contamination Investigation and Clean up. 155 p. MS, AQUATEST, Stavební geologie, Praha (in Czech)

7. Leitfaden (1983): *Leitfaden für die Beurteilung und Behandlung von Grundwasser-erunreiningungen durch leichtfüchtige Chlorkohlenwasserstoffe*. 104 p., Ministerium für Ernlährung, Landwirtschaft, Umwelt und Forsten Baden Würtermberg, Stuttgart

8. Øepa, P. (1991): Milovice II - Report on Hydrogeological investigation of Groundwater and Rock Contamination by the Soviet Army at the Milovice Military Training Base. 172 p., MS, AQUATEST, Stavební geologie, Praha (in Czech)

9. Seba Dynatronic (1971): Company brochure: Baunach Seba-Dynatronic GmbH

10. Švoma, J. (1981): The Karlovy Vary Airfield. Report on Contamination. 60 p., MS, AQUATEST, Stavební geologie, Praha (in Czech)

11. Švoma, J. (1993): Investigation and decontamination of soil and groundwater at former Soviet army bases in Czechoslovakia. F. Arendt, G. J. Annokkée, R. Bosman and W. J. van den Brink (eds.), *Contaminated Soil* 1993, p. 747 - 754, Kluwer Academic Publishers, printed in the Netherlands

12. Švoma, J. (1996): The Lipník-Všejany Pipeline. Report on Investigation and Clean up of Contamination caused by Soviet Army. p. 52, MS, AQUATEST, Stavební geologie, Praha (in Czech)

13. Švoma, J. (1996 b): Remediation of soil and groundwater in the Czech Republic, E.A. McBean et al. (eds.) *Remediation of Soil and Groundwater*, p. 251 - 270, Kluwer Academic Publishers, printed in the Netherlands

14. Švoma, J., Houzim, V. (1984): Protection of groundwater against oil pollution in the vicinity of airports, *Environ. Geol. Water Sci.* Vol. 6, No. 1, p. 21 - 30, Springer-Verlag New York, Inc.

REMEDIAL TECHNOLOGIES APPLIED AT SLIAC - VLKANOVA, A FORMER SOVIET MILITARY INSTALLATION

ELENA FATULOVA
Ministry of the Environment of the Slovak Republic
Nam. L. Stura 1
812 35 Bratislava

1. Introduction

Under the agreement concerning the withdrawal of Soviet troops from Czechoslovakia, preliminary assessments of environmental damage were carried out during the years 1990 - 1991.The assessments revealed that 14 out of the 18 former Soviet military installations in Slovakia were polluted by oil (hydrocarbon), oil derivatives and heavy metals. About 250 000 m3 of soil was polluted by hydrocarbon products, and contaminated groundwater was found over an area of about 20 km2.Overall ecological damage has been assessed to be about 931 million Slovak crowns (30 million USD). Of this amount more than half (621 million Slovak crowns) is allocated to the remediation of the Sliac - Vlkanova installation.

2. General information

Sliac - Vlkanova is located on the right bank of the Hron River. The river meanders through the flat surface of the southern part of the Sliac Basin. It was formed during the Baden period by downward tectonic movements that created a deep intervolcanic depression that filled with neogene sedimentary material. This fill lies on granites and Mesozoic rock at a depth ranging from 200 - 400 m below the surface. The neogene sedimentary fill itself is about 100 - 300 m thick, consisting of volcanic and volcanic sedimentary rock. From the hydrogeological point of view, this formation of alternating permeable and impermeable layers created conditions for irregular water-bearing layers with artesian piezometric levels. However, the top is assumed to be an impermeable base to quaternary sediments that represents a significant gravel aquifer.

Groundwater is bounded by the quaternary gravel aquifer. The thickness of the gravel layer is 5-15 m in the area of Vlkanova and 50 - 120 m in the area of the Sliac airport. The permeability of the gravel layers is low; the parameter of permeability k varies from 10-6 m.s-1 in the Sliac area to 10-4 m.s-1 in the Vlkanova area.

F. Fonnum et al. (eds.),
***Environmental Contamination and Remediation Practices at Former and Present Military Bases*, 299–303.**

The groundwater level is at a depth of 2 - 5 m, although primarily between 3 - 4 m Groundwater level fluctuation is 1 - 2,5 m, depending mainly on the rainfall and the depth of the surface water level.

During the last period of stay of the Soviet army in the Sliac - Vlkanova area and afterwards, many sources of pollution were identified, mainly underground and aboveground fuel tanks, pipelines, garages, fuel storage equipment, manipulation facilities, etc.

As the long-term investigations have revealed, previous activities of the air force and the army in the area of the Sliac - Vlkanova airbase have resulted in heavy pollution of soil and groundwater.

3. Soil contamination

Soil contamination was investigated using a very dense sampling net. More than 4500 analyses were carried out throughout the site. Soil samples were taken at 1 m depth intervals from the surface to the minimum groundwater level. Depth of sampling did not exceed 6 m.

Soil was found to be highly polluted by oil in at least 6 large, and numerous smaller areas (hot-spots) with the concentrations often far exceeding C - level of the Dutch list (1000 mg.kg-1). The maximum concentration, reaching 15,934 mg.kg-1, was found at three large polluted areas, and a intense sampling disclosed a number of other polluted locations also with very high concentrations (above C level) at the maximum depth drilled (4.5 - 6.0 m). While the horizontal borders are quite accurately known, the vertical extent of contamination can only be estimated.

The critically polluted area at Sliac - Vlkanova represents about 130 000 m2, and the total volume of soil contaminated by oil (over 500 mg.kg-1) was assessed to be 201,673 m3. Heavy metals that were found to occur repeatedly include mercury, copper, zinc, and cadmium. Mercury was the most often present in samples; however, its origin was not identified. Copper (up to 1280 mg/kg) and zinc (up to 8200 mg/kg) were found clustered in two areas only; cyanides in slightly higher concentrations (between B and C levels) were found in large quantities clustered in three areas. Soils at the site have not been surveyed for volatile hydrocarbons or chlorinated hydrocarbons (with the exception of PCB). The other compounds (PCB, PAH, F, BTEX) were not detected at elevated concentrations.

4. Groundwater contamination

The aquifer, formed mostly of a gravel layer between 7 - 100 m thick, was sampled in the uppermost shallow horizon of 2 to 14 m, primarily between 2 - 10 m. The wells used for sampling the shallow groundwater were equipped with continuous filters.

Free product in the groundwater level was found in 5 extensive areas. The initial thickness was locally 1 m. Currently, after sanitation pumping, the thickness of the oil layer still ranges from a few mm to several cm in some places. The extent of the area with a free oil layer in the groundwater is about 6,000 m2 with average thickness of about 1 cm, representing a volume of about 20 m3 of free product.The free product is a mixture of petrol (gas), kerosene and mineral oil.

The following important contaminants were found in the groundwater:

- non - polar extractable substances (NES),
- benzene (B), toluene (T), xylene (X),
- dichloroethene (DiCl), trichloroethene (TCE), tetrachloro-ethene (PCE)
- arsenic, copper

In addition, other chlorinated hydrocarbons including chlorobenzene, trichloromethane, tetrachloromethane were present.

5. Soil remediation

Remediation activities in the Sliac - Vlkanova area started in October 1992.Two basic remediation techniques were proposed for the soil and groundwater decontamination:

- Extraction and decontamination of the soil ex situ.
- Pumping and purification of the groundwater and repeated washing of the contaminated soil with already purified groundwater.

The soil was extracted prior to decontamination. The aim of the decontamination process was to reduce the content of oil products and non-polar extractable substances (NES) in the polluted soil to the values of 500 mg.kg-1 in the Sliac area, and to 200 mg.kg-1 at Vlkanova.

The decontamination of the soil ex situ has been performed in both of the main localities, Sliac and Vlkanova, by the Ropstop method, a composting process in which microorganisms and inorganic nutrients are added to promote microbial degradation. Despite a rather low content of carbon in the soil (less than 2 percent), a secondary organic substance did not have to be added.

A technological process of remediation is being carried out on the specially adapted decontamination surfaces, so-called ecoareas, located in the Sliac air - base and at Vlkanova, where the polluted soil has been piled up. The capacity of the ecoareas is 15 000 m3/year.

Due to the necessity for extraction and transportation of considerably large volumes of soil to a place where they can be cleaned up, the cost of the remediation process ex situ seems to be prohibitive. Therefore an amendment to the project has been made whereby the remediation process in situ will be carried out in the Vlkanova area.

Soil washing, based on a system of pumping and infiltration drains was widely used in connection with the remediation of groundwater. After being pumped out, the water was treated in two water purification plants and then taken away to infiltration trenches.

6. Groundwater remediation

For cleaning up the groundwater the following limits have been established by the Environmental District Office:

- NES	300 (g/l
- benzene	5 (g/l
- toluene	15 (g/l
- xylene	20 (g/l
- dichloroethene	10 (g/l
- trichloroethene	30 (g/l
- tetrachloroethene	10 (g/l

For groundwater remediation the following approaches have been applied:

- Pumping the groundwater out of wells followed by purification with technological equipment.
- Pumping the groundwater out of the drains and decontamination in water purification plants.

From the beginning of the remediation about 20 bores were pumped off in the Sliac area, and 15 boreholes at the Vlkanova site.In the Sliac locality, the water yield at the boreholes varied between 0,005 l.s-1 and 0,774 l.s-1. At the same time as the remedial pumping process was taking place, oil products were pumped from the groundwater level at the boreholes. In most of the samples of groundwater analysed, the NES concentrations did not exceed the value of 1,0 mg.l-l, and very often concentrations were less than 0,3 mg.l-1. Concentrations up to 10 mg.l-1 NES were exceptional.
The concentrations of chlorinated hydrocarbons in most of the samples varied from 0,0 to 0,2 mg.l-1, although some were as high as 4,6 mg.l-1.The maximum concentration of aromatic hydrocarbons (benzene, toluene and xylene) was 600 mg.l-1.In the water from the most polluted localities, the maximum concentration of aliphatic chlorinated hydrocarbons was 6,0 mg.l-1, and of aromatic chlorinated hydrocarbons was 600 mg.l-1. The concentrations of chlorinated hydrocarbons in water pumped from individual boreholes varied from 0,0 to 600 mg.l-1.

Polluted groundwater pumped from the surveying and sanitation boreholes was cleaned at the temporary purification plants. The water was pumped into a gravity separator (trap), where emulsified oil substances (products) were caught by a sorbent, Vapex, floating on the groundwater level. Then the water was repeatedly pumped off into the aeration towers where the dissolved oil products and chlorinated hydrocarbons were stripped, and consequently purified by the sorption process on the filters filled with an active coal. (Clean" water, as the result of whole procedure, was then drained off to a receiver, either the river Hron or a sewer.

Because the pollutants are primarily a combination of highly volatile substances, the efficiency of the remediation process varied from 50 to 90 per cent.

When the whole drainage system was completed, the sanitation pumping of the bores was reduced to only a few boreholes pumped off. From these the water will be cleaned further by means of temporary purification plats. The water withdrawn from the drainage system will be cleaned up in two water purification plats.

The entire drainage system was completed by the end of 1995. Seven sanitation drains were built together with two infiltration trenches adjacent to each drainage branch. The depth of individual drains varies from 4 to 5,6 m.

It is expected that the remediation of the Sliac - Vlkanova area will be finished within ten years.

References

Vojtasko I., Auxt A., Holubec M., Potucek M., (1996) Transfer of Know - how on Soil Remediation Techniques to Slovak Environmental Engineering Offices, Bratislava.

REMEDIATION OF POLLUTED ENVIRONMENT AT NAVAL PORTS OF THE BALTIC SEA

OLAVI TAMMEMÄE
Director General, Nature Protection Department
Ministry of the Environment, Republic of Estonia
Toompuiestee 24, EE 0100 Tallinn, Estonia

1. Localisation of naval port in the Baltik

Since Estonia, Latvia and Lithuania were situated at the western border of the former USSR which also was part of the sea border, many of the region's best ports were turned into support points for the military navy. This concerns the coasts of Estonia and Latvia - probably thanks to the neighbourhood of the Kaliningrad (Köningsberg) military port Lithuania remained untouched.

In Estonia, the largest and most polluted object within military possession was the town of Paldiski together with the Pakri Peninsula (incl. both the North and South Port) and the islands of Suur- (Large-) and Väike- (Small-)Pakri. The second largest object was the Tallinn Military Navy Base which consisted of three neighbouring ports - Miinisadam (the Mine Port), Hüdrograafia- (or Kabotaazi-)sadam (Hydrographical Port) and Vesilennukite sadam (Hydroplanes Port). In addition, the military factory No.7 had its own port, as well as the Baltic Base of the Admiralty Plants - the so-called Bekker Port.

The oil terminal of the military base was located at the very head of the Kopli Peninsula; the commanding and communication centres were located at Pirita-Kose, Laagri and Pääsküla (all within Tallinn), the coast guard rocket base was located at the head of the Viimsi Peninsula (in Rohuneeme). The island of Naissaar together with its Mine Storage also belonged to the Tallinn Navy Base. In addition to those mentioned, the port of the marine border guard was located within the Kopli Peninsula.

In Latvia, the former Soviet military navy bases include Riga, Ventspils and Liepaja). The Riga port was partly also used for cargo transport; Ventspils was the home base for marine border guard fleet, the pollution caused by which was virtually negligible as compared to the other mentioned ports.

The military navy bases of both Estonia and Latvia were connected with extensive rear services together with commanding and communication centres, fuel and ammunition depots; rocket bases defending the other objects; hydrographical services, training polygons, and units of the naval air force.

F. Fonnum et al. (eds.),
Environmental Contamination and Remediation Practices at Former and Present Military Bases, 305–311.

2. Pollution at naval ports

In the following, the presentation will focus on the ports of the military navy, as well as the pollution related to those and its remediation.

In all military ports of Estonia and Latvia, an inventory of pollution has been carried out and the preliminary investigations have been made. The type of pollution and its scope have been determined, the first assessment of the environmental risks has been made and remediation activities have been begun in areas with danger of further spreading of pollution. In most cases, problems are related to the pollution of water of the port water area, as well as of bottom sediments and the soil of the port with oil products, different chemicals and heavy metals - this in turn is the cause for secondary pollution of the port water.

Within the framework of phase I of the NATO/CCMS Pilot Study on the Environmental Aspects of Reusing Former Military Land, the typical profiles of different military objects, including military ports, were developed. According to the results, the most typical contamination profiles of navy facilities are the following:

- in accumulator and battery loading ports: acids and lyes, battery agents;
- in boiler houses: brown coal ash, brown coal/coal slack;
- in ammunition and weapon stores: anti-fat agents, explosives;
- in paints, varnish and lubricants stores: acids and lyes, anti-fat agents, anti-freeze, disinfectants;
- in sewage treatment facilities/sewage bittering facilities: detergents, crude oil, sludge;
- in tank farms: bitumen, cleaning agents in tank farms, crude oil;
- in wild dumps: ash, incineration residues, battery agents, disinfectants.

While comparing the above data with those of the already completed investigations, it is possible to focus the additional investigations at potential not-yet-identified components of pollution.

It is interesting to mention that the pollution analyses of the territory of military ports refer to quite similar contamination profiles, the differences relating to the share of different contaminants depending on the specific activities of the former navy port. For instance, in the case of submarine ports the share of pollution related to accumulators and their loading facilities is considerably higher than in other ports.

While in the case of different chemical substances it is possible to identify, collect/absorb and either destroy or pack and dispose of those, the case of oil products is different: according to the experience obtained to date, it is most common to either use biological remediation within the port territory (both *in situ* and *ex situ*) or transport the polluted soil to special isolated disposal sites. In order to prevent further contamination of surface water (and the port water area) during biological remediation activities, we have also practised the surrounding of the relevant area with "protective drainage" and additional treatment of water accumulated from such system.

So far, little attention has been paid to one of the most widespread types of pollution at the former military territories - unused buildings and facilities, as well as their remnants. The liquidation of such pollution is both costly and labour-intensive, the only option often being the destroying or blowing up of the remnants and the following clean-up of the territory.

Many water areas of the former navy ports are polluted with wrecks of sunk and scuttled vessels. In the Mine Port, 15 of those were found, in the Hydroplane Port - 2 ; in the North Port of Paldiski - 7, in the South Port of Paldiski - 2. To date, all of these wrecks have been lifted to the surface.

In the Liepaja harbour, 4 sunken submarines and 22 sunken military vessels have been found. In the Riga military port, 3 sunken vessels were found: 1 submarine, 1 mine trawler and 1 floating barrack.

Sunken vessels always bring about the risk of secondary pollution caused by the leaking of fuel from their fuel tanks or exploding of the explosives on board. As a rule, vessels which were scuttled on purpose during the Soviet Army's departure from our states did not carry ammunition.

The lifting up of recently sunken or scuttled vessels might sometimes even be economically profitable. For example, the Estonian Sea Inspectorate contracted a Norwegian private company to lift up sunken vessels from the former Estonian military ports mentioned above and in turn obtained the metal of the ships. By sales of this metal, the company compensated its expenditures and probably even made some profit. For the lift-up operation, the area around the wreck was surrounded with oil prevention booms in order to prevent the spreading of any fuel leakages.

The contamination of bottom sediments of the former navy ports is a serious problem both in Estonia and Latvia. As a rule, the polluted layer is the upper 1..1.5 metre layer of the bottom sediments and a wide selection of the table of chemical elements can be found there: Cu, Cd, Zn, Cr, Pb, Hg, As, Ni, as well as oil products. The pollution concentration typically exceeds the permitted ceiling levels of the soil to be dumped in the sea.

If not earlier, problems related to prevention of the spreading of secondary pollution from sediments and to disposal of the dredged polluted sediments need to be solved at the dredging of the port area. Otherwise there would be a serious threat of violating the international conventions aiming at the protection of the Baltic Sea, which our states are parties to, as well as of causing transboundary pollution.

At the dredging of the South Port at Paldiski, a protective dam was built between the low bant and the port water area, and the upper most polluted 1-metre-layer of bottom sediments was pumped there. Depending on the results of the subsequent investigations, the decision will be made as to the possibility of constructing the expansion of the port territory over this soil (in case the average pollution level of the soil concerning different components does not exceed the ceiling levels permitted for industrial areas), or alternative solutions would need to be found. At dredging activities, a special pump was used which enabled to extract the soil by pieces, and both the technologies used and the water quality in the vicinity was continuously checked. Due to that it can be said that additional pollution of the sea water was avoided while performing the dredging of the port water area. The construction of the dam was suitable for prevention of transport of the mud and silt polluted with heavy metals back into the sea and filtration of the water from within the soil back into the sea.

Although we do have data concerning the pollution level of bottom sediments of our ports, the database concerning the heavy metal content of marine wildlife, e.g. phyto- and zoobenthos, is very scarce or even missing.

Both Estonia and Latvia lack national financial means for the clean-up of the former military (incl. naval) sites. There are, however, possibilities to use the funds obtained from private

investors and from privatization of objects with the aim of additional investigation of the "historical" pollution, if necessary, and the following remediation activities.
For instance, the Estonian Act on the Use of Funds Obtained through Privatization (passed on April 3, 1996) stipulated the possibility to transfer 5% of the funds accumulated at privatization into the Environmental Fund. The procedure for using these funds is established with Governmental Regulation No.294 of November 19, 1996, according to which there are possibilities to provide means for "environmental investigations and analyses" at the privatized objects as well as for "remediation and sanitation of soil, ground water, and water bodies within polluted areas".

There are other legal means to "encourage" private investors to deal with environmental problems - e.g. the establishment of relevant permitted pollution levels which also enables the stipulation of restrictions to construction over too polluted soil.

Although within the framework of phase I of the NATO/CCMS Pilot Study mentioned above recommendations have been developed concerning the use of different remediation methods depending on the type of soil and pollution, problems at the implementation of these recommendations are often related to lack of practical experience and means for procurement of the necessary facilities.

Since the inventory of the former military sites as well as preliminary environmental assessments of those sites focussing on the threats/risks to the surrounding environment and public health have mostly been completed, both Estonia and Latvia have currently selected the option of focussing on monitoring and prevention of the spreading of existing pollution, and in parallel to this develop the measures for urgent remediation activities.

The complete remediation of the former military sites will, according to the privatization contract, already be responsibility of the new owners while the participation of the state in such activities will be performed through taxation measures in accordance with the relevant liability stipulations of the contract.

As an addition to the current paper, the "Temporary Environmental Quality Objectives for the Contaminants in Soil and Ground Water", as approved by Governmental Regulation No. 174 of April 11, 1995, of the Government of the Republic of Estonia, is given in Appendix 1.

Appendix 1

TEMPORARY ENVIRONMENTAL QUALITY OBJECTIVES FOR THE CONTAMINANTS IN SOIL AND GROUND WATER

Pollutant	Objectives for soil, mg/kg			Objectives for ground water, μg/l	
	Target value	Guidance value in living zone	Guidance value in industrial zone	Target value	Guidance value
Heavy Metals					
Mercury (Hg)	0.5	2	10	0.4	2
Cadmium (Cd)	1	5	20	1	10
Lead (Pb)	50	300	600	10	200
Zinc (Zn)	200	500	1500	50	5000
Arsenic (As)	20	30	50	5	100
Nickel (Ni)	50	150	500	10	200
Chromium (Cr)	100	300	800	10	200
Copper (Cu)	100	150	500	15	1000
Cobalt (Co)	20	50	300	5	300
Molybdenum (Mo)	10	20	200	5	70
Tin (Sn)	10	50	300	3	150
Barium (Ba)	500	750	2000	50	700
Other Inorganic Compounds					
Fluorides (as F^--ion, total)	450	1200	2000	1500	4000
Cyanides (as CN^--ion, free)	1	10	100	5	100
Cyanides (as CN^--ion, total)	5	50	500	100	200

Aromatic Hydrocarbons					
Benzene	0.05	0.5	5	0.2	5
Ethylbenzene	0.1	5	50	0.5	60
Toluene	0.1	3	30	0.5	50
Xylene	0.1	5	50	0.5	60
Phenols (individual compounds)	0.1	1	10	0.5	50
Chlorophenols (individual compounds)	0.05	0.5	5		
Aromatic hydrocarbons (total)	0.5	10	70	1	100
Oil products	100	500	5000	20	600
Polycyclic Aromatic Hydrocarbons (PAH)					
Benzo[a]pyrene	0.1	1	10	0.01	1
PAH (total)	5	20	200	0.2	10
Halogenated Hydrocarbons					
Aliphatic chlorinated and aromatic hydrocarbons (individual compounds)	0.1	5	0	1	70
Polychlorinated biphenyls (PCB, total)	0.1		10	0.1	1
Amines					
Aromatic amines (aniline, xylidines) (total)	5	10	50	0.1	5
Aliphatic amines (total)	50	300	700	1	20
Pesticides					
Organochlorine pesticides (individual compounds)	0.1	0.5		0.05	1
Organochlorine pesticides (total)	0.2	1	10	.1	2
Pesticides (total)	0.5	5	20	0.3	5

Explanation:

1. The objectives in the list are either target or guidance values. **The target value** for a pollutant in the environment indicates the concentration which is considered harmless for human health and ecosystems and which is set as a goal of consistent and systematic efforts of the society. **The guidance value** indicates the concentration which, when exceeded, would cause unacceptable health or environmental risk at the specific location. In order to take decisions concerning the possibility of further use or the necessary treatment method, investigations should be carried out at the risky site and/or region.

2. Values for the **groups of substances** (e. g. cyanides, phenols) should be considered as maximum values in the given group, unless indicated otherwise. In case of necessity, more strict requirements for individual compounds in the group may be established, depending on their risk.

3. If the guidance value is exceeded in an industrial zone, the establishment of new enterprises and the expansion of existing enterprises should be avoided at the specific site.

A PROPOSAL FOR A DECISION FLOW CHART FOR THE SELECTION OF TECHNOLOGIES FOR THE REHABILITATION OF POLLUTED AQUIFERS

TERESA E. LEITÃO
LNEC - National Laboratory for Civil Engineering
Av. do Brasil, 101 - 1799 Lisboa Codex PORTUGAL
Phone: (+351 1) 848 21 31 ext. 2802; Fax: (+351 1) 847 38 45
E-mail: tleitao@lnec.pt

Abstract

Sustainable use of groundwater requires the design of an appropriate protection and management programme that has three main objectives: maintenance of unpolluted groundwater, prevention of future pollution problems, and restoration of polluted groundwater. This paper focuses on the third objective.

A decision flow chart for the selection of the most appropriate technologies for the rehabilitation of polluted aquifers is presented.

Appropriate criteria for the selection of technologies for cleaning up polluted sites are established. The criteria are determined by the type of groundwater pollution problem (type of contaminants and their concentration) and on hydrogeochemical and biological conditions (hydraulic conductivity, pH, adsorption properties of the matrix, presence of micro-organisms, etc.).

Measures to contain and minimise the spreading of pollution, and also the physical, chemical and biological technologies available for the rehabilitation of polluted aquifers are described.

This study was undertaken in the Laboratório Nacional de Engenharia Civil for the author's Ph.D. thesis which was presented to the Faculty of Science of the Lisbon University, Portugal.

1. Introduction

Groundwater is a vital resource in securing a base supply of water in most countries.

In general, groundwater benefits from the natural protections provided by soil cover, by its depth under the ground and by the slow rate of recharge and water flow. On the other hand, the slow movement of groundwater, while favourable to its protection, works against rehabilitation.

The long term nature of groundwater pollution problems and the intrinsic vulnerability to contamination of some groundwater systems combine to make groundwater an endangered resource. Sources of pollution include inappropriate

F. Fonnum et al. (eds.),
***Environmental Contamination and Remediation Practices at Former and Present Military Bases*, 313–321.**

agricultural activities (a diffuse source), industrial and urban activities and salt water intrusion caused by overpumping of coastal aquifers.

An awareness of this problem has lead different countries to introduce ameliorative policies. The European Community has established a "Fifth Environment Action Programme *Towards Sustainability*" in which the Council requests that any action programme should have the following objectives:

- maintain the quality of unpolluted groundwater;
- prevent further pollution;
- restore, where appropriate, groundwater.

2. Main Approaches to Groundwater Rehabilitation

Restoration of polluted groundwater to acceptable quality standards is a process which involves, in most cases, the application of several methods to maximise the performance of the overall treatment process. Groundwater rehabilitation can be accomplished mainly through three different approaches: (1) removal of groundwater, usually followed by further treatment; (2) *in situ* treatment; and (3) water clean-up in a restricted area around an abstraction well.

Groundwater containment and restoration measures include the following (Figure 1.): (1) physical and hydraulic containment of contaminated groundwater; (2) extraction of groundwater (pump-and-treat procedure); (3) *in situ* chemical treatment; and (4) *in situ* bioremediation.

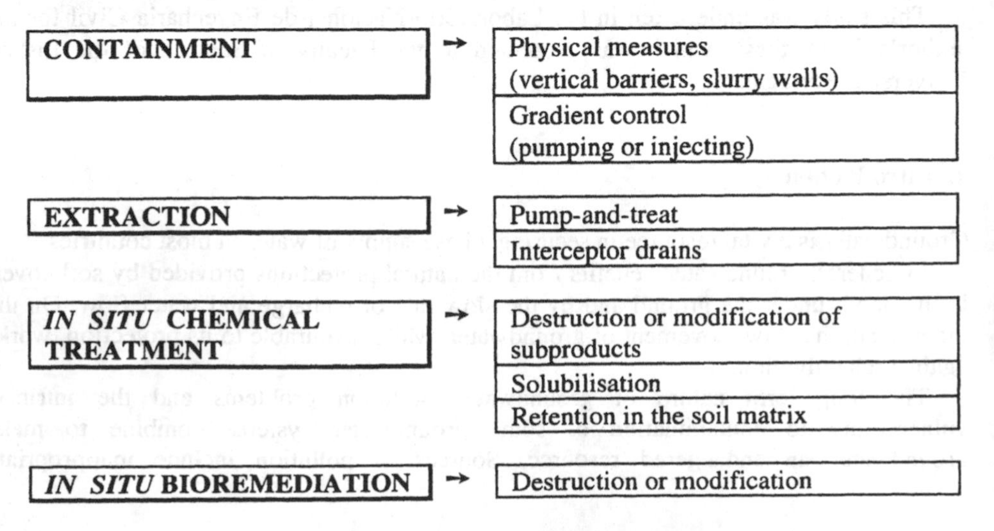

Figure 1. - Main groundwater containment and restoration options

The containment of groundwater pollution is an immediate concern after the detection of a contamination problem. Therefore, containment processes are used in most of the rehabilitation programmes with the purpose of minimising the migration of contaminants located at a specific site, and of protecting the surrounding areas.

Containment processes can be achieved either through the use of hydraulic or physical barriers. Examples of the former are the use of a combination of pumping and injection wells capable of creating a hydraulic barrier. The latter may be slurry walls and grout curtains. In this type of process the restoration of groundwater quality through natural processes may be possible in some cases.

The second main set of restoration processes are **extraction** processes capable of removing mobile contaminants from the aquifer. They aim to achieve the migration of pollutants toward pumping systems or interceptor drains from which they can be extracted. This procedure aims at hydraulic control of the contaminated area with the abstraction of contaminated water. The water is subsequently treated to remove both dissolved and free floating pollutants (pump-and-treat).

***In situ* chemical** rehabilitation is based on the control of chemical processes in groundwater in order to stimulate restoration. Three different approaches are used: (1) the destruction or modification of contaminants to safe sub-products; (2) the solubilisation of contaminants from the soil matrix into the water, for subsequent pumping; and (3) retention of contaminants in the soil matrix. This last option is used to diminish, temporarily, the concentration of pollutants in water, allowing its abstraction at a higher quality.

***In situ* bioremediation** works on the basic principle of promoting naturally occurring microbiological processes. This is achieved by introducing essential nutrients (e.g. nitrogen and phosphorous) and appropriate electron acceptors (e.g. oxygen or hydrogen peroxide). The ultimate goal of this technology is to convert organic wastes into biomass and harmless by-products of microbial metabolism, such as carbon dioxide, methane and inorganic salts.

3. Aquifer Characterisation

The characterisation of local conditions at a specific site provides the basic background information for the selection and design of the most appropriate system for restoration. This implies the previous characterisation and interpretation of flow and transport in the aquifer.

The following factors must be characterised: (1) the source of contamination; (2) aquifer hydrogeology and hydrochemistry (through aquifer monitoring); (3) contaminants present (identification and quantification). The cleanup levels required and the most appropriate rehabilitation technologies are decided on the basis of this information.

The information gathered from groundwater monitoring, sampling and study of contaminant behaviour gives an understanding of the aquifer hydrogeology, groundwater flow paths, as well as contaminant behaviour, concentration and distribution.

4. Selection of the Appropriate Technology

The main technical considerations in the selection of a restoration technology are:

hydraulic conductivity, water pH, concentration and type of elements present in the water, soil adsorption potential, and the presence of micro-organisms in the soil.

The characterisation of these parameters allows the elimination of technologies on the basis of the type of pollutants and/or hydrogeological conditions. Figure 2. presents a proposal for a simplistic decision flow chart for the selection of a rehabilitation technology. The flow chart was constructed on the basis of a quantification of some of the parameters referred to [6].

Hydraulic conductivity plays a key role in determining the possibility of using restoration processes. In general, for hydraulic conductivity values lower than 10^{-8} m/s it is not possible to use *in situ* techniques nor pumping strategies. This is due to the very slow circulation of fluids in the media impeding abstraction of water, as well as *in situ* treatment. In these cases containment procedures are advisable, using physical barriers.

For values of hydraulic conductivity with an order of magnitude lower than 10^{-6} m/s (between 10^{-8} to 10^{-6} m/s) rehabilitation is only possible for volatile compounds. They can be eliminated by the injection of air into the saturated zone (biosparging) [7].

Theoretically, extraction (pump-and-treat) technologies are applicable for all media with hydraulic conductivity values higher than 10^{-6} m/s. These technologies are suitable when contaminants are mainly present in solution (low adsorption rate), with volatile pollutants, and for the abstraction of light non-aqueous phase liquids (LNAPLs). The technologies are not applicable in the case of dense nonaqueous-phase liquids (DNAPLs).

As a result of sorption, ion-exchange, chemical precipitation and biotransformation processes, contaminants may be retained in the soil matrix. To increase contaminant solubility and mobility, some retardation effects, like sorption and ion-exchange, can be reduced. This approach can be used in conjunction with the pump-and-treat procedure, increasing the concentration of pollutants present in the water to be treated.

Examples of this procedure include particle transport, co-solvation and phase shifting. In these cases, changes in the media - caused by, for example, raw chemical discharge to the subsurface or changes in pH and Eh - can ionise neutral compounds, reverse precipitation reactions, produce complexes with other chemical species, and limit bacterial activity [4].

Bioremediation is one of the most used technologies for the rehabilitation of aquifers contaminated with organic compounds.

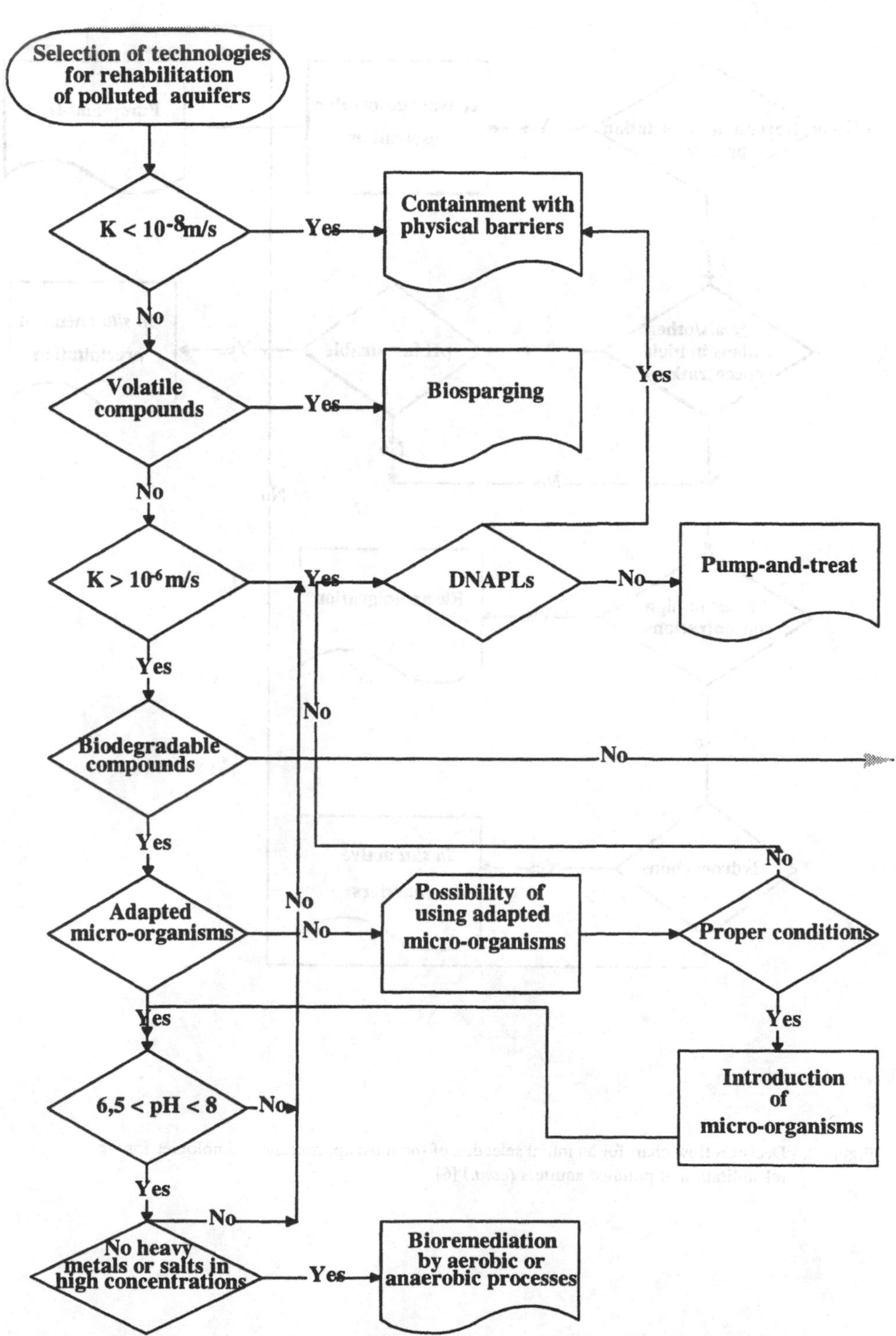

Figure 2. - Decision flow chart for an initial selection of the most appropriate technologies for the rehabilitation of polluted aquifers [6].

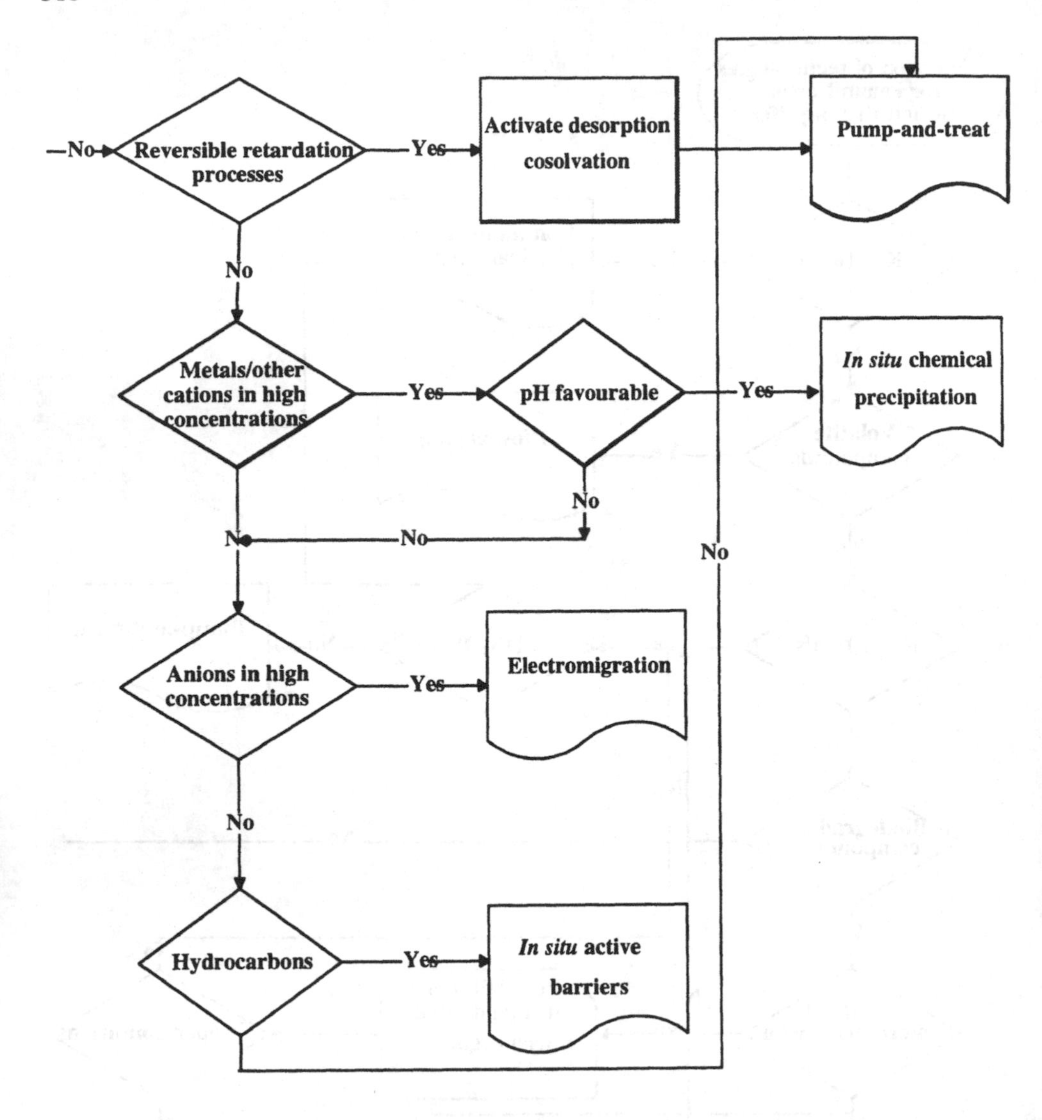

Figure 2. - Decision flow chart for an initial selection of the most appropriate technologies for the rehabilitation of polluted aquifers *(cont.)* [6].

In order to permit biological degradation of organic pollutants, appropriate conditions for microbiological activity in the subsurface must be achieved. These are redox conditions, pH, medium permeability and presence of adapted micro-organisms.

Redox and pH conditions must be in a range that allow bacterial growth and avoid toxic concentrations of inhibiting agents, such as heavy metals. Most subsurface bacteria prefer a pH range of 6.5 to 8, but they may adapt to conditions slightly beyond this range [1].

The presence of adapted micro-organisms is normally not a problem, although some recalcitrant contaminants may be untouched by existing bacteria. The micro-organisms in some uncontaminated soils might require little or no acclimatisation before they can degrade xenobiotics. On the other hand, some pollutants require long periods of acclimatisation before biodegradation can occur [5].

Permeability is a very important parameter for *in situ* biorestoration due to its influence on flow conditions. For the projects reviewed by Staps [10], the K value varied between 10^{-2} and 10^{-5} m/s, but mainly in the range 10^{-3} - 10^{-5} m/s. Generally, a K value of 10^{-5} m/s is regarded as being the minimum permeability for successful application of *in situ* biorestoration [10]. Materials with low permeability, whether present in layers or in lenses, may hold significant levels of contaminants but receive little of the injected solution. Therefore, most successful applications of *in situ* biorestoration have usually been in shallow sandy or highly fractured aquifers [8].

Another important question concerning biodecomposition is the existence of a minimum substrate concentration; below that minimum level bacteria cannot utilise the targeted substrate. The substrate concentration may be so small that the rate of energy extraction from the substrate's consumption is not sufficient to maintain the bacterial activity; if this happens, the bacteria will decay. Bouwer *et al.* [2] report that the use of a pollutant as a primary substrate for microbial metabolism is restricted to those cases of ground water contamination where highly degradable organic compounds are present in concentrations above 10 mg/l.

Finally, micro-organisms might be inhibited or killed by high concentrations of the pollutants themselves [5]. Biodegradation occurs therefore only in the more diluted part of the contaminated site.

Therefore, most remediation projects in which enhanced biorestoration has been applied have started with the removal of heavily contaminated soils. Biorestoration is therefore used in conjunction with other remediation procedures. In this type of procedure pumping systems are usually first installed to remove free product floating on the groundwater. Then *in situ* biorestoration enhancement measures are initiated to degrade the more diluted portions of the plume.

In some cases existing bacteria might leave contamination untouched. The addition of micro-organisms acclimatised to degrade a specific pollutant can then be used. This technology, although in use for more than a decade, has not been convincing until recently, due, in particular, to potential inoculation problems.

For the removal of non-biodegradable pollutants - i.e. most inorganics like metals and some salts - other technologies are available. Processes like *in situ* chemical precipitation of iron and manganese are currently used (Vyredox, Hallberg and Martinell, [3]). In this process ferrous ion (which is soluble in water) is oxidised to ferric (which is insoluble), before water is pumped into a well. This method, which uses iron-oxidising bacteria and aeration wells, achieves a high degree of oxidation around the well [3].

In situ electromigration is potentially applicable to any electrically charged species, whether organic or inorganic, present in groundwater. In this process, an electrical potential is applied to metallic or carbon electrodes that have been placed in wells in the contaminated aquifer [9]. This can be used as a pre-concentration process for soluble species such as SO_4^{2-}, NO_3^-, Cl^-, and Na^+ for subsequent removal by extraction methods.

The use of permeable barriers is an alternative treatment process. Permeable barriers allow the passage of water, but prevent the migration of certain types of pollutants. This can be accomplished by extraction, stabilisation and degradation of groundwater contaminants. These technologies include adsorption, ion-exchange, chemical precipitation, and biological stabilisation.

The practical application of these technologies is only known at field scale. They have been used to achieve precipitation and adsorption of inorganic pollutants, and also for organic contaminants removal.

Some of the technologies presented have only yet been tested at field scale applications.

5. Conclusions

In this paper the main technologies available for aquifer remediation are summarised.

A proposed decision flow chart is presented for the initial selection of the most appropriate technologies for the rehabilitation of a polluted aquifer. The aim of the flow chart is to present the most important factors that need to be considered in deciding the best remediation technologies available.

The flow chart presented is solely a generalised first approach to the problem. Each case needs a deeper analysis and characterisation of specific constraints like, for example, changes in hydraulic conductivity or final quality standards incompatible with a candidate technology.

Despite the expansion of polluted aquifer rehabilitation technologies, general understanding of and faith in some of these technologies is still a matter for debate.

ACKNOWLEDGEMENTS

We would like to express out gratitude for the assistance and financial support of Laboratório Nacional de Engenharia Civil, and the Lisbon Faculty of Sciences.

We would also like to thank Seán Quigley for his help in reviewing the English version of the paper.

6. References

1. Barbee, G.C. (1994) *Fate of Chlorinated Aliphatic Hydrocarbons in the Vadose Zone and Ground Water.* Ground Water Monitoring Review, Winter 1994, pp. 129-140.
2. Bouwer, E., Mercer, J., Kavanaugh, M. and DiGiano, F. (1988) *Coping with Groundwater Contamination.* Journal of the Water Pollution Control Federation, Vol. 60, n° 8, pp. 1415-1427.
3. Hallberg, R.O. and Martinell, R. (1976) *Vyredox - In Situ Purification of Ground Water.* Ground Water, Vol. 14, n° 2, pp. 88-93.
4. Keely, J.F. (1989) *Performance Evaluations of Pump-and-Treat Remediations.* U.S. EPA Office of Research and Development: EPA/540/4-89/005, Oklahoma, 1989, 19 pp.
5. Lee, M.D., Thomas, J.M., Borden, R.C., Bedient, P.B., Ward, C.H. and Wilson, J.T. (1988) *Biorestoration of Aquifers Contaminated with Organic Compounds.* CRC Critical Reviews in Environmental Control, Vol. 18, n° 1, pp. 29-89.
6. Leitão. T.E. (1996) *Metodologias para a Reabilitação de Aquíferos Poluídos.* Ph.D. Dissertation Thesis, Lisbon Faculty of Sciences, 1996, 489 pp.
7. MacDonald, J.A. and Rittmann, B.E. (1993) *Performance Standards for In Situ Bioremediation.* Environmental Science and Technology, Vol. 27, n° 10, pp. 1974-1979.
8. Rainwater, K., Mayfield, M.P., Heintz, C. and Claborn, B.J. (1993) *Enhanced In Situ Biodegradation of Diesel Fuel by Cyclic Vertical Water Table Movement: Preliminary Studies.* Water Environment Research, Vol. 65, n° 6, pp. 717-725.
9. Runnells, D.D. and Wahli, C. (1993) *In Situ Electromigration as a Method for Removing Sulphate, Metals, and Other Contaminants from Ground Water.* Ground Water Monitoring Review, Winter 1993, pp. 121-129.
10. Staps, J.J.M. (1991) *International Evaluation of In Situ Biorestoration of Contaminated Soil and Groundwater, in* "Demonstration of Remedial Action Technologies for Contaminated Land and Groundwater", Ed. F. Olfenbuttel, Vol. 2, Part 2, NATO/CCMS, Report n° EPA/600/R-93/012c, pp. 741-766.

INVESTIGATION AND REMEDIATION OF FORMER MILITARY FUEL STORAGE OF VALČIŪNAI (LITHUANIA)

N. ŠEIRYS
Hydrogeological Company GROTA
Eišiškių pl. 26, Vilnius, Lithuania

1. Introduction

The supply of potable water in Lithuania completely depends on groundwater. Hence it is important to evaluate and minimise any negative impact on this resource from underground pollution. At present, the economic situation in Lithuania restricts the remediation of polluted areas. Consequently, many of the sites will not be cleaned up in the foreseeable future. Meanwhile, however, contaminants continue to migrate into the environment. The most that can be achieved at present is the monitoring of these sites.

The former military fuel storage centre at Valčiūnai is the first military site in Lithuania where remediation of subsurface pollution was undertaken.

2. Site Description

Valčiūnai Oil Product Base is located 14 km south of Vilnius, the capital of Lithuania. For 30 years, light oil products and rocket fuels were stored at this former military fuel depot. The total volume of the underground storage was some 33,100 m^3. The base's area of 82 hectares is divided into two parts: a western one for light oil product storage (tank volume c. 27,500 m^3) and an eastern one formerly used for rocket fuel storage (tank volume c. 3,200 m^3). The two storage area separated by a 200-300 meter wide forest belt (Fig.1). The rocket fuel storage area contains both rocket fuel and fuel oxidizing agents at separate locations. The total volume of the tanks with the oxidant is approximately 2,400 m^3.

The first site investigations at the Valčiūnai Oil Product Base were carried out in the summer of 1993, prior to its being taken over by Lithuanian government. The presence of hazardous compounds in the natural subsurface environment was demonstrated in numerous locations.

F. Fonnum et al. (eds.),
***Environmental Contamination and Remediation Practices at Former and Present Military Bases*, 323–328.**

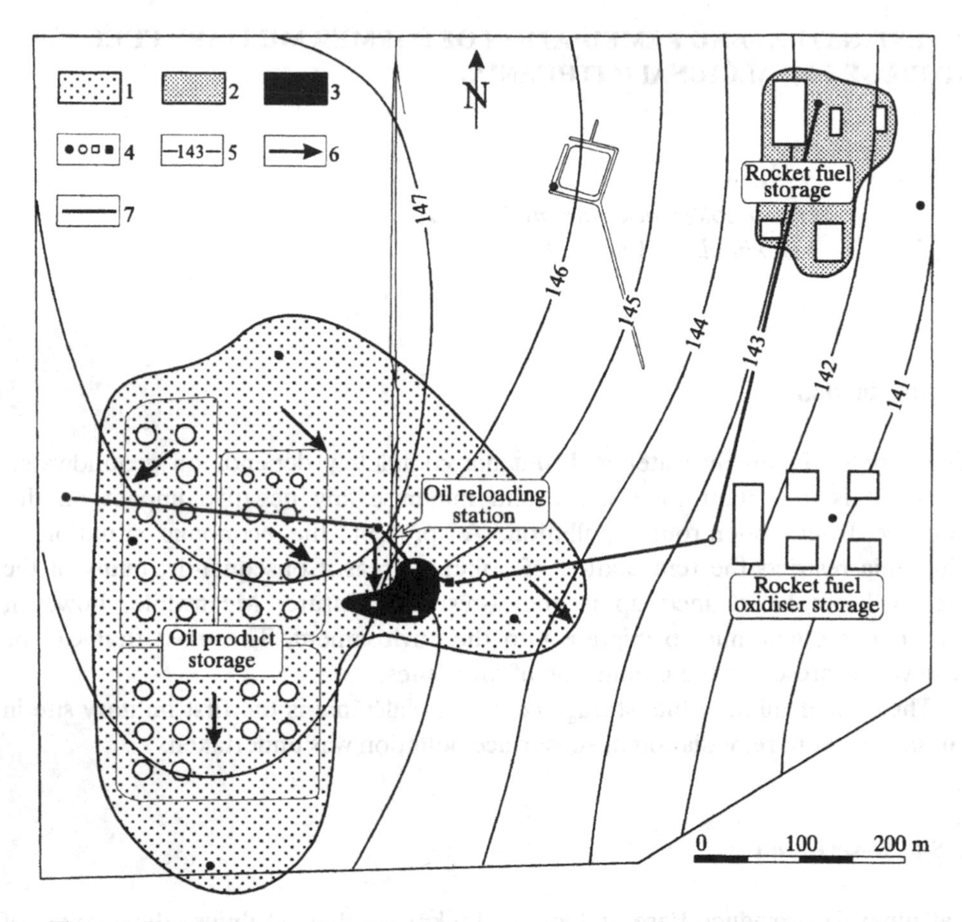

Figure 1. Distribution of pollutants in shallow groundwater in the Valčiūnai Oil Product Base

1 - 3 - areas of groundwater contamination: 1 - by aromatic hydrocarbons (BTEX), 2 - by chlorinated hydrocarbons (TCE), 3 - free phase oil layer on the groundwater table; 4 - monitoring, probe, production and infiltration well; 5 - isoline of groundwater level, elevation above sea level (m ASL); 6 - contaminant migration pathway; 7 - geological cross-section line.

The sources of the pollution pose a major risk to the quality of surface- and ground water. Firstly, pollutants occur in groundwater both as a dissolved aqueous plume and as a pure product. These pollutants migrate laterally and disperse transversely; hence the area of pollution in the shallow aquifer gradually widens. The polluted area lies at the watershed of two rivers. Shallow groundwater flowing from the base's territory drains into one of these rivers. Secondly, there is a risk of downward migration into the deeper freshwater aquifers. In this respect, the geographical position of the base is also unfavourable. It is located within the third sanitary protection zone of the groundwater reservoir supplying water to Vilnius, a zone in which chemical pollution is prohibited. Therefore, it is vital to determine the actual and potential impact of pollution on groundwater quality in the aquifer.

Regular hydrogeo-ecological studies in the area of the Valčiūnai Oil Base began in the spring of 1994. A total of 70 boreholes have been drilled at the site and a monitoring network for regular observation of the evolution of the extent of underground pollution created. This network comprises a total of 13 observation wells installed for observation of groundwater quality and level in shallow, intermediate and deep (productive) aquifers. For observations of the downgradient migration of the pollutant plume, 10 boreholes were instaled in the shallow aquifer with depths varying from 6.3 to 18.6 m. Active remediation of the site has been in progress since January 1997.

3. Subsurface pollution

Light non-aqueous phase liquids (NAPL) have contaminated the groundwater system beneath the western part of Valčiūnai Oil Product Base, whereas dense NAPL (DNAPL) contamination occurs beneath the northeastern part. The subsurface pollution is illustrated in Figs. 1 and 2.

The expansion of the pollutants in the groundwater system under the base is uneven. The rate of pollutants migration and free phase oil products layer accumulation are controlled by the lithology of the layers and also depend on the nature of the pollutants.

Under the western and central parts of the base, the soil and shallow groundwater are polluted over an area of more than 15 ha. Total concentrations of hydrocarbons in the subsoil at depths of 5-15 m range from 17 to 611 mg/kg of wet soil. The total concentration of aromatic hydrocarbons (BTEX) in the shallow groundwater ranges from 14 to 72 mg/l.

During the drilling of boreholes in the rocket fuel storage area a strong odour was noted in the vadose zone at depths of 2.0 - 6.5 m. Beneath the tanks, shallow groundwater is polluted by trichloroethene (TCE) and aromatic hydrocarbons in an area of 1.0-1.3 ha. Total concentration of BTEX in shallow groundwater reaches 70 μg/l, whereas TCE reaches c. 15 mg/l. In the lower, intermediate aquifer at a depth of 33-36 m the concentration of TCE in groundwater is 93 μg/l.

A ravine was mapped in the southern part of the oil reloading station. A layer of coarse, gravel-like gravely sand covering an area of 3,000 - 3,500 m^2 with a thickness of 3-4 m occurs at a depths of 6-8 m. It acts as an area of accumulation of free phase hydrocarbon layer. Due to conditions in this area, leaking oil products have been accumulating there for almost 30 years and reached a thickness of up to 2 m, at about 5.5 to 7.5 m below land surface. Further migration of the oil phase to a larger area is limited by the lithology of the site, although dissolved hydrocarbons are likely to migrate further. In order to alleviate the pollution in the western part of the base, active removal of accumulated oil products from the subsurface is being undertaken.

4. Oil product removal

The existing free phase oil plume is the main source of subsurface pollution in the central part of Valčiūnai Oil Product Base. Hazardous organic compounds are dissolved and transported away from the pollution source via groundwater flow. The primary goal of the site clean-up was to remove the free phase oil product layer from the subsurface and to reduce the extent of the polluted area.

For product removal, a single pump system producing a mixture of hydrocarbons and water was employed. The remediation network is composed of four production/infiltration wells and a 45 m^3 volume oil separation tank.

Three wells are installed in the zone of oil product accumulation; they can be used for extraction or infiltration purposes at different stages of the remediation operation. The fourth well is installed outside the boundary of oil product spreading. It is used for infiltration of separated water into the subsurface. Prior to remediation, the top of the oil layer was at a depth of 5.80 - 6.0 m BGL. The thickness of this layer was varying from 2.04 to 2.85 m in different wells. Oil thickness in the wells proportionally reflects the real thickness of the oil layer accumulated on the shallow groundwater table. The absolute real thickness of oil layer on groundwater is approximately 0.5-0.7 m lower than the apparent thickness in the wells.

Repeating cycles of oil-water mixture removal take about 7 hours per day. Only one production well is being operated. The pumped yield corresponds to the

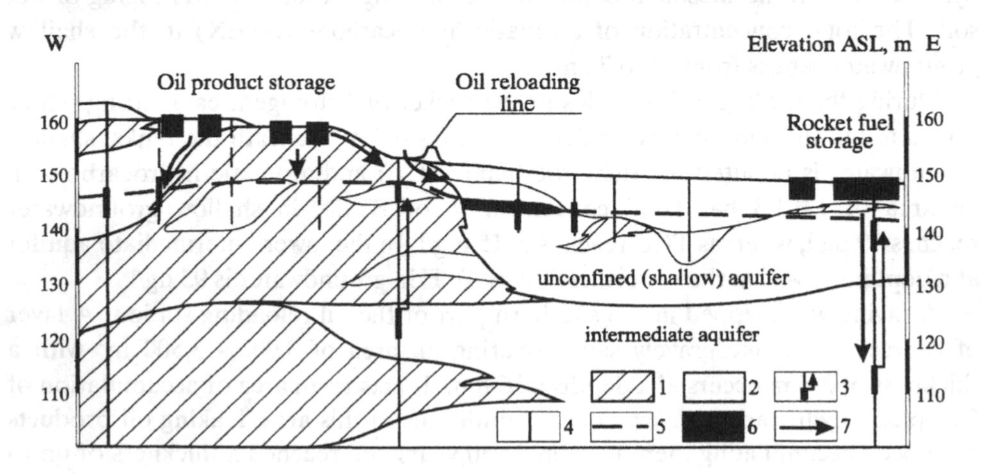

Figure 2. Geological cross - section and antropogenic load of the Base territory

1 - low permeability layer; 2 - aquifer; 3 - groundwater sampling point and groundwater level of intermediate aquifer; 4 - probe borehole; 5 - groundwater level of shallow aquifer; 6 - free phase oil layer on the groundwater surface; 7 - contaminant migration pathway.

capacity of infiltration well to accept the separated water, *i.e.* c. 1 m^3/h. The amount of oil product in the extracted mixture ranges from 10 to 36 percent. At an optimal yield of 0.8 - 1.2 m^3/h, the oil comprises 22 - 32 percent. More than 250 m^3 of combustible yellow hydrocarbon products have been pumped out during the initial 9 months of remediation. The mixture contains about 60% petrol, with the remainder being diesel fuel and kerosene.

It is difficult to estimate the time required to remove the oil layer from the subsurface. The thickness of the oil layer in operating production wells decreased by 0.77-1.03 m (from 2.04-2.85 m to 1.27-1.82 m) over the initial 9 month period. The thickness of oil layer in all production wells thus decreased approximately 40 percent.

Information is lacking concerning the amount of spilled and distributed oil product in the subsurface. The product that remains in the unsaturated (vadose) zone is an important source of contributing to contamination in the dissolved and free phase oil plume. The product is leached by groundwater as the water table rises. Changes of oil product thickness in the wells confirm that, at present, accumulation of oil is continuing.

It has been estimated that 400 - 500 m^3 of oil product could potentially be removed from the subsurface within a two years period.

5. Conclusions

There are two polluted areas in the subsurface of the Valčiūnai Oil Product Base. The first area consists of soil and shallow groundwater polluted with hydrocarbons (LNAPL) in the western part under and around the oil storage tank complex. The second area lies in the north-eastern part of the base under the rocket fuel tanks, where soil is polluted with chlorinated hydrocarbons (TCE).

The geological structure and hydrogeological conditions at the site vary, both laterally and vertically. Moraine till occurs immediately beneath the surface at the western part of the base, and dips under sand-gravel deposits in the central and eastern parts where a shallow aquifer occurs. In the eastern part of the area, where till layers have been washed out, there exists hydraulic connection with the deeper intermoraine aquifer supplying the Pagiriai well-field, the largest potable water supply alimenting the city of Vilnius.

Oil products have leaked from the storage tanks and accumulated on the water table below the central part of the base. Free phase oil covers an area of 3,000 - 3,500 m^2, with a maximum thickness of approximately 2 m, beneath the rocket fuel storage area, the soil and shallow groundwater are polluted by trichloroethene. This chemical (in its free phase) is heavier than water and therefore migrates to deeper aquifers.

Local accumulation of oil product is controlled by the lithology and structure of aquifer sediments. Oil products have also dissolved in the groundwater and are now migrating with groundwater flow. Thus the size of the polluted area is thus increasing. In order to hinder the further migration of oil products, remediation of the subsurface has started. In the initial nine months of a pump-treat-recirculate system, more than 250 m^3 of liquid oil products were extracted.

The extraction of liquid oil products accumulated on the shallow groundwater table will probably continue for some two years, by which time about 400 - 500 m^3 of oil product should have been removed. It has now become obvious that area after that, will still not be possible to call the pollution source "clean". Subsequently, for the extraction of oil adsorbed on sediments and in pore spaces, a vacuum-extraction technique and bioremediation will be introduced for the next stages of remediation process.

6. References

1. Paukštys, B. and Belickas, V. (1996) Detection of groundwater contamination at former military sites in Lithuania, *Regional Approaches to Water Pollution in the Environment (NATO ASI Series, 2. Environment - Vol. 20), 71 - 90.*
2. Kruger Consult in Partnership with the Baltic Consulting Group. (1995) Inventory of Damage and Cost Estimate of Former Military Sites in Lithuania, Final Report, Vilnius.
3. *Šeirys, N. (1995)* Studies of soil and groundwater pollution in the Valčiūnai fuel storage, *Geologijos akiračiai 1,* 13 - 16 [in Lithuanian].
4. Palmer, P.L. (1996) Vacuum-enhanced recovery, *In Situ Treatment Technology, 149 - 194.*

DUMPED AMMUNITION IN MINE SHAFTS

Hydrogeochemical evaluation of two mine shafts at dalkarsberg.
A repository for decommissioned ammunition and other explosives

J.G. HOLMÉN and U. QVARFORT
Uppsala University, Institute of Earth Sciences
Norbyvägen 18B, S-752 36, Uppsala, Sweden.

B. Liljedahl
The National Defence Research Establishment,
Department of NBC Defence,S-901 82 Umeå, Sweden

Abstract

Ammunition and explosives have been dumped at Dalkarsberg in two water-filled mine shafts, the Central shaft and the Bäckaskog shaft. The future hydrogeological status of the repository has been evaluated, based on field investigations and mathematical modeling. The field investigations consisted of borings and chemical analysis of water from the shafts. The flow and transport in the tunnel system and in the fractured rock mass were simulated by use of three-dimensional models. The heterogeneous properties of the rock mass were represented by a stochastic continuum.

The models predict that all the water in these two shafts, as well as all transported pollutants, will discharge to the surface water system in the very near vicinity of the uppermost part of the Bäckaskog shaft. All pollutants will end up in a small stream located close to the shafts.

We conclude that the explosives (TNT and lead acid) as well as metals present in the mineshafts as a result of diffusion from the dumped material will be completely released after about 2 000 to 20 000 years. Because of a relatively high dilution in the shafts, the amount of released components in the water will be very low. The main conclusion is therefore to leave the dumped material in the shafts and install a monitoring system for future water analysis.

1. Introduction

After the Second World War, a huge amount of ammunition was destroyed or dumped at various sites. The Swedish Armed Forces used lakes, the sea or a few former mining shafts for dumping. Potential environmental hazards now exist from the presence of ammunition and explosives, together with municipal sludge and household wastes, in two

F. Fonnum et al. (eds.),
Environmental Contamination and Remediation Practices at Former and Present Military Bases, 329–342.

mineshafts at Dalkarsberg. The migration of substances such as TNT from ammunition, as well as components from the municipal waste, is of current concern.

The purpose of the study is to evaluate the future hydrogeochemical status of the repository for explosives at the Central and Bäckaskog shafts of the Dalkarsberg mine. The evaluation is based on field investigations, chemical analysis of water from the shafts, and mathematical modeling. The mathematical flow modeling provides a theoretically study of the groundwater system at the Dalkarsberg mine, and thereby assess the flow pattern and estimate the size of the flow and transportation time for the groundwater inside and in the vicinity of the Bäckaskog and Central shafts. This part of the study was based on a system analysis approach. The results of the modeling were then used in combination with leaching calculations to estimate the release of the TNT and other elements from the dumped material.

2. Basic Site Description

The investigation focuses on an abandoned iron mine, Dalkarsberg, which is located in the province of Närke, central Sweden. The ore occurs in a granitic gneiss host rock. The mine was originally worked as early as 1559. During the period 1819 to 1925 the production was 3 000 000 tons of iron. Mining at Dalkarsberg ceased at the end of the 1930s, when the price for iron dropped. Figure 2.1 gives an overview of the mines based on old mine maps.

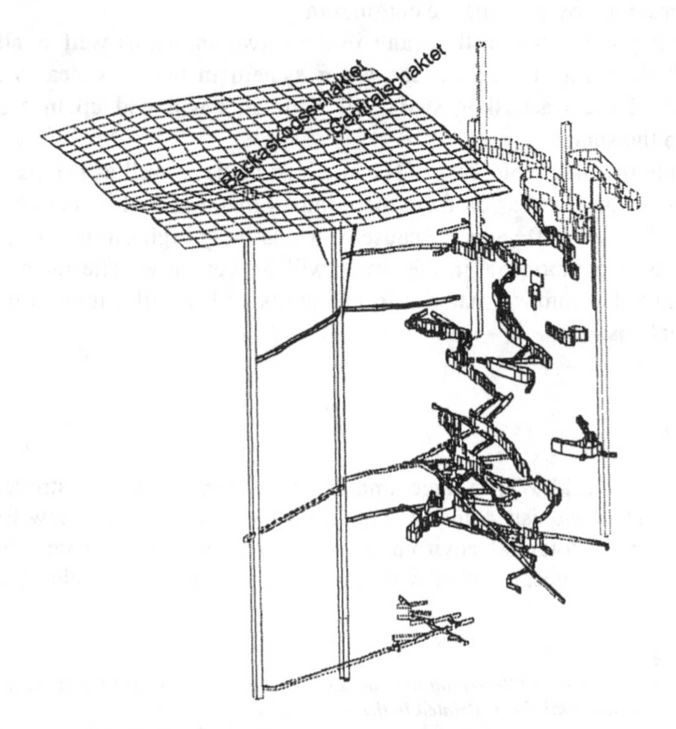

Figure 2.1 Mining map of the Dalkarsberg mine

During the period 1955 – 1968 the mineshafts was used for dumping ammunition and explosives together with household wastes and sludge's. According to the ammunition, the shafts contains about 7 tons of gun ammunition, 143 tons of handgranates, 35 tons of land and personal mines, chemicals and 1.7 millions of detonators. Together with the household waste and sludge the shafts are today filled up to the surface. The approximate chemical composition of the dumped material is given I Table 2.1

TABLE 2.1. Chemical Composition of the dumped material. From [8], [10]

Kind of material	Amount ton	Chemical substance
Gun ammunition,	0.4	Gun powder
Gun ammunition 7.5-21	7.5	Gun powder, TNT
Explosives	383	TNT, lead acid
Handgranates	143	TNT
Picrine acid	35	Picrine acid
Gun powder	0.2	Gun powder
Detonators	3	TNT, lead acid, Al
Capsule material	330	Cu, Pb, Zn
Municipal wastes	?	?

3. Field investigations and chemical analysis

In the summer of 1997 water was sampled from the shafts. The sampling were done by diamond drillings at different angles from the surface and into different depths in the shafts. By this method is was possible to get water samples from 20, 50 and 100 meters respectively below the surface. According to the used drilling technique is was not possible to reach any deeper samples. A vertical boring through the dumped material and into the shafts was not possible according to safety reasons

The samples were analyzed for metals by using ICP-MS and ICP-AES by the SGAB, Laboratory, Luleå, Sweden. The organic components were detected using GC-techniques according to the EPA-programs by the Institute for Applied Environmental Studies, (ITM), University of Stockholm, Sweden. FOA, Department of NBC Defence, Umeå, Sweden, made the TNT analysis.

The results of the chemical analysis show no high contents of investigated elements either for metals or organic components. All the water samples were suitable for drinking. The results were confusing as most of the samples were taking inside the waste material and were believed to have high concentrations of especially nitrogen and picrine acid.

4. Establishment of the flow models

4.1 DEFINITION OF MODELS

4.1.1 Original flow equation

The formal model is a three dimensional mathematical description of the studied flow system. The basic flow equation is a differential equation developed from Darcy´s law [1]. The development of a three dimensional flow equation from Darcy´s law is well known, see for example [2] or [3]. We assume that the groundwater has a constant temperature and a constant density. The modeling is carried out under steady state condition.

4.1.2 Size and lay-out of the models

Three different models have been established, a regional model that represents a large area surrounding the Dalkarsberg mine and two local models that represent smaller areas surrounding the studied mine.

Regional model. The purpose of the regional modeling was to estimate the regional groundwater flow pattern, and thereby derive suitable boundary conditions for the smaller local models. The local models are located inside the regional model. For the regional modeling we used the computer code MODFLOW [7]. This is a three-dimensional finite difference model. The regional modeling was carried out as a sensitivity analysis, as regards the size of the model. The size of the regional model was increased both horizontally and vertically. For every new size of model, we compared: (i) the calculated head at positions representing the shafts of the Dalkarsberg mine, to (ii) the head calculated at the same positions by smaller models. The regional model was made so large that no significant change in head took place at the studied positions when the regional model was further increased in size. The head values, in the vicinity of the Dalkarsberg mine, that we obtained for this large regional model were used as boundary conditions for the smaller local models. The regional model extends over an horizontal area of about 50 km^2.

Local models. The local modeling was performed by the use of the computer code GEOAN [4]. This is a three dimensional finite difference model, capable of modeling the interaction between the groundwater surface, the ground surface and the groundwater recharge, hence, calculating recharge and discharge areas etc. It is also a model specially designed for use of the stochastic continuum approach. The local models surrounding the Dalkarsberg mine are of the following size:

Model 1, Horizontal: 1200m x 2600m, vertical: 600m. *Model 2,* Horizontal: 700m x 750m, vertical: 600m. The purpose of Model 1 was to estimate the flow pattern of the groundwater in the rock mass, assuming that no mine exists. We have established Model 1 as a comparison to the actual situation, which is represented by Model 2. In Model 2 we have included a generalized and limited layout of the tunnel system of the Dalkarsberg mine. The purpose of Model 2 was to estimate the flow pattern of the groundwater in the rock mass as well as estimate the flow pattern of the groundwater in the tunnels.

4.1.3 Boundary conditions

The outer boundaries of the regional model were located along topographic water divides (no flow boundaries) and along significant streams and lakes (specified head boundaries), the bottom of the model was defined as a no flow boundary. The local models were located inside the regional model. The outer boundaries of the local models, including the bottom of the models, were assigned the specified head boundary condition, the head values were given by the regional model.

4.1.4 Hydraulic conductivity

The rock mass is a fractured crystalline rock with heterogeneous properties. In the local models, the heterogeneous hydraulic properties of the rock mass were represented by a stochastic continuum. Many authors, e.g. [4], [5] and [4], discuss the stochastic continuum approach. For the regional model a homogeneous conductivity was defined, an effective value representing the heterogeneous conductivity was used.

Conductivity of rock mass, results from field investigations. The conductivity of the rock mass has been investigated by the use of packer tests in a 100m deep bore hole, located about 100m from the Bäckaskog shaft. The tests were carried out as double packer tests, with a test section length of 3m. The bore hole was also tested as one section. The tests show that the conductivity varies and can be represented by a statistical log-normal distribution. The conductivity of the whole bore hole, tested as one section, is larger than the geometric mean of the values of the 3m sections. Hence, the conductivity of the rock mass is heterogeneous and scale dependent. An effective value, applicable at large scales, have been estimated to 1.8×10^{-6} m/s.

Conductivity of rock mass, as defined in local models. Based on the field investigations and a method consistent with the stochastic continuum approach, given in [4], the following two distributions have been defined for the conductivity (K) of the rock mass in the local models. *Model 1.* Block size 50x50x50m, Log-normal distribution, Geometric mean of K= 1.38×10^{-7} m/s, Standard deviation of eLog K= 1.10 *Model 2.* Block size 25x25x25m, Log-normal distribution, Geometric mean of K= 1.25×10^{-7} m/s, Standard deviation of eLog K= 1.35.

Uncertainty in the defined conductivity of the rock mass. Large uncertainties are included in the field investigations and in the evaluation of the conductivity as well as in the method for establishing models of the heterogeneous properties of the rock mass. Therefore, an uncertainty in the conductivity has been introduced in the local models. We assume that the geometric mean of the log-normal distributions, given above, can vary within one order of magnitude. The assumed deviation from the best estimate, given above, is log-normal distributed with a standard deviation of 0.25 orders of magnitude.

Conductivity of tunnels and shafts. For tunnels and shafts that are very permeable, not drained and in equilibrium with the surrounding groundwater system, there is a maximum theoretical flow that depends on the permeability of the surrounding rock mass only. In the models, the conductivity of the tunnels is defined in a way that the flow of the tunnels will represent the flow of a tunnel that is empty (the maximum theoretical flow).

4.1.5 Effective porosity (chinematic porosity)

In the models the effective porosity of the rock mass is assumed to be approximately normal, distributed and correlated to the conductivity of the rock mass. The effective porosity is defined as approximately normal distributed with a mean of 0.005, a standard deviation of 0.0018, a maximum of 0.01 and a minimum of 0.001; the correlation coefficient between the effective porosity and the logarithms of the conductivity is set to 0.9. For the tunnels the effective porosity is set to one. Hence, as regards the flow, the tunnels are assumed to be empty.

The regional and local models have been assigned conductivity based on the results from the field investigation. The models have been calibrated by varying the groundwater recharge until the calculated head values and the size of the calculated recharge and discharge areas, are in accordance with estimated and field conditions. In the calibrated models, the potential recharge is set to 50 mm/year. This recharge value produces groundwater heads that represent average head values for a normal year, (See Figure 4.1)

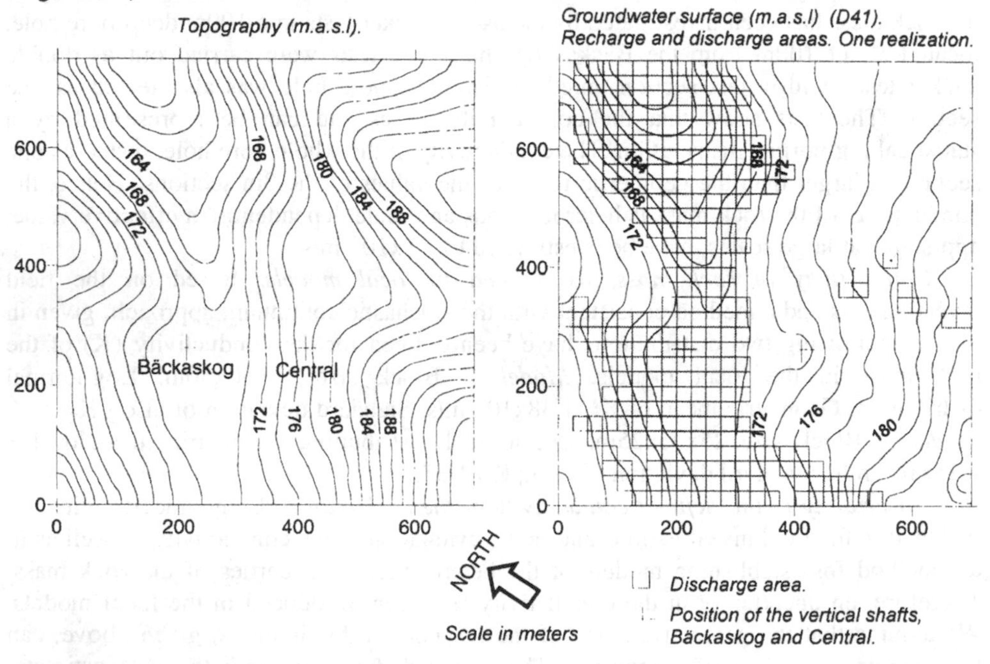

Figure 4.1 Topography and position of the shafts. Calculated groundwater surface and discharge areas.

5. Results of the flow modeling

5.1 FLOW PATTERN

5.1.1 General

The Dalkarsberg mine is located at a slope towards a valley. At the bottom of the valley a stream is flowing towards a lake. The general flow pattern of the groundwater is from the surrounding heights towards the stream. The modeling predicts that the Bäckaskog shaft is located in a discharge area, the Central shaft and the rest of the tunnel system are situated in recharge areas. The Central shaft and the rest of the tunnel system are connected to the Bäckaskog shaft through several different tunnels. As the Bäckaskog shaft is a discharge area, the water in the tunnel system will flow towards this shaft. During periods with heavy rainfall, it is possible that both the Bäckaskog and the Central shafts will be discharge areas, and for such a situation the water in the tunnel system will flow towards both shafts. In the vicinity of the tunnels, the groundwater flow in the rock mass is directed towards the tunnels (see Figures 4.1 and 5.1).

5.1.2 Flow pattern in shafts and tunnels

The modeling predicts that all the water that occur in the tunnel system will discharge at the uppermost part of the Bäckaskog shaft (see Figure 5.1).

Flow pattern in the Central shaft. In the upper 135m of the Central shaft, the water is flowing downwards, to the first tunnel that connects the Central and Bäckaskog shafts. Deeper down in the shaft the flow is directed upwards to tunnels that connect the two shafts at different depths. When the flow reaches a connecting tunnel, it will follow the tunnel to the Bäckaskog shaft.

Flow pattern in the Bäckaskog shaft. In the Bäckaskog shaft, the water is flowing upwards, to the uppermost part of the shaft and to the ground surface.

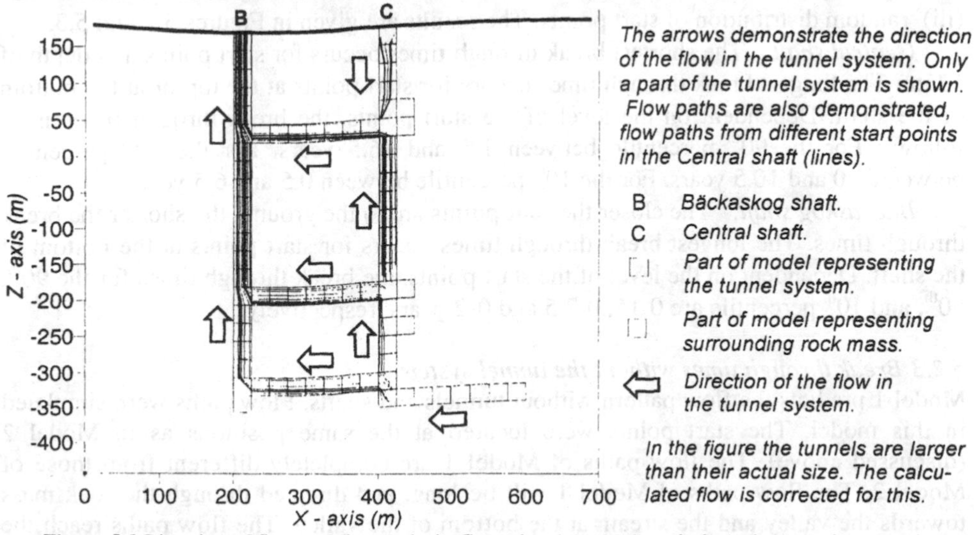
B
C
150
100
50
0
-50
-100
-150
-200
-250
-300
-350
-400
Z - axis (m)
0
100
200
300
400
500
600
700
X - axis (m)
The arrows demonstrate the direction of the flow in the tunnel system. Only a part of the tunnel system is shown.
Flow paths are also demonstrated, flow paths from different start points in the Central shaft (lines).
B Bäckaskog shaft.
C Central shaft.
Part of model representing the tunnel system.
Part of model representing surrounding rock mass.
Direction of the flow in the tunnel system.
In the figure the tunnels are larger than their actual size. The calculated flow is corrected for this.

Figure 5.1 Direction of flow and flow paths in Central and Bäckaskog shafts and connecting tunnels.

5.2 THE TRANSPORT AND THE BREAK THROUGH TIME OF A POLLUTANT IN THE SHAFTS

5.2.1 General

Parts of the waste, deposited in the shafts, might dissolve in the groundwater. Dissolved components will be transported by the groundwater towards the ground surface. As previously stated, the modeling predicts that all the water of the tunnel system will discharge at the upper part of the Bäckaskog shaft. Hence, the probability is very small for long flow paths, from the tunnels, through the rock mass, and to distant discharge areas. The discharge of the dissolved components will take place at the very upper parts of the Bäckaskog shaft and in the very near vicinity of the Bäckaskog shaft. From there, the dissolved components will flow with the surface water, or the groundwater close to the surface, towards the stream that flows in the valley. Mixing with surface water will take place during this transport and in the stream.

In this study the *break through time* is defined as the time it takes for a pollutant, dissolved in the groundwater, to flow with the average water velocity (advection) from its starting point to the ground surface (end point). In the calculations we have not included diffusion, dispersion or chemical reactions.

5.2.2 Break through time, Central and Bäckaskog shafts

Model 2 was used to calculate break through times. Flow paths were simulated from different levels in the shafts and the start points were randomly distributed in the shafts. All simulated flow paths had their end points at the uppermost part of the Bäckaskog shaft. The shafts were divided into sections of 25m vertical length.

Statistical analyses were conducted for the flow paths that start within each section. In the models the variation in calculated break through time depend on the: (i) heterogeneity of the rock mass, (ii) uncertainty in the conductivity of the rock mass and (iii) random distribution of start points. The results are given in Figures 5.2 and 5.3.

Central shaft. The shortest break through times occurs for start points at a depth of 135m. The longest break through times occurs for start points at the top or at the bottom of the shaft. Dependent on the level of the start points, the break through times is as follows: For the 90th percentile between 1.5 and 16.5 years. For the 50th percentile between 1.0 and 10.5 years. For the 10th percentile between 0.5 and 6.5 years.

Bäckaskog shaft. The closer the start points are to the ground, the shorter the break through times. The longest break through times occurs for start points at the bottom of the shaft. Dependent on the level of the start points, the break through times for the 90th, 50th, and 10th percentile are 0-15, 0-7.5 and 0-2 years, respectively.

5.2.3 Break through times without the tunnel system

Model 1 predicts the flow pattern without tunnels and shafts. Flow paths were simulated in this model. The start points were located at the same positions as in Model 2 (discussed above). The flow paths of Model 1 are completely different from those of Model 2. The flow paths of Model 1 will be long, and directed through the rock mass towards the valley and the stream at the bottom of the valley. The flow paths reach the ground, at the bottom of the valley, along a distance of about one-kilometer. The break through times of the flow paths depends on level of start points and varies between a 0 and more than 500 years. In Model 1 (without tunnels) as well as in Model 2 (with tunnels) all the polluted water will end up in the stream in the valley.

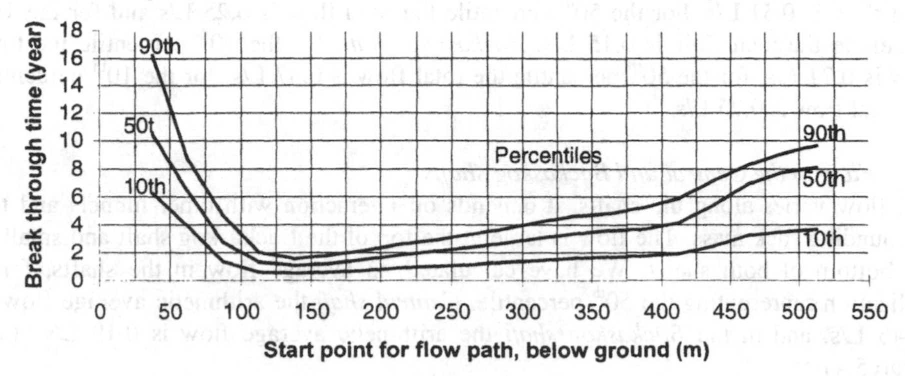

Figure 5.2 Break through times in Central shaft versus depth.

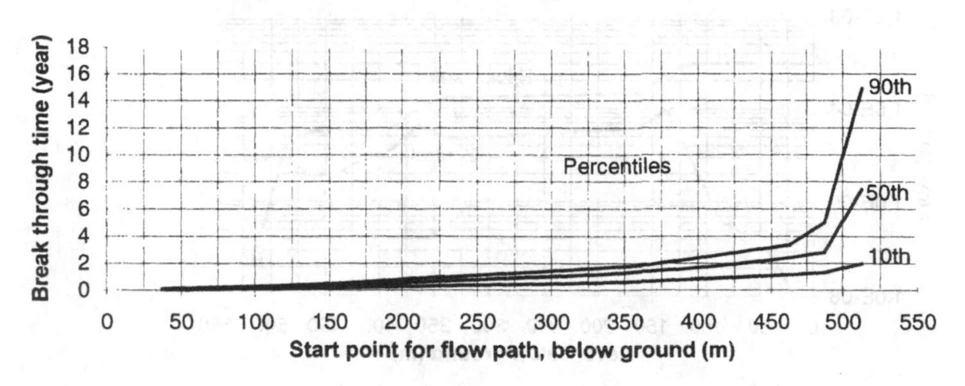

Figure 5.3 Break through times in Bäckaskog shaft versus depth.

5.3 SIZE OF FLOW IN THE SHAFTS

5.3.1 General

The tunnel system is not a system of connected tubes. The flow in the tunnels interacts with the flow in the surrounding rock mass and is consequently different at different sections. The flow in the tunnels and the shafts can be calculated based on different concepts. At one section we will have a flow given in [Length^3/time] and a velocity given in [Length/time], at another section the flow and velocity will expressed otherwise. The total flow that enters and leaves a given part of the tunnel system can be calculated based on a mass balance taken over the envelope of the studied part of the tunnel. We will call the flow that enters a studied part of a tunnel *the total flow* and it is given in [Length^3/time]. In the models the variation in calculated flow depend on (i) the heterogeneity of the rock mass and (ii) the uncertainty in the conductivity of the rock mass.

5.3.2 Total flow in the Central and Bäckaskog shafts

The total flow has been calculated based on a mass balance taken over the envelope of the studied shafts (L/s is Liters per second). *Central shaft:* for the 90th percentile the total flow is 0.51 L/s. For the 50th percentile the total flow is 0.25 L/s and for the 10th percentile the total flow is 0.15 L/s. *Bäckaskog shaft:* for the 90th percentile the total flow is 0.71 L/s, for the 50th percentile the total flow is 0.37 L/s, for the 10th percentile the total flow is 0.25 L/s.

5.3.3 Flow in the Central and Bäckaskog shafts

The flow varies along the shafts, it depends on interaction with other tunnels and the surrounding rock mass. The flow is large at the top of the Bäckaskog shaft and small at the bottom of both shafts. We have calculated an average flow in the shafts, for a realization representing the 50th percentile. *Central shaft* the arithmetic average flow is 0.045 L/s, and in the *Bäckaskog shaft* the arithmetic average flow is 0.19 L/s (see Figure5.4)

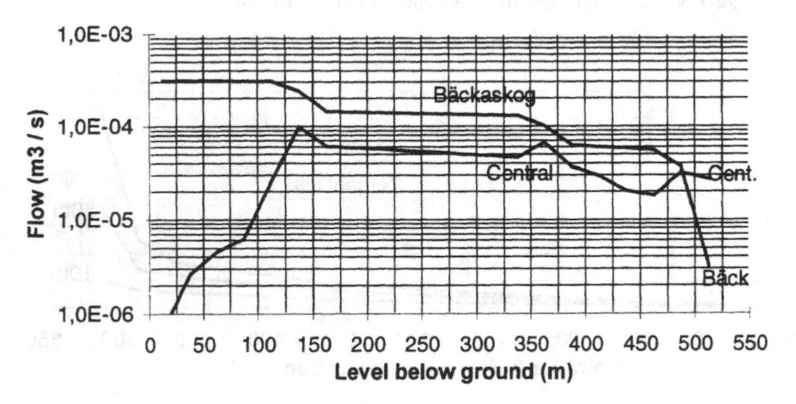

Figure 5.4 Flow versus depth, in Bäckaskog and Central shaft.

5.3.4 Renewal time of the water in the shafts

With renewal time, we mean the time it takes to naturally replace the water at a studied section of the tunnel system. Based on the previously discussed break through times and flows, we estimate renewal times. We note that the renewal times are short at the top of the Bäckaskog shaft and long at the bottom of both shafts. *Central shaft:* with a probability of 90% all the water above a depth of 140m will be renewed in less than 16 years, with a probability of 90% all the water between a depth of 100m and 350m will be renewed in less than 3 years. *Bäckaskog shaft:* with a probability of 90% all the water above a depth of 240m will be renewed in less than 1 year, with a probability of 90% all the water above a depth of 440m will be renewed in less than 3 years.

5.4 MIXING AND DILUTION IN THE SHAFTS

5.4.1 General

If a dissolved component occurs at a certain concentration at a certain depth in a shaft, this concentration will changed as the water in the shaft interact with the surrounding groundwater and mix with the in-flowing water. In the models the mixing/dilution will take place as the flow in the tunnel system increases towards the upper part of the Bäckaskog shaft and as we assume that the water at every depth will form a homogeneous mixture. Based on the variation of the flow in the shafts, it is possible to calculate the mixing/dilution that has taken place for a given concentration at a given depth, when this water reaches the uppermost part of the Bäckaskog shaft. For example, a certain concentration at a depth of 480m in both shaft, will be mixed/diluted 10 times before this water reaches the uppermost part of the Bäckaskog shaft.

5.4.2 Mixing and dilution in the Central and Bäckaskog shafts

The mixing/dilution varies along and between the shafts. It depends on interaction with other tunnels and the surrounding rock mass. We have calculated an average value with regard to the whole of the shafts, for a realization representing the 50th percentile. In the *Central shaft* the arithmetic average value is 39 times, and in the *Bäckaskog shaft* the arithmetic average value is 7 times.

5.5 UNCERTAINTY IN THE RESULTS

5.5.1 General

The models have been calibrated by varying the groundwater recharge. The most important parameters that influence the calibration and the predicted results are the (i) present groundwater heads (ii) conductivity of the rock mass – size and heterogeneity, (iii) topography, (iv) size and the continuity of the tunnel system and (v) the generalizations included in the established mathematical model. When considering the results from a mathematical modeling, one should remember that a model is a generalized description of the studied system and based on a limited amount of data. We have reliable data only for parameter (iii).

As regards the other parameters, the effects of the generalizations and the uncertainty in the input data will make the results uncertain as well. Therefore, the results of a modeling should be regarded as estimations.

6. Estimation of the status and release from the deposed waste

The chemical analysis of the water samples from the shafts did not indicate any release of contaminants from the dumped material. According to the groundwater- model there are relatively high water flows in the shafts, which to some extent can be responsible for a relatively high dilution. This means that all the soluble parts from the dumped material have been washed out from the shafts, and passed the upper parts of the Bäckaskogs shaft. Cf. Cap 5.2.1.

The explosive materials, lead acid and TNT occur in capsules and are placed into boxes before dumping. The release of agents from these parts of the dumped material is a combination of geochemical processes as well as fluid dynamics. The chemical aspects of the release is generally described as the results of several reactions, such as dissolution and desorption, in combination with a transport by diffusion through the boundary layer (capsule) to the free water.

By using a combination of diffusion and dissolution it may be possible to calculate the release and transport of the contaminants from the dumped material through the shafts. For this reason we use the following model, mainly from [11]

$$L = (2Dt)^{½} \quad \text{Einstein – Smoluehowskis ekv.} \tag{1}$$

L= distance from the solid (cm)
D= diffusion coefficient (m^2/s)
t= time (s)

The time can then be calculated

$$t = h/v \tag{2}$$

h = high of the solid (m)
v = flow (m/s)

The amount of the components can be calculated

$$m = LO(C/2)v \tag{3}$$

m = The amount of the components kg/s
L = Diffusion distance (m)
O = The circumference of the component (m)
C = Solubility (kg/m^3)
The concentration of the components from the shafts can be calculated as:

$$C_{(out)} = m/F \tag{4}$$

$C_{(out)}$ = The concentration in the water from the mine
m = The amount of the components kg/s
F = Total flow from the shafts (m^3/s)

If we presume that the area of the shafts are 9 m^2 and if the diffusion constant for TNT is 10^{-5} cm^2/s and the solubility is 0.1 kg/m^3 it will take, about 2000 – 20 000 years for a complete release. For the lead acid a similar calculation gives a time of about 1000-year. This will presume that the release from the capsules to some extent must be controlled by corrosion [9], [10], [11].

7. Summary and Conclusions

Field investigations and chemical analysis of watersamples from the upper 150m level in the shafts revealed no trace of explosives or its degradation products. Findings of explosives in the upper 50 m part of the Central shaft showed no corrosion or other impact and were still active. The lack of TNT and lead acid in the water is thus explained by the fact that the ammunition has not started leaching. Estimations show that due to diffusion all TNT and lead acid will be released after 2000 – 20 000 year.
However, this does not explain the lack of trace from the extensive amount of picrine acid that was dumped in both shafts. This also applies for e.g. nitrogen and chloride that would be expected from the municipal sludge and household waste in the shafts.

The model predicts, however, that all water in the mine system will discharge to the surface water in the very near vicinity of the uppermost part of the Bäckaskog shaft. A stream will then drain the discharge area to a lake in the vicinity of the mining area. Due to the local topography in combination with the design of the mines the models predict a somewhat unexpected flow pattern. Unlike a very slow groundwater flow that is commonly found in abandoned, hydrological stabile mine in Sweden, the groundwater flow is relatively fast and concentrated to the tunnels and shafts. No long flow paths of potential polluted water are thus expected in to the surrounding bedrock.

The water above the 150 m level fulfil the Swedish drinking water quality standards. This unexpected high quality of the water might be explained by the high flow indicated by the model. The model predicts a break through time at the Central shaft between 1.5 and 16.5 year. In the Bäckaskog shaft the time varies between less than one year up to 15 years. This means that all the soluble components, as picrine acid and nitrogen/chloride has been washed out from the system.

In a long time perspective, over 1000 years, the concentration of TNT in the disharge area at the Bäckaskog shaft might be expected in a magnitude of 0.001 mg/L. This is at USEPA restriction level. The main conclusion will be a recommendation to seal the shaft to prevent further dumping and to install a monitoring system at the Bäckaskog shaft for future water analysis. A plan for water treatment system in case of lowered water quality should be completed.

8. Acknowledgments

This study has been conducted as a part of the research work between the Institute of Earth Sciences, Uppsala University and The National Defense Research Establishment, Department of NBC Defense. The calculation of the amount of explosives and some of the background formulas for calculation has been given by T. Carlsson and S. Lamnevik , FOA. The project has been funded by The Swedish Armed Forces.

9. References

1. Darcy, H. 1856: "Les Fontaines Publiques de la Ville de Dijon" Dalmont, Paris, France.
2. Bear, J. And Verruijt, A. 1987: "Modeling groundwater flow and pollution". D. Reidel publishing company, Dordrecht, Holland. ISBN 1-55608-014-X
3. Strack, O., 1989: "Groundwater Mechanics" Prentice-Hall Inc., Englewood Cliffs, New Jersey, 07632, USA, ISBN 0-13-365412-5
4. Holmén, J. G. 1997: "On the flow of groundwater in closed tunnels. Generic hydrogeological modeling of nuclear waste repository, SFL 3-5", 286 pp, Doctoral Dissertation Uppsala University, Uppsala, Sweden, ISBN 91-506-1231-X.
5. Matheron, G. 1967: "Eléments pour une théorie des milieux poreux." Masson, Paris, France
6. Neuman, S. P. 1987: "Stochastic continuum representation of fractured rock permeability as an alternative REV and fracture network concepts", In: Farmer, I.W. et al (eds.) Proc. 28th U.S. Symp. Rock. Mech., 533-561, Balkema, Rotterdam.
7. McDonald M. G. and Harbaugh A. W. 1988: "A modular three-dimensional finite-difference ground-water flow model", MODFLOW manual, U.S. Geological Survey, FederalCenter, Box 25425, Denver, CO 80225, USA.
8. Wiman, L. 1995: "Försvarets Miljöfarliga Lämningar". Lägesrapport från Försvarsmakten. Bilaga till skrivelse dnr. HKV 21210:79607.
9. Selim, H.M. and Iskandar. I.K. 1994: "Sorption-Desorption and Transport of TNT and RDX in Soils". CRREL. Report 94-7
10. "Encyploclopedia of Explosives and Related Items". PATR
11. Carlsson T & Lamnevik S, 1997: "Flödesberäkningar Dalkrsberg". Försvarets Forskningsanstalt. Avd för Vapen och Skydd. Stockholm. In Swedish
12. CRC. 1995: "Handbook of Chemistry and Physics". 75th Ed.

NUCLEAR ACCIDENTS AND RADIONUCLIDE TRANSPORT

M. KROSSHAVN AND F. FONNUM
Norwegian Defence Research Establishment,
Division for Environmental Toxicology,
P.O.Box 25, N-2007 Kjeller, Norway

1. Characteristics of ionising radiation

The ionising radiation will occur in many different forms of energy levels such as α, β, γ, and neutrons. The unit for an absorbed dose is 1 Gray equivalent to the classical 100 rad. The relative biological effects of the different types of radiation differ. For röntgen, β and γ-rays the relative biological effect is 1, for neutrons it may vary from 5-20 and for α-rays it is 20-40. The unit for the biological effective dose in man is Sievert equivalent to the classical 100 rem. One Sievert is 1 Gray x Relative Biological Effect, for a population exposed to a radiation hazard the term used is Collective Effective Dose. If 1 million people are exposed to 0.001 Sv, the collective effective dose is $0.001 \times 10^6 = 1000$ man-Sv.

The most important isotopes distributed to nature from military sources are ^{90}Sr, ^{131}I, ^{134}Cs, ^{137}Cs, ^{235}U and ^{239}Pu. In addition we are exposed to naturally occurring isotopes such as ^{40}K, ^{14}C and ^{3}H. Most people will also be exposed to radon and its daughters from the ground (Table 1). The effect on humans and their effective half life of several important isotopes is given in Table 1.

TABLE 1. The properties of some important radionuclides

Isotopes	Type of radiation	Critical organ	Competitive element	Half-lives		
				Physical	Biological	Effective
^{90}Sr	β, γ	Bone	Ca	28.5 y	50 y	20 y
^{131}I	β, γ	Thyroid	I	8 d	138 d	7 d
^{134}Cs	β, γ	Cytoplasm	K	2 y	20-100 d	70 d
^{137}Cs	β, γ	Cytoplasm	K	30.1 y	20-100 d	70 d
^{235}U	α, γ			4×10^4y	50y	50 d
^{239}Pu	α, γ	Lungs		2×10^4y	50y	50y
^{240}Pu	α, γ	Lungs		6×10^3y	50y	50y
^{40}K	β, γ	Cytoplasm		1×10^9		
^{238}U	α, γ	Cytoplasm		4×10^9y		

F. Fonnum et al. (eds.),
Environmental Contamination and Remediation Practices at Former and Present Military Bases, 343–353.

2. Military sources of radioactive material

The most important sources of radioactive material used by the military are derived from energy production, nuclear weapons, phosphorescence material and medical diagnosis and treatment. The most important radioactive contamination is due to atmospheric nuclear tests which were conducted in the 1950´s and the 1960´s. Large amounts of radioactive particles were transported into the atmosphere. The particles in the troposphere were transported with winds and deposited on a local and regional scale depending on the wind direction. The radioactive particles in the stratosphere had long residence times, up to 3 years, before being deposited on a global scale. Immediately after release the short-lived ^{131}I was the most important isotope. The major long-lived radioactive contaminants from global fallout was ^{137}Cs, ^{90}Sr and $^{239-240}Pu$.

The production of the fissile material ^{239}Pu for nuclear weapons production has caused discharges of large amounts of radioactive contamination to rivers and seas particularly during the 1950´s and 1960´s. The waste water from chemical separation of fission products, amounted to several million cubic meters which was discharged directly to the ground causing widespread contamination. Today such waste is stored in steel tanks until it can be stored as solids in a permanent repository [10].

Nuclear reactors are widely used by the military in nuclear ships, particularly submarines, spacecrafts and lighthouses. The ship nuclear reactors are small (100 megawatt) compared to the civilian nuclear power plants (Several 1000 megawatt). The main causes for accidents with nuclear reactors are loss of cooling, fire or sea water penetrating into the ship/submarine reactor. Nuclear submarines are fuelled with highly enriched uranium Depending on the reactor design the enrichment of ^{235}U varies from 20% to more than 90%. In a submarine reactor 99% of the radioactivity is in the fuel rods of the reactor and the remaining activity is due to activation products of the reactor compartment.

In the Arctic there are more than 200 nuclear submarines with 400 nuclear reactors. Some are still in service, but many are moored near the shore awaiting decommissioning. There are also at least 21000 spent nuclear fuel assemblies which are partly onboard ships and partly stored on land. The reactors each contain about 160 fuel elements. Under normal service the fuel assemblies has to be removed every seven years and twice during the life-time of a submarine [10].

At present a major concern is the slow rate of decomissioning of the Russian submarines in the Arctic. In Russia the submarine is prepared for prolonged water-borne storage with the fuel rods in the reactor. After at least 3 years the fuel rods are cooled and can be removed. They are then stored temporarily in water until redisposed. The reactor compartment is then removed together with the neighbouring compartments and stored on a barge.

There has been several accidents with nuclear submarines and 6 submarines are lost in the sea. Among many accident scenarios, only core heat up event which is caused by loss of coolant or power, and criticality events due to accidents or irregular defuelling procedures are of serious concern [11]. The risk of release of radioactivity from the Russian submarine Komsomolets has been carefully studied. Radionuclide releases are estimated to be about 1012 Bq annually. Considering that the submarine has sunk at 1700 m depth the effective dose to fish is negligible because of the dilution effects.

TABLE 2. Known Accidents in nuclear ships

Countries	Cause	No	Comment
US	leakage of sea-water into submarine	3	One submarine lost
US	Defects of weapon system	1	One submarine lost
US	Leak in primary circuits	1	
Ru	Failure in nuclear power plant	12	5 loss of cooling
Ru	Fires and explosion	10	2 submarines lost
Ru	Loss of propulsion	4	
Ru	Defects in weapon system	3	
Ru	Leakage of water	2	2 submarines lost
Ru	Other causes	1	

A Danish group have collected information from 55 accidents of which 22 were regarded as serious. Thereof were 3 Western and 19 were Soviet/Russian. The number of ship reactors operating in number of years are 12000 ship reactor x year. The possibility of an accident is therefore $5x10^{-3}$ per ship reactor year or the possibility of serious accident is $2x10^{-3}$ per ship reactor year [9]

Another potential hazard are the radionuclide thermoelectric generators which have been used to power light houses in remote locations. These reactors contain ^{90}Sr in a solid strontium-titanate compound which is insoluble in fresh water and sea-water. The reactor have been extensively tested against heat, mechanical damage and chemical agents. No accidents are known

An example of the devastating effects of an accident in a civilian nuclear power plant is Chernobyl. The Chernobyl accident started early in the morning April 26, 1986, at the Chernobyl Nuclear Power Plant. A steam explosion initiated several events and other explosions leading to a collapse of the reactor core. The roof was blown off due to the explosions and exposed the reactor core. During ten days of fire concrete, graphite and fuel debris were ejected into the atmosphere. Most of the released ^{90}Sr and Pu-isotopes was deposited in the vicinity of Chernobyl. A part of the radioactive releases was deposited in Belarus and Ukraine [1]. The radioactive releases were also transported with the wind to the Baltic states, Finland, Sweden, Norway, Archangels, Kola Peninsula, Western part of Russia and to the Arctic.

It is interesting to note that the radioactive pollution in Norway from Chernobyl is in many places 10-50 fold larger than from the atmospheric nuclear weapons testing. Releases due to an accident with a nuclear ship reactor which is 100 times smaller, would be much less.

Spacecraft normally generate their electric power from solar panels. But in some cases with a need for extended energy as in Radar Ocean Reconnaissance Satellites, it has been necessary to resort to nuclear power. All together 31 such satellites have been launched and 5 accidents on reentry are registered. The most known is the accident with Cosmos 954 which contained 20-50 kg of highly enriched ^{235}U which was scattered over 80 000 km2 in Canada. Most of the material was recovered [9].

Radioluminescensce is produced by radiation of special material. Important military uses are instrument panels and fire escape. It has also been used in marker lights at remote airport-runways. Earlier ^{226}Ra was the primary isotope used, but has now been replaced by the weaker ^{3}H and ^{147}Pm [10].

Depleted uranium is a by-product from the uranium enrichment industry. It is used as penetrators in armour-piecing and counter-weight in aircraft. In principle the radiation from depleted uranium, (238U), is insignificant [10].

The occupational exposure to monitored workers 1985-1989 showed that workers involved in the total nuclear fuel cycle was exposed to 2500 manSv with an average of 2.9 mSv per person. The highest exposure was obtained during mining and milling with 4.4 and 6.3 mSv per person respectively. Medical applications are responsible for 1000 manSv with an average of 0.5 mSv, whereas defence applications are responsible for 250 manSv with an average of 0.7 mSv per person.

Collective dose committed to the world population during a 50 year period from 1945 to 1992 was 850 million manSv. The collective dose due to natural sources was 650 million manSv, medical exposure accounted for 165 million manSv and atmospheric nuclear weapons tests 30 million manSv [14].

3. Transport pathways of radionuclides in the environment

A number of natural radionuclides are present in soil from weathering of rocks and decay of organic material. ^{40}K is the main naturally occurring isotope from weathering of rocks and ^{14}C originate from cosmic radiation of nitrogen in the atmosphere. ^{14}C is present in soil as a result of incorporation of CO_2 into organic matter and subsequent decay. The other important natural radionuclides present in soil is ^{87}Rb, uranium and thorium-isotopes, and their daughter products particularly ^{210}Pb. The ^{40}K levels in soil is normally at least 1-2 order of magnitudes higher than the levels of artificial radionuclides from global fallout [3].

Following discharges and accidental releases, radioactive contaminants enter the biogeochemical cycle of elements in the environment. The most mobile transport-phase is atmospheric circulation, while ocean circulation is less efficient and much slower. Radioactive contamination deposited in the terrestrial environment is mainly transported with run-off from the catchment areas to rivers and further to the sea. After deposition external radiation will be the primary contributor to the dose.

Fallout, from atmospheric nuclear tests and the Chernobyl accident, was transported by atmospheric circulation and deposited in terrestrial and aquatic environments. River transport of radioactive contaminants effects large human populations by contaminating drinking and irrigation water, and also fish stocks.

3.1. SOIL

In order to understand the transport of radionuclides in the environment it is necessary to study their chemical properties. Radioactive caesium and strontium have the same chemical properties as the stable nuclide of the respective elements. Strontium appears to have similar chemical properties to calcium, while the properties of caesium is similar to the properties of potassium.

Deposited ^{90}Sr is easily dissolved in soil water as positive charged ions and can to some extent be adsorbed by inorganic colloids and humic substances. ^{90}Sr undergoes cation exchange in soil and becomes available for uptake in plants as does calcium. When calcium levels in soil are low, higher levels of ^{90}Sr can be expected in plants. This is due to competition for ion uptake from soil into roots. In soil and run-off ^{137}Cs occur as cation and tend to be sorbed to particles, inorganic colloids and humic substances. In soils with high content of inorganic colloids and particular clay minerals caesium appear to be less available for transport and uptake in biota. In some soil where clay is absent, a larger part of the present ^{137}Cs is transferred to vegetation.

Plutonium compounds appear to be less soluble in water than caesium. Plutonium can occur in several different oxidation states depending upon the chemical complexes these elements form in soil. Much less is known about biological uptake of plutonium than for caesium or strontium.

Radioactive contaminants are transported from the catchment area via run-off to river systems which may transport the contaminants all the way to the sea. In the temperate and alpine Arctic zones most of the transport takes place during flood seasons, particularly during spring flooding. The transport from catchment areas to the sea is complex and dependent upon different factors including soil properties, climate, hydrological conditions, variable chemical regimes throughout the transport process and the chemical properties of the radionuclides themselves.

Radioactive contamination is transferred to vegetation by direct deposition from the atmosphere and uptake from soil and water. Animals are contaminated by inhalation, (primarily of ^{131}I), and by the consumption of water and food.

3.2. RIVERS

Radionuclides stored in soil can be considered as a source which continuously provides radioactive contamination into river water. Radioactive contamination from global and Chernobyl fallout in the catchment areas continue to contaminate rivers. These transport processes are very complex. Radionuclides can be dissolved and remain as simple ions in water, undergo chemical reactions with other ions or molecules, be sorbed to organic and inorganic colloids, (which are large and complex molecules), and particles. Caesium and plutonium will sorb easily to particles. A large fraction of Cs and Pu will therefore be removed from the water by the settling of particles to the bottom sediments. The transport of particles within the river will depend upon the quantity of particles, river flow and resuspension of sediments.

Strontium is more conservative than caesium and plutonium and will, to lesser extent, be sorbed to particles and transferred to the bottom sediments. Strontium tends to stay dissolved in the water and therefore will more easily be transported over long distances.

Radionuclides sorbed in the bottom sediments can be transferred back into the river water by desorption of radionuclides from sediments, by resuspension and by erosion of sediments. Radionuclides in the sediments can be transferred to vegetation or to biological organisms living in the sediments, and in this way enter the food chain. Radionuclides can also be transferred to plankton and vegetation directly from water. They are transferred to fish by the consumption of plankton and through direct uptake from the water by the gills.

4. Scandinavia

Studies of global fallout from atmospheric nuclear tests were conducted in Norway during the years 1957-1959. The levels of ^{90}Sr in soil increased 3 times during this period in the Oslo region while the levels around Bergen on the Norwegian East coast increased even more [7]. The increment of ^{137}Cs and ^{90}Sr levels in soil was mainly due to wet deposition of fallout by precipitation, but also to some extent by dry deposition. The annual transfer from soil to run-off water was 1-5% of the total ^{90}Sr content, and 0.5-3.8% of the ^{137}Cs content. The highest removal occurred after heavy rain-fall which increased the run-off from the catchment area and the transport of radionuclides to the rivers [7].

The initial biological uptake of ^{137}Cs and ^{90}Sr in plants was observed to be much more efficient from directly deposited material, than via root uptake from soil. In the following years active root uptake of ^{137}Cs and ^{90}Sr in vegetation was found to become more significant. The transfer of ^{137}Cs and ^{90}Sr from soil to vegetation and enrichment in terrestrial and aquatic food chains is evident [4].

Large areas of Scandinavia was contaminated with ^{137}Cs after the Chernobyl accident in late April 1986. In Sweden the deposition from global fallout amounted to 1.25 PBq ^{137}Cs and 0.83 PBq ^{90}Sr. The deposition of Chernobyl fallout was 4.25 PBq ^{137}Cs, which is factor of 3.4 higher than global fallout. The fallout was deposited in mountainous regions, agricultural areas, forests, streams and lakes [12]. A substantial part of the mountainous areas were covered with snow of various thickness, but also snow free ridges. On the ridges the deposition was 40-60 kBq/m2, while snow covered areas received up to 20 kBq/m2. The average ^{137}Cs concentration in the soil was 39.2 kBq/m2 with maximum level of 97.9 kBq/m2 [2, 4].

Most of ^{137}Cs was retained in living plants, organic litter, and the upper part of the humus layer. In podsol soils 90 - 99% of ^{137}Cs was found in the upper 4 cm of the humus layer. In brunsolic soils 61-76% of ^{137}Cs was found in the upper part consisting of mineral layer mixed with organic matter. The vertical migration of ^{137}Cs was very low in the studied types of soil. [2, 4].

The ^{137}Cs activity was about 4 times higher in grass growing in podsol soil than in grass at brunsolic soil. The ^{137}Cs activity in leaves and fruit bodies of higher plants was related to uptake via the root system. Higher plants growing in nutrient poor soil were observed to have a higher uptake of ^{137}Cs than the same species growing in nutrient rich soil. These areas received the same amount of ^{137}Cs deposition [2, 4]. About 97.3% of ^{137}Cs appeared to be retained in soil, while 2.4% was transferred to biota. Only 0.3% of the ^{137}Cs pool was found in the recycling biomass like leaves, fruits and needles. The ^{137}Cs activity in fungi was much higher than in other vegetation and the ^{137}Cs levels in grazing animals depends on the occurrence of fungi during grazing season [2]. The ^{137}Cs levels in sheep and reindeers after the grazing season are still, (1997), exceeding the limits acceptable for consumption in some areas. After the initial reduction of ^{137}Cs levels shortly after the deposition, no significant decrease in the ^{137}Cs levels of moose, blueberry and birch have been observed [13].

Between 1964 and 1985 the ^{90}Sr activity concentrations in Finnish rivers decreased from 100 Bq/m3 to 10 Bq/m3. In 1986 a slight increase of ^{90}Sr activity levels in river water was evident in the most effected areas following the Chernobyl fallout. Between 1960 to 1980 strontium was released from catchment areas into rivers and about 25-50% was removed with run-off. The ^{90}Sr transported by rivers influenced the Baltic Sea, particularly during the 1960´s and 1970´s. Between 1950-1994 the river transport of ^{90}Sr accounted for 24% of the total ^{90}Sr contribution to the Baltic Sea in this period [12].

Following the Chernobyl accident, ^{137}Cs activity in Finnish rivers increased from around 1-5 Bq/m3 to 100-1,000 Bq/m3. Levels up to 20,000 Bq/m3 were determined in small streams immediately after the Chernobyl fallout. About 40% of the initial deposited ^{137}Cs on mires were transported with snow melt to streams during 1986. In the following years about 30% of the remaining ^{137}Cs fraction was transported from the water saturated part of mires. The annual release from dry parts of mires accounted for only 2% of the total ^{137}Cs content. No transport of ^{137}Cs via ridges and moraine slopes to ground water was detected. The reduction of ^{137}Cs levels in forest soils appeared to be due to the physical decay. Caesium was also released from marshlands where the annual release amounted to a few percents which is unusually high [12, 13].

The river transport of global fallout related ^{137}Cs amounted to 1% of the total contribution to the Baltic Sea for the period 1950-1985 and increased dramatically following the Chernobyl accident. This was mainly caused by deposition on water body surfaces, snow covered areas and wet marshlands immediately before the snow melt and spring flooding. The river transport of ^{137}Cs was also unusually high as a result of increased mobility of ^{137}Cs due to the high proportion of marshlands in the catchment areas. The latter is expected to influence river transport rates of ^{137}Cs in the following decades. The river transport of ^{137}Cs into the Baltic Sea is one of the highest registered. The ^{137}Cs levels in fish from the Baltic Sea were about 10-100 Bq/kg in 1989-92. These levels are about the highest levels registered in marine fish. The average ^{137}Cs levels in fish from the Black Sea was 3 Bq/kg and even lower in Arctic seas at less than 1 Bq/kg. Nevertheless, these levels are much lower than the concentration levels in fresh water fish in the rivers and lakes in Fenno Scandinavia, which were as high as 10,000 Bq/kg in 1989-1992 [12].

5. Russia

The largest Russian rivers are Ob, Yenisey and Lena. These rivers are located transitively in the tundra and the forest tundra zones of the Eurasian territory. Ob and Yenisey rivers discharge into the Kara Sea, while Lena river discharge into the Laptev Sea. The main source of radioactive contamination of the catchment areas of Ob (2.990.000 km^2), Yenisey (2.580.000 km^2) and Lena (2.490.000 km^2) is global fallout. The total deposition in these catchment areas is 20.0 PBq ^{137}Cs and 12.2 PBq ^{90}Sr. The Ob and

Yenisey rivers and catchment areas are also influenced by discharges from nuclear plants and reprocessing plants in the Urals, and accidental atmospheric releases from these nuclear facilities [12, 15].

The ^{137}Cs removal from the catchment areas via run-off about 2-3% which was much less than the removal rate of ^{90}Sr in the period 1960's-1990's. About 10% of ^{90}Sr has been removed from the Ob and Yenisey catchment areas during this period. The physical half-life of ^{90}Sr is 28.5 years which means that more than 50% of the initial deposited ^{90}Sr has decayed since the 1960's. The ^{90}Sr migration from the catchment area to the river depends on variations in the climatic and hydrological conditions which influence the river flow. About 50-70% of the annual ^{90}Sr flux occurs normally during 2-4 months of spring flooding [6]. The reduction of ^{137}Cs and ^{90}Sr in the terrestrial environment is due mainly to physical decay. Even though the transport of radioactive contamination from the catchment area generally decrease, the release from soil to run-off will continue to contaminate the river systems, particularly during spring-flooding for decades.

The ^{90}Sr flux to the Kara Sea from the Ob and Yenisey catchment areas and the discharges from nuclear facilities amounted to 1.4 PBq during the period 1962-1993. Of which almost 30% is related to the discharges from nuclear facilities. The river transport of ^{90}Sr exceeds the direct deposition of ^{90}Sr on the Kara Sea following the atmospheric nuclear tests [15, 16].

6. Belarus and Ukraine

Regions in Belarus and Ukraine received large amounts of radioactive contamination following the Chernobyl accident. The contaminated area is drained by the Dnieper river and its major tributaries: Pripyat, Sozh, Besed and Iput. The catchment area of Pripyat river extends a territory of 121.000 km^2, while the Dnieper catchment area covers 504.000 km^2. Pripyat and Dnieper rivers are located in Belarus and Ukraine. Most of the catchment areas are extensively tilled and contain high proportions of boggy areas. Afforestation is also characteristic in many regions.

About 71.6 PBq ^{137}Cs and 8.1 PBq ^{90}Sr was released during the Chernobyl accident [8]. The deposition of ^{137}Cs and ^{90}Sr in Belarus was 50 PBq and 5.6 PBq respectively of which 24.5 PBq ^{137}Cs and 2.96 PBq ^{90}Sr was deposited in the catchment area of Pripyat and Dnieper rivers. The total contaminated area covers more than 15.282 km^2. The contaminated areas are divided into regions depending on the contamination level: ^{137}Cs: 37-185 kBq/m^2; ^{137}Cs: 185-555 kBq/m^2; ^{137}Cs > 555 kBq/m^2; ^{239}Pu > 3.7 kBq/m^2 [12].
In the Chernobyl 30 km exclusion zone, very high contamination levels were detected in fish of about 20-30 kBq/kg ^{137}Cs and 50 kBq/kg ^{90}Sr, shortly after the accident, although levels up to 150 kBq/kg ^{137}Cs were registered. During 1986 levels of ^{137}Cs in fish from the Kiev reservoir increased up to an average level of almost 2 kBq/kg in 1987. This had decreased 10-15 times by 1996 [12].

Transport of radioactive contaminants via river run-off is the most important transport process removing Chernobyl related fallout from the contaminated areas. Suspended particles in rivers contributes significantly to ^{137}Cs migration. Almost 40% of the ^{137}Cs amount removed by rivers was associated with suspended materials. Unlike

^{137}Cs, ^{90}Sr tends to migrate from the catchment area in a dissolved state, up to 92-98%. Only 1 to 8% of the total ^{90}Sr activity was found to be attached to finely dispersed material [12].

Six artificial reservoirs are located along the Dnieper river downstream the confluence with Pripyat river. For the period 1986-1995, the Pripyat and Dnieper transport into the Kiev reservoir amounted to 1% of ^{137}Cs and 6% of ^{90}Sr deposited in the Pripyat-Dnieper catchment areas. Run-off from the upper Pripyat and Dnieper basins and the Chernobyl exclusion zone have been the primary contamination sources following the initial direct deposition of fallout. During the same period the physical decay accounted for 16% of each of the radionuclides. The transport of radionuclides from soil with river run-off is not an efficient for removal from the but is a significant secondary contamination source of rivers for decades [12].

The reservoirs are characterised by low water velocities and high sedimentation rates. ^{137}Cs attached to suspended sediments is deposited in the bottom sediments. As a result of sedimentation, bioaccumulation and adsorption processes, only 1% of ^{137}Cs entering the Dnieper reservoir system reach the Black Sea. In contrast most of ^{90}Sr occurs in the dissolved state and about 40% passes through the Dnieper cascade system to the Black Sea [12].

7. Uptake of radionuclides in biota

Both after the atmospheric nuclear explosions and after the Chernobyl accident the uptake of radionuclides have been studied extensively. The uptake depends both on the type of isotopes and the elements they are competing with (Table 1). Traditionally we have in Norway always found a high uptake of radioactivity in reindeer. During wintertime when reindeer consume moss and lichen, the levels are very high, several kBq/kg body weight. In summer when the diet changes to grass, the level decreases rapidly. Sheep which during summer eat mushrooms, also reaches a high level of radioactivity. The levels in sheep decreases significantly after grazing for a few weeks. For several types of domestic animals the radioactivity can be reduced by administrating the chelating agent Berliner Blue. Trout in mountain lakes also reaches a high dose of caesium and the biological half-life is several months. Immediately after release of activity from an nuclear explosion or from a nuclear reactor, the ^{131}I concentration is a high. Cows grazing on contaminated grass have a high uptake of ^{131}I which is rapidly transferred to the milk. The physical half-life of ^{131}I is 8 days and the precaution is to keep cows indoor until the levels in grass is acceptable for consumption, but also to keep the milk in storage for a few weeks. Consumption of potassium iodide tablets will also reduce the radioactive iodide uptake by a dilution effect.

8. Medical effects of radiation

8.1. ACCUTE EFFECTS

The acute and early effects of radiation occur from cell death. A dose of 2.5-5 Gy will cause death from damage to the haemapoetic system within 4-6 weeks. The effect is due to the death of the bone marrow stem cells and the consequences is a loss of red and white blood cells and platelets. The death is due to an infection caused by the impairment of the immune system or to bleeding or anaemia from depression of platelets.

It is possible in the case of low doses to treat the infection with antibiotics until the immune system recovers. The blood cells decrease in number from the day of exposure and can reach a low level already at day 10. Death usually occurs 30 days after irradiation, but it may take up to 60 days. Today it is possible to increase and extend the synthesis of the blood cells by using cytokines and this may extend the lethal radioactive dose to 5-7 Gy.

At a higher dose of 5-10 Gy the gastrointestinal syndrome will also occur, resulting in death 3-10 days after exposure. The symptoms include noisia, vomiting, diarrhoea and loss of water. The symptoms and subsequent death are due to damage of epithelium cells in the gastrointestinal tract. At present there is no treatment for this effect.

An even higher dose of 50-100 Gy will cause death by cerebrovascular damage within two days. The symptoms are as described before including disorientation and loss of co-ordination of muscular activity.

8.2. LATE EFFECTS

The late effects of radiation results mainly from cell damage and not cell death. Carcinogenic and hereditary damage results from stochastic effects of radiation and are the principal risk to health from low level of radiation. Radiation can cause cancer in any part of the human body. Prior to the appearance of a tumour there is a latency period of several years, the length of which depend on the type of cancer.

Leukaemia is the most frequently induced cancer after radiation and latency period is 2-5 years, while for other cancers ten or more years is normal. There is also a further delay before the transformed cell begin to develop a tumour. Various factors influence the probability of cancer development following exposure to radiation.

From Japanese atomic bomb survivors, 92228 people, a total of 3435 died of cancer, but only 357 of these cases were due to radiation. There are 144 dead of leukaemia and 81 of these where due to radiation [5]. Most of the leukaemia appeared 5-10 years after the explosion where most other cancer types appeared 10-30 years afterwards.

There are 11 studies of underground mining exposed to high concentrations of radon and radon daughters. The carcinogenic agents is in this case the short lived daughters of radon: ^{218}Po and ^{214}Po. The excess lung cancer risk from exposure to radon is about 1-3 per 10 000 working months [5].

It is important to notice that there is an increase of cancer due to radiation, but cancer develops several years after exposure. At present all estimations of cancer assume a linear dose-effects relationship. This does not take into account that radiation damage is continuously repaired. This has significant consequences when considering population effective dose since small doses of radioactivity multiplied by large populations will give an appreciable man-sv dose. ICRP adopted in 1991 a probability of 0.04 fatal cancers pr man-Sv for workers in atomic installations and 0.05 fatal cancers for the general population.

Since DNA is the target for radiation, it is expected that radiation will lead to changes in a single gene or in chromosome aberrations. Experiments on human cells in vitro and observations of the Japanese survivors suggests that humans are not particularly sensitive to chromosonal aberrations and gene mutations by radiation.

References

1. AMAP (1997) *Arctic Pollution Issues: A state of the Arctic Environment Report.* AMAP, Arctic Monitoring Assessment Programme, Oslo, Norway.
2. Blytt, L. D., Garmo, T. H., Haugen, L. E. and Olsen, R. A. (1993) The distribution of radiocesium in a sub-alpine birch forest in southern Norway. *In: Environmental Radioactivity in the Arctic and Antarctic.* Eds: P. Strand and E. Holm. Scientific Committee of the Environmental Radioactivity in the Arctic and Antarctic, Norwegian Radiation Protection Authority, Østerås, Norway, pp 279-282.
3. Brady, N. C. (1984) *The Nature and Properties of Soils.* 9th ed. Macmillan Publishing Company Inc, USA, pp 750.
4. Bretten, S., Gaare, E., Skogland, T. and Steinnes, E. (1993) Investigations of radiocaesium in alpine ecosystems in Norway following the Chernobyl accident. *In: Environmental Radioactivity in the Arctic and Antarctic.* Eds: P. Strand and E. Holm. Scientific Committee of the Environmental Radioactivity in the Arctic and Antarctic, Norwegian Radiation Protection Authority, Østerås, Norway, pp 291-295.
5. Casarett, L. J. and Doull, J. (1986) *Toxicology. The Basic Science of Poisons.* 3. ed Eds: C. D. Klaassen, M. O. Amdur and J. Doull. Macmillan Publishing Company, New York, pp 974
6. Chumichev, V. B. (1995) Sr-90 discharge with main rivers of Russia into the Arctic Ocean during 1961-1990. *In: Environmental Radioactivity in the Arctic.* Eds P. Strand and A. Cooke. Scientific Committee of the Environmental Radioactivity in the Arctic and Antarctic, Norwegian Radiation Protection Authority, Østerås, Norway, pp 79-83.
7. Lund, L., Michelsen, O. B. and Wik, T. (1962) *A study of Sr-90 and Cs-137 in Norway 1957-1979.* FFI report K-253. Norwegian Defence Research Establishment, Kjeller, Norway, pp 129.
8. Lyutsko, A. M., Rolevich, E. V. E, and Chernov, V. I. (1990) *To Live after the Chernobyl Disaster.* Vysheishayashkola Publ., Minsk, pp 109.
9. NACC. (1995) *Cross-Border Environmental Problems Emanating from Defence-Related Installations and Activities. Phase I: 1993-1995, Volume 1: Radioactive Contamination.* NATO/CCMS Report No 204, Brussels, Belgium.
10. NACC. (1998) *Cross-Border Environmental Problems Emanating from Defence-Related Installations and Activities. Phase II: 1995-1998, Volume 3: Management of Defence-Related Radioactive Waste.* NATO/CCMS Report No 226, Brussels, Belgium.
11. NACC. (1998) *Cross-Border Environmental Problems Emanating from Defence-Related Installations and Activities. Phase II 1995-1998, Volume 4: Environmental Risk Assessment For Two Defence-Related Problems.* NATO/CCMS Report No 227, Brussels, Belgium.
12. NACC. (1998) *Cross-Border Environmental Problems Emanating from Defence-Related Installations and Activities. Phase II: 1995-1998, Volume 2: Radioactive Contamination of Rivers and Transport Through Rivers, Deltas and Estuaries to the Sea.* NATO/CCMS Report No 225, Brussels, Belgium.
13. Nylén, T. (1996) *Uptake, Turnover and Transport of Radiocaesium in Boreal Forest Ecosystems.* Dr Thesis. National Defence Research Establishment, Umeå, Sweden.
14. UNSCEAR. (1993) *Sources and Effects of Ionizing Radiation.* United Nations, New York, pp 922
15. Vakulovsky, S. M., Kryshev, I. I., Nikitin, A. I., Savitsky, Y. V., Malyshev, S. V. and Tertyshnik, E. G. (1995) Radioactive contamination of the Yenisey River. *J. Environ. Rad.* **29,** 225-236.
16. Vakulovsky, S. M. (1997) Unpublished data.

MIGRATION PROCESSES OF RADIONUCLIDES IN TERRESTRIAL ECOSYSTEMS OF UKRAINE

N.V. GOLOVKO, A.N. ROZKO, T.I. KOROMYSLICHENKO
State Scientific Center of Environmental Radiogeochemistry, National Academy of Sciences and Ministry of Extraordinary Situations of Ukraine, Kyiv, Ukraine, 34 Palladin av., Kyiv-252142

1. Introduction

The measures directed at eliminating the consequences of the Chernobyl accident were successful only after a thorough study of the behaviour of technogenic radionuclides in the environment. These measures (technical, medical, biological, agro-industrial, etc.) were based upon a knowledge of environmental tolerance for technogenic contamination, and of environmental capacity for self-restoration [1].

Estimation of the radio-ecological situation in contaminated ecosystems required precise knowledge of the state of the Chernobyl radionuclides, the main mechanisms of their mobilisation, their transformation in the environment, and the key parameters determining these mechanisms.The model describing radionuclides' behaviour in ecosystems includes several parameters: the organic component of soil, radioactive contamination, meteorological factors, peculiarities of landscape, and the mineral component of soil (Figure 1). According to the current view of the behaviour of radionuclides in soil, their redistribution is presented by a scheme in which radionuclides within the "hot particle" are transferred to the soil solution, then to the reversibly bound form, and then to the irreversibly bound form as a result of transformation of the fuel matrix. The irreversibly bound form is strong fixation of radionuclides. From our point of view it consists of natural organic substances - humic acids; natural minerals - clay minerals; complexes of soil - organic acids with clay minerals; natural biopolymers - microorganisms and products of their metabolism; biocomplexes of microorganisms with clay minerals (Figure 2).

F. Fonnum et al. (eds.),
***Environmental Contamination and Remediation Practices at Former and Present Military Bases*, 355–361.**

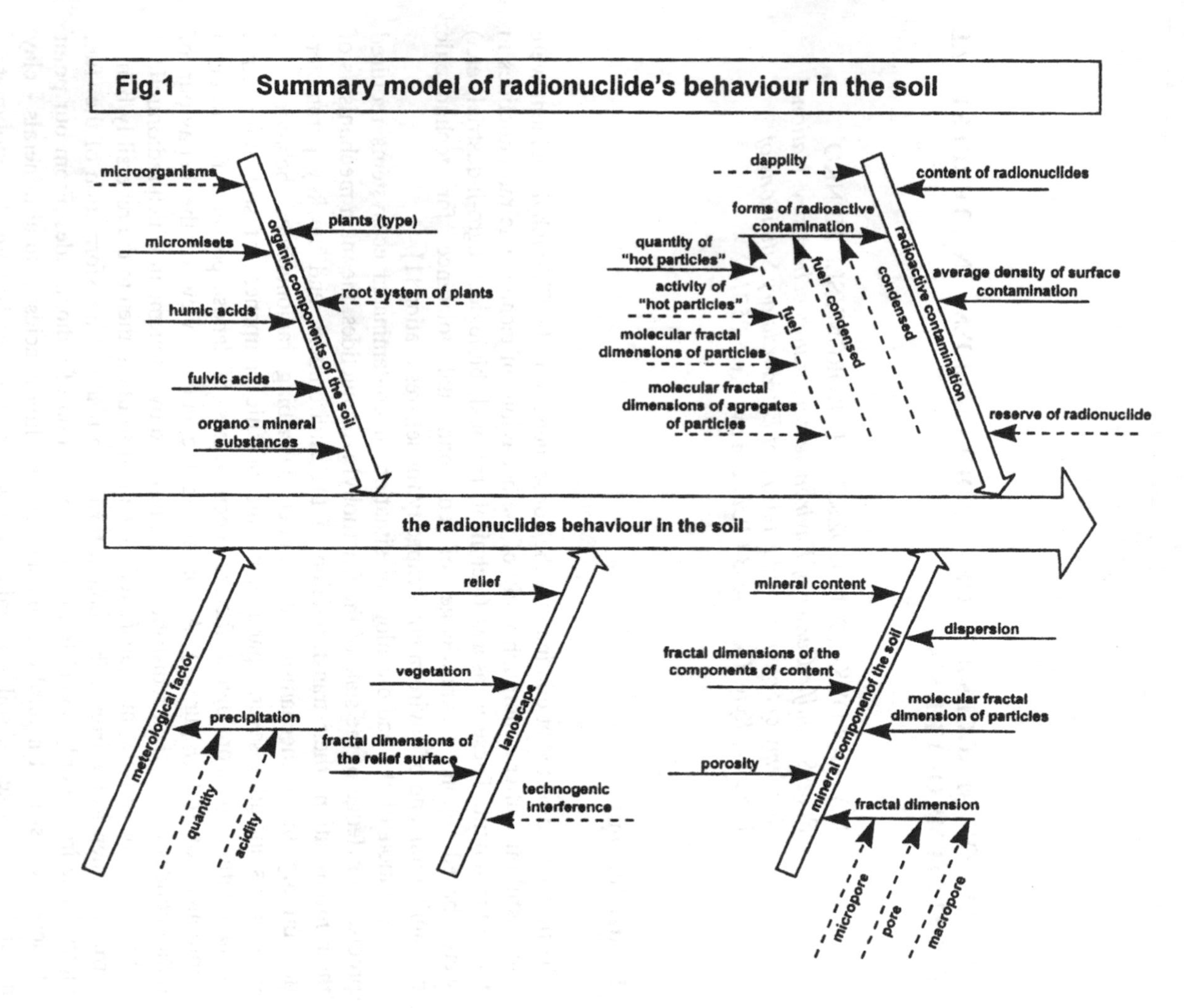

Fig.1 Summary model of radionuclide's behaviour in the soil

2. Migration processes associated with oil products

Among the thousands of technogenic organic substances that are pollutants of soils and water, oil products are specially important. Oil products act as transfer agents by binding radionuclides. By analyzing the content of oil products in soils, as well as in water, one can predict the ecological situation in different regions contaminated by radionuclides. Interaction of the organic component of soils with technogenic pollutants plays an important part when considering the ecological capacity of a system, and its ability to undergo decontamination.

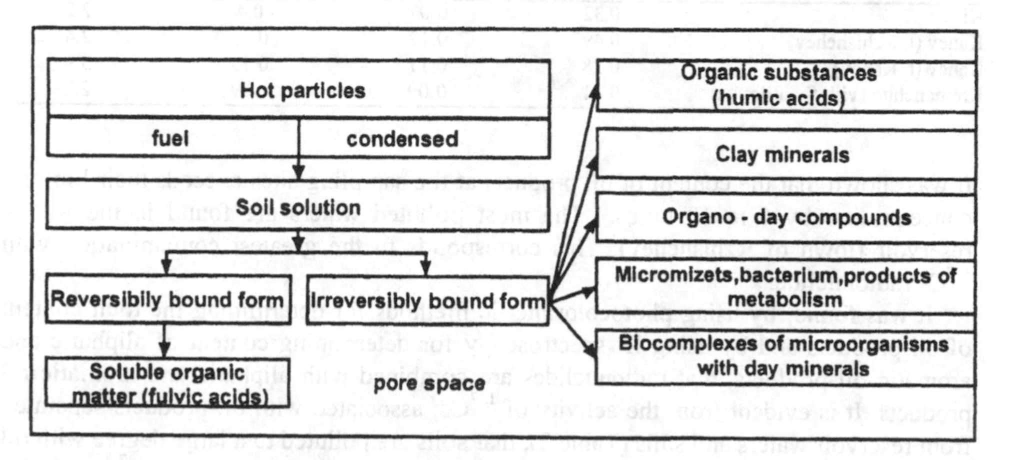

FIGURE 2. Scheme of redistribution of radionuclides

The goal of this work is to determine the influence of technogenic organic substances, natural organic substances, and organo-mineral compounds on the redistribution of radionuclides. Geochemical criteria for ecosystems stability of Ukrainian Polissya can then be obtained, and used to assess technogenic contamination and to develop methods for rehabilitation.

To investigate the behaviour of radionuclides bound with oil products, the following samples were tested:

- waters of the Kiev, Kanev and Kremenchug reservoirs;
- soil enriched with oil products;
- soil from the zone nearest to the Chernobyl NPP.

To investigate the behaviour of radionuclides bound with natural organic substances, the following samples were tested:

- contaminated soils from the 30-km zone - Chistogalovka, Novoselky;
- contaminated soils from Narodychi region.

In the course of the experiment, the following methods were used: chemical methods of extraction, selection, and separation of humic substances and oil products; gel-chromatography; visible and infrared (IR) spectrophotometric methods, and x-ray diffraction.

The content of radionuclides combined with oil products in the Kiev, Kanev and Kremenchug reservoirs is represented in Table 1.

TABLE 1. The content of radionuclides combined with oil-products

Reservoirs	Oil products Concentration, mg/l	^{137}Cs activity, combined with oil products	^{137}Cs activity total, Bq/l	^{137}Cs activity specific, Bq/kg E5
Kiev	0.32	0.07	0.4	2.2
Kanev (t. Rzhishchev)	0.49	0.12	0.15	2.4
Kanev (t. Kanev)	0.28	0.11	0.15	3.9
Kremenchug (vill. Sviatilovka)	0.41	0.09	n. av.	2.2

It was shown that the content of oil products at the sampling sites exceeds their limiting concentration by 6 to 10 times. The most polluted waters are found in the Kanev reservoir (town of Rzhishchev). This corresponds to the greatest contamination with ^{137}Cs radionuclide.

It was found, by using photocolorimetric methods for determining the total content of oil products and by using IR spectroscopy for determining content of aliphatic and aromatic oil products, that radionuclides are combined with aliphatic and aromatic oil products. It is evident from the activity of ^{137}Cs, associated with oil products separated from reservoir waters and soils (Table 2), that soils are polluted to a large degree with oil products as well as with the associated radiocaesium. For soils the value of ^{137}Cs activity reaches 5.4 Bq/kg; this differs significantly from the 0.12 Bq/kg activity observed in the most polluted waters of the Kanev reservoir in the town of Rzhishchev.

TABLE 2. Content of aliphatic and aromatic oil products, separated from soils

Sample	Aliphatic oil products mg/kg dry soil	Aromatic oil products mg/kg dry soil	Cs-137 activity associated with oil products Bq/kg
PP-91	116.7+0.62	3.0+0.62	5.4
P-1	39.2	-	5.2

Consequently, the discovery of the bonding between the radionuclide ^{137}Cs with aromatic oil products in aqueous solution points to the possibility of a relatively rapid transfer of radionuclides into water flows, including percolation (filtration) into porous media. Radiocaesium combined with oil products is not subject to sorption in suspension and is not removed from the water during its movement. Oil products like radionuclides are industrial pollutants and for this reason, the migration process of radiocaesium exhibits technogenic process characteristics.

3. Migration processes associated with humus substances.

The redistribution of radionuclides and natural organic substances along the vertical profile of the selected samples was obtained by studying one of the stages of the scheme showing radionuclide behaviour (Figure 2).

The activity of peat and podzolic soils sampled in Novoselki, Chistogalovka settlement and Narodychi region is shown in the Table 3-6.

TABLE 3. Activity of samples, Novoselki settlement, peat soil

Interval, cm	^{90}Sr, Bq/kg	^{137}Cs, Bq/kg
0-1	1.37×10^4	3.76×10^3
2-3	1.04×10^4	8.88×10^2
3-4	7.55×10^4	2.14×10^2
5-6	1.44×10^3	2.03×10^2

TABLE 4. Activity of samples, Chistogalovka settlement

Interval, cm	Peat soil Basic activity, Bq/kg	Podzolic soil Basic activity, Bq/kg
0-1	1.08×10^6	2.27×10^5
1-2	7.73×10^5	1.13×10^5
2-3	2.03×10^5	7.75×10^4
3-4	3.84×10^4	7.4×10^4
4-5	1.78×10^4	3.1×10^4
5-7.5	4.4×10^3	2.85×10^4
7.5-10	4.6×10^3	2.5×10^3
10-15	4.4×10^3	1.27×10^3
15-18	2.45×10^3	1.2×10^3

TABLE 5. Activity of samples, Narodychi region, peat soil

Interval, cm	^{90}Sr, Bq/kg	^{137}Cs, Bq/kg
0-1	3.2×10^3	3.97×10^3
1-2	2.4×10^4	4.78×10^3
2-3	3.2×10^3	3.81×10^3
3-4	3.8×10^3	3.73×10^3
4-5	1.4×10^3	6.27×10^3
5-6	2.7×10^3	4.12×10^3
6-7	2.5×10^3	4.34×10^3

TABLE 6. Activity of samples, Narodychi region, podzolic soil

Interval, cm	^{134}Cs, Bq/kg	^{137}Cs, Bq/kg
0-2	9.26×10^2	0.43×10^2
2-4	-	0.53×10^2
4-6	-	0.17×10^2
6-10	-	3.53×10^2
10-14	-	1.86×10^2
14-20	-	1.24×10^2

It was found that the samples from Chistogalovka settlement, especially in peat soil, are the most active. The fixed forms of radionuclides and fulvic acids that belong to mobile forms of the main dose-creating radionuclides were obtained on the basis of conformity to natural laws of redistribution of radionuclides between humic acids (Table 7).

TABLE 7. Content of ^{137}Cs in percentage, combined with humus substances in peat and podzolic soils

Interval, cm	Peat soil, Novoselki	Peat soil, Narodychi	Peat soil, Chistogalovka	Podzolic soil, Narodychi	Podzolic soil, Chistogalovka
0-1	31.1	No determ.	0.06	0.05	0.07
1-2	No determ.	84.4	0.03	0.40	0.05
2-3	27.4	84.4	-	No determ.	0.08
3-4	10.7	31.9	0.20	No determ.	0.03
4-5	No determ.	16.5	-	No determ.	0.23
5-6	1.4	35.2	2.7	No determ.	0.12
6-7	No determ.	66.3	-	No determ.	0.68
7-8	No determ.	3.0	-	No determ.	No determ.

Studying the migration processes of Cs-137 and Sr-90 showed that, in comparison with the total activity, the percentage of extraction of radioactive caesium with humic substances, which are not bound with clay minerals, varies with the sample. This value is high only for peat soils in Narodychi region, and varies from 66,3 to 84,4 %. For Novoselki settlement it varies from 10,7 to 31,1% and is very small in Chistogalovka settlement for the contaminated areas in the 30-km zone - especially for podzolic soils (0,05-2,7). This may be explained by very slow velocity of transformation of fuel particles in this region. Owing to this fact the highest activity is concentrated in "hot particles". By separating organic substances on fixed and mobile forms of radionuclides it was shown that 48% is bound to humic acids and 37% is bound to fulvic acids. Using IR-spectrometry and gel-chromatography it was found that activity of Cs-137 is accumulated in the upper (0-7 cm) layer and is correlated with humic acids, enriched with carboxyl groups. Strontium is present in soil mostly in the form of intricate complexes of unspecific organic substances and humic acids, rather than in the form of individual compounds. The obtained results will allow the establishment of quantitative dependence of different forms of radionuclide migration on content of organic components [2].

Ten years after the accident, a trend is apparent in which there is a natural transfer of technogenic radionuclides from geochemically mobile forms to less mobile forms, and a decline in surface radioactive contamination density due to radioactive decay and radionuclide migration away from contaminated areas. In the course of radionuclide transformation in soils, some radionuclides – mostly radiocaesium – form immobile entities that are not taken up by vegetation. Owing to this phenomenon, considerable reduction of radiocaesium biomass contamination of the surface ecosystem may be expected over the next 10-20 years.

References

1. Sobotovich, E. V. (1996) Radioactive Contamination of the Ukrainian Territory, International Symposium *Technological Civilization Impact on the Environment* 22-26 April 1996, 60.
2. Stevenson, F. J. (1994) Humus chemistry: genesis, composition, reactions, John Wiley & Sons, Inc.

INDEX

9 780792 352488